S0-EJQ-639

Antitumor Compounds of Natural Origin: Chemistry and Biochemistry

Volume II

Editor

Adorjan Aszalos, Ph.D.

Head, Biochemistry/Chemotherapy
National Cancer Institute
Frederick Cancer Research Center
Frederick, Maryland

CRC Press, Inc.
Boca Raton, Florida

Library of Congress Cataloging in Publication Data

Main entry under title:

Antitumor compounds of natural origin.

Bibliography: p.
Includes index.
1. Tumors—Chemotherapy. 2. Antineoplastic agents. 3. Natural products. I. Aszalos, A. A.
RC271.C5A68 616.99'4061 80-26903
ISBN 0-8493-5520-6 (v. 1)
ISBN 0-8493-5521-4 (v. 2)

This book represents information obtained from authentic and highly regarded sources. Reprinted material is quoted with permission, and sources are indicated. A wide variety of references are listed. Every reasonable effort has been made to give reliable data and information, but the editor and the publisher cannot assume responsibility for the validity of all materials or for the consequences of their use.

Direct all inquiries to CRC Press, Inc., 2000 N.W. 24th Street, Boca Raton, Florida 33431.

International Standard Book Number 0-8493-5520-6 (Volume I)
International Standard Book Number 0-8493-5521-4 (Volume II)

Library of Congress Card Number 80-26903
Printed in the United States

PREFACE

The importance of cancer chemotherapy is apparent to all who work in the cancer field. Several books published on topics of cancer chemotherapy, especially the volumes of "Methods in Cancer Research", deal with various aspects of this field. It is our feeling at this time, that a book which deals with the fundamental chemical and biochemical aspects of cancer chemotherapy would be a valuable addition to the existing books, especially for researchers. Recent insights into the biochemical, nutritional, and cell cycle differences of the cancerous cell can be related now to mode of action and to chemical structures better than before. Therefore, by illuminating these relationships and pointing out possible new research avenues, we hope, researchers will be stimulated in their efforts to develop more selective antitumor agents.

The 30 or so clinically applied cancer chemotherapeutic agents used today were obtained through chemical synthesis or by isolation from natural resources. Because of the magnitude of these topics, this book will deal only with products obtained from natural resources, for example, those from marine or Earth plants and animals and from fermentation media. Topics for the book were selected in order of current interest, but certain topics often discussed recently were omitted. However all isolated natural agents known to have in vitro or in vivo anticancer activity were collected up to July 1979 and are listed in the first chapter according to a chemical classification system, and in alphabetical order within each class. This tabulation should serve researchers who are looking for similar compounds within a certain class and those who want to relate some biological activities to compounds of different chemical classes.

A relatively large number of interesting, new compounds were isolated from marine sources, but as pointed out by the Hegyelis, in their up-to-date Chapter, the applied screening method needs some revision.

New insights were gained recently into the mode of action of immunostimulants, especially those of macrophage activators. It could be demonstrated that when macrophages become tumoricidal due to the action of these stimulators, the biological changes of the macrophages can be connected to certain cell membrane changes called "cis-membrane effects". These effects are measurable now by physicochemical methods. Immunostimulators are derived from various sources, such as bacterial cell membranes or shark liver. Screening for these agents could be done now on an extended basis. Speaking of screening, the current general in vitro prescreening method for antitumor agents by the NCI is based on cytotoxicity measurements and it is hoped that it will be replaced by more selective methods.

New research directions are indicated for the compound classes of enzymes and proteins. Enzyme cancer chemotherapy is advanced to the point where definite research objectives can be formulated. Development of derivatives of therapeutically useful enzymes to obtain extended half-life and to avoid immunogenecity of these proteins, could be persued and would yield valuable tools for the clinicians, as pointed out by Drs. Abell and Uren. Antitumor proteins with their effect on RNA and DNA metabolism require specific receptors on the cell membrane in order to exert their biochemical effects. One could use these compounds to identify cell membranes receptors as the first chemotherapeutic event in the recognition of tumor cells from their normal counterparts, as is recommended by Drs. Montgomery, Shepherd, and Vandre.

Aureolic acid type compounds and the ansamycins represent a very interesting group of antibiotics as is well detailed by Drs. Skarbek, Speedie, and Sethi. Some of these compounds are used in the clinic "successfully". There is ample room, however, for improvement in designing new compounds of these types, perhaps compounds with improved pharmacodinamic properties and better target selectivity. A great variety of compounds included in the chapter "Quinones and Other Carbocyclic Antitumor Antibiotics" by Dr. Eckardt, I contend, will stimulate the speculation of chemists and biochemists alike.

Vitamin A seems to influence the differentiation state of epithelial cells. Related observations were reported as early as 1926. These observations were followed by investigations on the influence of Vitamin A and its derivatives on skin, mammary gland, lung, colon, and bladder cancers as detailed in Chapter 4, Volume II, by Drs. Frolik and Roller. Along this line very useful in vitro test systems were developed to test the relative potency of Vitamin A derivatives. These in vitro test systems, like the one using fibroblast cells transformed by sarcoma growth factor, or the one employing mouse mammary glands exposed in vitro to 7,12-dimethylbenz(a)anthracene, or the one which determines ornithin decarboxylase activity in mouse prostate and hamster tracheal organ cultures, were useful to demonstrate that retinoic acid can reverse malignant transformations. The correlations among these cell cultures and their relation to in vivo systems seems to be very good. It would be interesting to test these systems for screening other antineoplastic agents, like natural products. Perhaps these in vitro systems would lead to more selective agents than the present ones.

Retinoids exert their effect via the immune system in some cases. Since the interaction of terpenoids with membranes is a known fact, could it be that this mechanism is similar to that which is known for the P_3 factor of BCG and for certain polyene antibiotics in relation to macrophage activation? It would be interesting indeed to relate the mode of action of these substances.

Polyene antibiotics were classified according to their effect on membranes into two groups. One group with interesting biological activities was shown to exert a variety of effects on different cells by Drs. Brajtburg, Medoff, and Kobayashi. One of the most interesting actions of these compounds is on cells of the immuno system. These include polyclonal B cell activation, increase of tumoricidal capability of macrophages, and action on some T cells resulting in impairment of the normal suppressor regulation of immune responses. Studies with in vivo models indicate that polyenes potentiate the uptake of antitumor agents into tumor cells as well as stimulate host-mediated immunomechanism. It is likely that these exciting results will put some polyene antibiotics among the clinically useful antitumo agents, as is already indicated by the conducted phase one studies.

Terpenoids were treated in Chapter 6, Volume II, according to chemical classes in an excellent and comprehensive review by Drs. Misra and Pandey. While there are very few among these terpenoids considered for clinical use, the well tabulated structure activity relations in this chapter clearly indicate possible future research directions.

We hope that the above paragraphs will convince the reader to continue reading the following chapters and that through these chapters we will have contributed to his ideas, experimental designs, and scientific dreams. These dreams are instrumental in developing better cancer chemotherapeutic agents at least as much as are scientific rationales. Finally I would like to thank all the contributors whose dedication and excellent work in cancer chemotherapy made this book possible.

Adorjan Aszalos

THE EDITOR

Adorjan Aszalos, Ph.D., recently joined the Bureau of Drugs, Food and Drug Administration, in Washington, D.C. Previously, he was the Head of Biochemistry, Chemotherapy, Frederick Cancer Research Center, Frederick, Maryland. He also held positions at the Squibb Institute for Medical Research and at Princeton University.

Dr. Aszalos graduated from Technical University of Budapest, Hungary with B.S. and M.S. degrees in chemical engineering and biochemistry. He received his Ph.D. in bioorganic chemistry from Technical University of Vienna, Austria in 1961. Subsequently, he was Post-Doctoral Fellow at Rutgers University.

Dr. Aszalos is a member of the American Chemical Society, Interscience Foundation, and New York Academy of Sciences. In the latter society, he served as Vice Chairman of the Biophysics Section in 1973 to 1975. Dr. Aszalos received, among other awards the Austrian Industrial Research Award and the Army Post-Doctoral Research Award.

Dr. Aszalos has presented over 30 lectures at National and International meetings and published over 80 research papers and several review articles and chapters. His current major interest is antibiotics and enzymes in chemotherapy.

CONTRIBUTORS

Creed W. Abell, Ph.D.
Professor and Director of Biochemistry
Department of Human Biological Chemistry and Genetics
University of Texas Medical Branch
Galveston, Texas

Janos Bérdy, Ph.D.
Senior Research Fellow
Antibiotic Section, Institute of Drug Research
Budapest, Hungary

Janina Brajtburg, Ph.D.
Research Assistant Professor
Division of Infectious Diseases
Washington University School of Medicine
St. Louis, Missouri

Klaus Eckardt, Dr. Sc.
Director of Laboratory
Department of Chemistry and Antibiotics
Central Institute for Microbiology and Experimental Therapy
East Germany

Charles A. Frolik
Laboratory of Chemoprevention
National Cancer Institute
Bethesda, Maryland

Andrew F. Hegyeli, D.V.M., Ph.D.
Program Director, Carcinogenesis
National Cancer Institute
Bethesda, Maryland

Ruth J. Hegyeli, Ph.D.
Foreign Relations Director
Heart Lung Institute
National Institute of Health
Bethesda, Maryland

George S. Kobayashi, Ph.D.
Professor, Associate Director
Department of Medicine
Washington University
Barnes Hospital
St. Louis, Missouri

Gerald Medoff, M.D.
Professor, Head, Infectious Disease Divison
Departments of Medicine, Microbiology and Immunology
Washington University School of Medicine
St. Louis, Missouri

Renuka Misra, Ph.D.
Associate Professor
Department of Chemistry
University of Toronto
Canada

Rex Montgomery, Ph.D., D.Sc.
Professor
Department of Biochemistry
University of Iowa
Iowa City, Iowa

Ramesh C. Pandey, Ph.D.
Scientist
Frederick Cancer Research Center
Frederick, Maryland

Peter P. Roller, Ph.D.
Research Chemist
Analytical Chemistry Section
National Cancer Institute
Bethesda, Maryland

V. Sagar Sethi
Research Associate Professor
Hematology/Oncology Section
Bowman Gray School of Medicine
Winston-Salem, North Carolina

Virginia L. Shepherd, Ph.D.
Assistant Research Scientist
Department of Biochemistry
University of Iowa
Iowa City, Iowa

Jerry Daniel Skarbek, Ph.D.
University of Maryland
Baltimore, Maryland

Marilyn K. Speedie, Ph.D.
Assistant Professor of Pharmacognosy
Department of Medicinal Chemistry and Pharmacognosy
University of Maryland
Baltimore, Maryland

Jack R. Uren, Ph.D.
Scientist
Sidney Farber Cancer Institute
Division of Pharmacology
Boston, Massachusetts

Dale Vandré, Ph.D.
Department of Biochemistry
University of Iowa
Iowa City, Iowa

To
Ildiko, Attila, and Rita

TABLE OF CONTENTS

Volume I

Volume II

Chapter 1

MARINE ANTITUMOR AGENTS

Andrew F. Hegyeli and Ruth Johnsson Hegyeli

TABLE OF CONTENTS

I. INTRODUCTION

Cancer existed on this planet long before modern civilization and industrialization. The earliest evidence is found in animals thought to have lived some 50 million years ago. For instance, tumors have been identified in the remains of two Cretaceous dinosaurs and one Pleistocene cave bear.[1,2] Beyond establishing the fact that cancer goes back to prehistoric times, these isolated findings offer little information about its prevalence in these populations of animals. Cancer is predominantly a disease of old age. In the wild, very few animals survive to old age. Even then, most become preys for stronger animals, thus eliminating any evidence. For these reasons, even today, cancer is rarely detected in free-living animals. The assessment of the extent of cancer in prehistoric times is further limited by the fact that even in animals who did survive to old age, few remains have been preserved to our time, and most of these remains are confined to bones. In view of these odds, it is surprising that any evidence of cancer in prehistoric animals has been uncovered at all. The earliest reported diagnoses of human malignancies are three cases of osteogenic sarcoma found among the thousands of mummies preserved from the Third (around 2800 B.C.) and Fourth Dynasties in Egypt.[3,4] Many subsequent accounts in humans followed.

The earliest known records of attempts to control tumor growth stem from the same time period as the first records of definite diagnoses of cancer in humans. Marine antitumor agents were some of the first to be attempted. For example, a Chinese physician who lived about 2700 B.C. is reported to have experimented with about 100 drugs a day, using 3000 condemned criminals as research subjects. On the basis of his experimental results, he introduced the first recorded chemotherapy for tumors — the use of seaweeds in the treatment of swellings and tumors in all parts of the body, particularly of the neck glands. It is now known that seaweeds contain iodine, among other chemicals. Hence, the Chinese were using iodine to successfully treat tumors of the thyroid gland (now called goiters) some 4700 years ago; this treatment is still being used.[5] A second marine antitumor treatment also used for thousands of years in traditional Chinese medicine is clams[6,7]

Although there are many records from antiquity and from the Middle Ages of a variety of agents being used in the treatment of tumors, a systematic search for antitumor agents did not get started until the beginning of this century, when the founder of modern chemotherapy, Paul Ehrlich, turned his attention to this problem.[8] In this modern search for antitumor substances, a number of investigators are again focusing their search on the sea, the origin of life on this planet. The sea, with its millions of species, represents the richest source of animals and plants. This vast potential source of antitumor agents remains largely unexplored. Nevertheless, in recent years a number of potential therapeutic agents have been isolated from the marine fauna and flora. To date, only one of these investigations has progressed to the stage of successful synthesis of a drug actually used in cancer treatment. An arabinosyl nucleoside isolated from the Caribbean sponge *Crypotethya crypta* by Bergmann and co-workers[9-13] is now utilized in the treatment of myeloid leukemia.[14-17] A symposium held in 1960 on the biochemistry and pharmacology of compounds derived from marine organisms gave new emphasis to the search for marine antitumor agents.[18] Four subsequent symposia on food and drugs from the sea[19] and Gordon Research Conferences devoted to marine natural products reflect the renewed scientific interest in this area of research.

Most marine antitumor agents reported in the literature are crude extracts. However, in recent years the systematic search for new cytotoxic substances from the sea[20-22] has resulted in the isolation and chemical identification of a number of new compounds in pure form. These substances show a wide range of chemical structures and molecular weights. In most cases, the evaluation of the anticancer potential of crude extracts from different sea organisms has been carried out by in vitro cytotoxicity tests in malignant cell cultures. These approaches have resulted in the isolation of many toxic substances in pure form, since the sea is very rich in toxic and venomous animals and plants.[23,24]

These new compounds need further study in order to confirm their potential antitumor activity. The differential action of these agents on malignant cells needs to be determined in parallel in vitro investigations using malignant as well as normal cells. Also, in vivo animal tests need to be carried out to carefully evaluate systemic toxicity and antineoplastic activity. A review of the literature reveals that in many cases, research on new "antitumor" compounds of marine origin has been published without any reference to the determination of the biological activity. Often, these new compounds have been derived from biologically active crude extracts without subsequent monitoring. This should be accomplished through purification guided by in vitro or in vivo toxicity or antineoplastic systems to verify retention of the antitumor activity of the final purified fractions. Thus, detailed biological studies of these compounds are needed before any consideration can be given as to their potential usefulness in cancer chemotherapy. Progress in isolating new compounds from the sea in pure form is presented by Scheuer,[25,26] Ruggieri,[27] and Faulkner and Anderson.[28] They describe the structures of many compounds isolated from marine organisms. Baker and Murphy[29] list 504 compounds isolated during the period from June 1969 to the end of 1973 alone.

In this chapter, recent information on sources, isolation, biological tests, and chemical characterization of potential antitumor agents have been approached in a holistic manner, relating each to the other in a systematic way. This will make it possible for interested investigators to get an appreciation for the depth of inquiry in a given area and the challenge and problems that remain.

In considering approaches to the study of potential anticancer agents from marine fauna and flora, it is important to keep in mind that the mechanism of action of some of these substances may not always be a direct cytotoxic effect discernable in in vitro systems. Rather, the antitumor effect may be an indirect action, mediated through the host's self-defense mechanisms and only be assessable through in vivo test systems. The hypothetical "immunologic surveillance system" and/or the nonspecific protective systems in humans may be important in preventing 80% of the world population from developing cancer, while exposed to the same environmental and occupational carcinogenic hazards that contribute to the other 20% developing malignant disease. Marine invertebrates may be a particularly good choice for the study of these self-defense mechanisms for several reasons. In the first place, marine invertebrates and lower vertebrates very rarely have neoplasms.[30-32] Secondly, invertebrates do not produce antibodies and their defense mechanisms are based on phagocytosis by leucocytes, aided by a low molecular weight, nonproteinaceous substance.[33,34] These low molecular weight substances may have a high level of cytospecificity for malignant as opposed to normal cells. Thus, they may provide models for the synthesis of novel anticancer compounds. Thirdly, these organisms would be an easily accessible source for producing large quantitites of substances for test and evaluation, should one or more active agents be identified from this source. This subject will be addressed more fully later in this chapter.

In preparing this review, an attempt has been made to provide an extensive up-to-date coverage of all available international information on antitumor agents from the sea. First, the computerized data bases were searched to locate all recent primary sources. *Toxline, Cancerline, Medline,* Chemical Biological Abstracts and reference articles and books were consulted. *Cancerproj,* the off-line bibliographical citation lists of *Medlar II,* provided information of 53 ongoing programs under the general heading of "marine-derived antitumor agents". The principal investigators of these programs were then contacted for bibliographies of their pertinent publications. The authors of this chapter would like to express their thanks especially to Drs. Angeles, Erickson, Norton, Pettit, Prendergast, Rugieri, Ryoyama, Rinehart, Schmeer, Schmitz, Shimizu, van der Helm, and Weinheimer for providing reprints and preprints of publications on their ongoing research on antineoplastic compounds of marine origin.

II. CYTOTOXIC ANTICANCER SUBSTANCES FROM THE SEA

Isolation of cytotoxic antitumor substances from marine organisms has been reported in many books and review articles. The historical review of *Pharmacologically Active Agents from the Sea* by Emerson and Taft in 1945[35] was followed by reviews by Nigrelli,[36,37] Chapman,[38] Jackson,[39] Schwimmer and Schwimmer,[40] Russell,[41] Lewin,[42] Kreig,[43] Burkholder,[44] Freudenthal,[45] Gullion,[46] Harshbarger,[32,33] Baslow,[47,48] Li,[6] Ruggieri,[27] Boolootian,[49] and Pettit.[20-22] These reviews, as well as articles by Nigrelli,[18] Li,[50] Schmeer,[51] Hegyeli,[52] Norton,[53,54] Quinn,[55] and many other authors, have generated interest in the field of anticancer substances from the sea. As a result, many new investigators have been attracted to this field of research. Progress has been greatly accelerated, particularly in recent years, and hundreds of potential antitumor agents of marine origin have been isolated and are currently under study throughout the world.

The approaches used by the investigators for determining the biological activity of the extracts are many and varied. The National Cancer Institute (NCI) has developed a number of test systems which have been refined over time. NCI protocols, incorporating the latest approaches for screening natural products, were published in 1972.[56] In these new protocols, the number of in vivo tests has been reduced to five, compared to 24 tests in earlier protocols. The five systems presently recommended by NCI for in vivo tests are the L-1210 (LE) lymphoid and P388 (PS) lymphocytic mouse leukemias; B16 mouse melanoma; mouse Lewis lung carcinoma (LL); and rat Walker 256 carcinosarcoma (W256). The in vitro test recommended for routine screening is the KB culture line derived from human epidermoid carcinoma of the nasopharynx. According to the natural product flow chart of NCI's testing program, the crude extracts are first tested in the PS system, using BDF_1 mice. The results are expressed as a percentage of the control survival time, and the natural product is considered "active" if the T/C* (test vs. control) percentage ratio is equal to or greater than 130, and if the two lowest active dose levels are nontoxic. The purification and isolation of the crude extract are then usually monitored by in vitro tests. According to the 1972 NCI protocol, a natural product reaches a confirmed level of activity in the KB cell culture test if the concentration of the test material which induces 50% growth retardation (ED_{50}), is equal to or lower than 20 μg/mℓ culture medium.

The systematic evaluation of extracts of marine invertebrates, vertebrates, and sea flora by Pettit and co-workers at the University of Arizona,[20] Weinheimer and co-workers at the University of Oklahoma,[57] Norton and Moore and co-workers at the University of Hawaii,[53-55] and many others has generated more than 10,000 extracts from marine organisms for tests in NCI's screening program. About 10% of these

* $$T/C\% = \frac{\text{median survival time of test animals}}{\text{median survival time of control animals}} \times 100$$

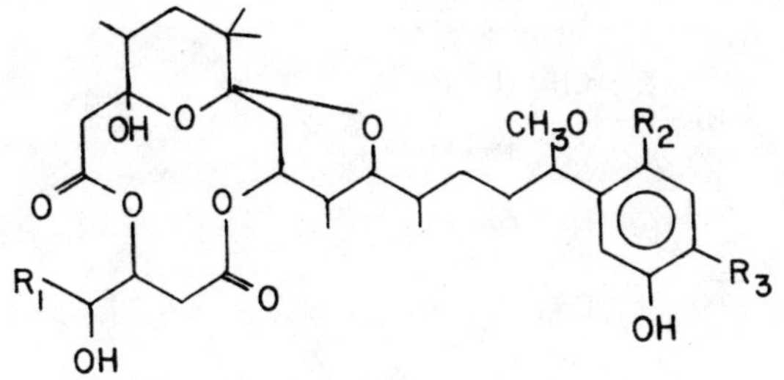

I $R_1 = \overset{31}{CH_3}, R_2 = R_3 = H$

II $R_1 = R_2 = R_3 = H$

III $R_1 = R_3 = H, \quad R_2 = Br$

IV $R_1 = H, R_2 = R_3 = Br$

V $R_1 = \overset{31}{CH_3}, R_2 = R_3 = Br$

FIGURE 1. Dibromoaplysiatoxin (I), oscillatoxin-A (II), 21-bromooscillatoxin (III), 19,21-dibromooscillatoxin (IV), and 19-bromooscillatoxin (V).

extracts were found active in one or more of the NCI test systems. The organisms used by the investigators for the preparation of the antitumor extracts belong to different phyla of the plant and animal kingdoms. In order to facilitate orientation and future reference by scientists interested in gaining systematic and reliable information regarding this area of research, the marine substances discussed in this chapter are presented according to the taxonomic classification of the organism, which is the source of the agent. The authors hope that this methodologic approach will be a helpful benchmark for the future.

A. Marine Flora

1. Phyla of Toxic Marine Algae

Most toxic algae belong to three phyla: Dinoflagellates, phytoflagellates and blue-green algae. These organisms vary in size from minute unicellular forms of a few microns in diameter, to large seaweeds reaching many meters in length. Some of these algae are toxic enough to be lethal to livestock, fish, waterfowl, and humans.[58] Currently, there are relatively few scientists working in this area of research. However, some promising leads are beginning to emerge.

Shiomi and co-workers isolated a glycoprotein from the red alga *Agardhiella tenera*.[59] This compound agglutinates L5178Y mouse leukemia cells but is inactive against L-1210 cells. Its major active component is a glycoprotein with 2.7% glucose content. An acetylated glucan was isolated by Shibata and co-workers[60] from the lichen *Glyrophora esculenta*. This agent is active against S-180 tumors in mice.[61,62] Hechendorf and Shimizo identified a similar polysaccharide from another lichen, *Umbilicaria mammulata*.[63] Myderse and co-workers studied a variety of antitumor substances from toxic marine algae. Figure 1 shows the chemical structure of five of these substances:

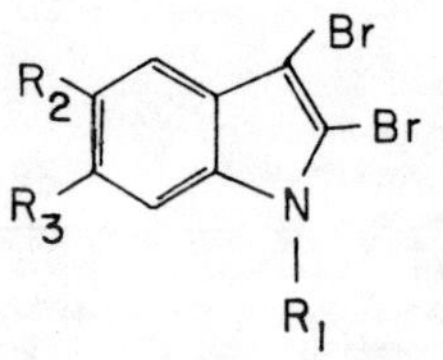

I	R_1	R_2	R_3
II	CH_3	H	Br
III	CH_3	Br	H
IV	H	Br	Br
V	CH_3	Br	Br

FIGURE 2. Molecular structures of brominated indols.

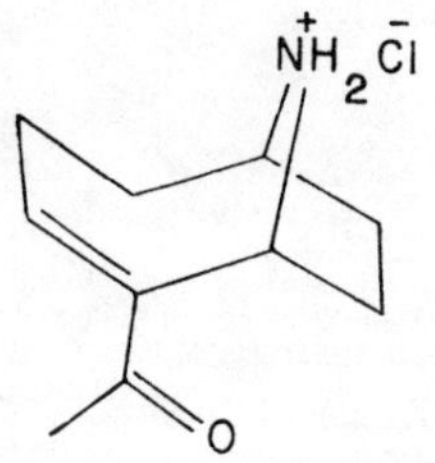

FIGURE 3. Anatoxin-A hydrochloride.

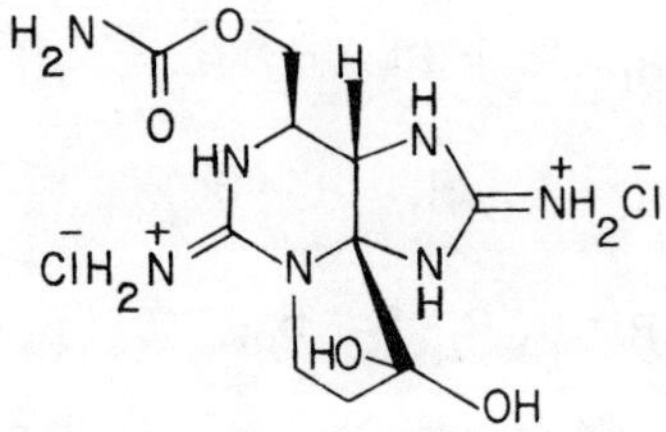

FIGURE 4. Saxitoxin dihydrochloride.

Dibromoaplysiatoxin(I) from the blue-green alga *Lyngbya gracilis*[64] and oxcillatoxin-A(II), 21-bromooscillatoxin(III), 19, 21-dibromooscillatoxin(IV) and 19-bromo-aplysiatoxin(V) from a mixture of *Oscillatoria nigroviridis* and *Schizotrix calciola.*[65] The anticancer activity and toxicity of dibromoaplysiatoxin is comparable to that of oscillatoxin-A.

Dibromoaplysiatoxin has a T/C value of 186% at a dose of 1.8 mg per mouse. However, it is dermatonecrotic, and it is believed that this toxin is responsible for the sporadic outbreaks of contact dermatitis among swimmers in Hawaiian waters. This toxin was first isolated from the digestive tract of the sea hare, *Stylocheilus longicauda*[66] which feeds on algae. Kato and Scheuer[66,67] elucidated the chemical structure of dibromoaplysiatoxin (Figure 1).

Another group of chemicals, brominated indols, were isolated from the alga *Laurencia brongniartii* by Rinehart and co-workers.[68] Figure 2 shows the chemical structure of the four different brominated compounds. Of the four compounds, only III was active against L-1210 tumor cells in tissue culture with an ED_{50} value of 3.6 $\mu g/m\ell$.

An alkaloid, anatoxin-A was discovered in the algal bloom of *Anabaena flos-aquae* by Starvic and Gorham[69] and the detailed chemical structure of this alkaloid was established by Devlin and associates.[70] Figure 3 shows the structure of anatoxin-A. Still another type of compound, saxitoxin, was extracted from the bloom of the blue-green alga *Aphanizomenon flos-aquae* by Sawyers and associates.[71] Saxitoxin was previously found in toxic mussels and clams by Schantz and co-workers.[72] The structure of saxitoxin dihydrochloride is illustrated in Figure 4.

In addition, Erickson and his associates are working on the isolation of a variety of antitumor agents from toxic marine algae. To date, they have isolated a number of halogenated sesquiterpenoids,[73] polyhalogenated indols,[74] and 15-halogenated compounds[75] from the marine alga *Laurencia nidifica.* However, to date none of these compounds have shown significant antitumor activity in the currently used biological test systems.

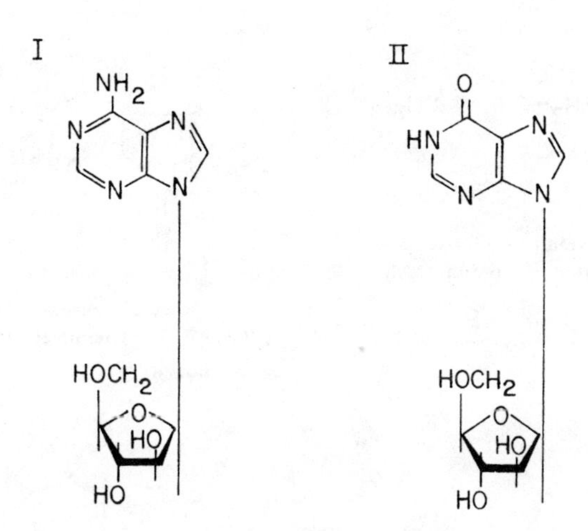

FIGURE 5. Chemical structure of (I) 9-β-D-arabinofuranosyladenine(ara-A), and (II) 9β-D-arabinofuranosylhypoxanthine(ara-Hx).

B. Marine Fauna

Several factors make the marine fauna especially attractive for anticancer studies. First, there is an abundance of marine animals. Secondly, as already mentioned, cancer is relatively rare in marine invertebrates and lower vertebrates. For example, very few of the 200 or more species of sharks have been found to have cancer.[22]

In 1965, in collaboration with the National Cancer Institute, several research groups began a systematic and world-wide search for anticancer substances from marine animals.[76] As a result, many new agents have been found. Most are crude or purified extracts and the specific components responsible for the anticancer activity still remain to be identified. Nevertheless, a number of substances have been isolated in pure form. In some cases, the chemical structure and even the absolute configuration of the molecule have been determined. The aspect of these investigations which is the most neglected to date is the biological activity. In most cases, even where the chemical information is quite sophisticated, biological test data on these chemicals are still rudimentary. Usually, it is limited to in vitro toxicity data in malignant cell cultures, or the substances have not yet been tested.

1. Phylum Porifera (Sponges)

There are more than a million species of marine invertebrates. Among these animals, the sponges are the most primitive multicellular organisms. As was already mentioned, the arabinosyl nucleosides isolated from the Caribbean sponge *Cryptotethya crypta* by Bergman and co-workers[9-13] led to the synthesis of adenine arabinoside(ara-A) and cytosine arabinoside (ara-C).[77-79] These chemicals have antiviral and anticancer activity with broad spectrum activity against DNA and RNA tumor viruses.[79] Ara-A is currently used with some success in the treatment of one type of cancer, myeloid leukemia.

Recent studies indicate that ara-A selectively inhibits viral DNA synthesis and that the 5′-triphosphate derivative of ara-A can inhibit the mammalian cell DNA polymerase and ribonuclease enzyme activity. Ara-A is rapidly converted in animals, bacteria, and mammalian cell cultures to a less active degradation product, 9-D-arabinofuranosylhypoxanthine(ara-Hx). Inhibitors of adenosine deaminase effectively block the deamination of ara-A to ara-H and may therefore enhance the biological activity of this anticancer drug. Figure 5 shows the structures of ara-A(I) and ara-Hx(II).

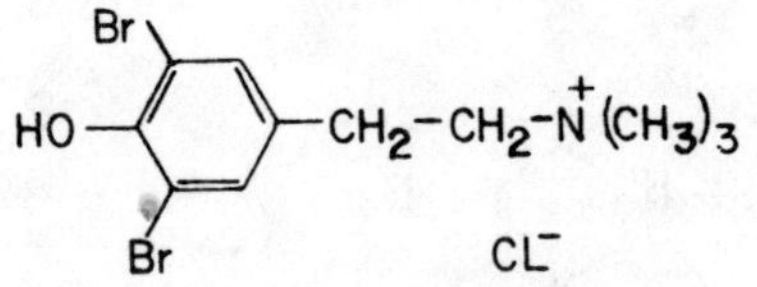

FIGURE 6. [2(3,5-Dibromo-4-hydroxyphenyl) ethyltrimethylammonium chloride].

R = H

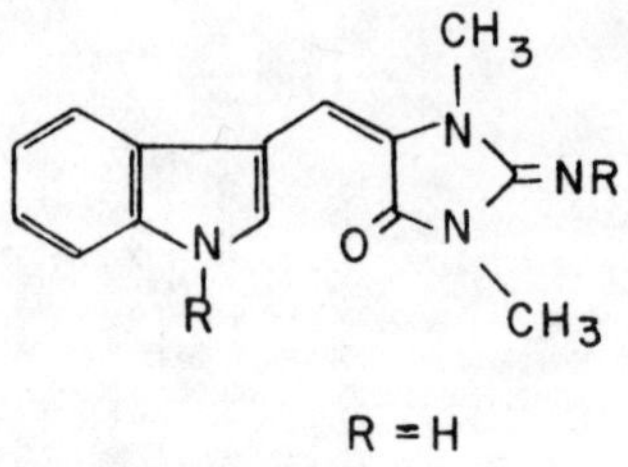

FIGURE 7. Chemical structure of aplysi nopsin.

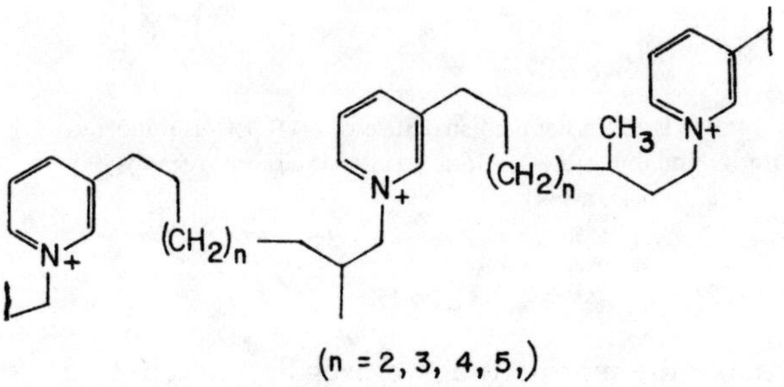

(n = 2, 3, 4, 5,)

FIGURE 8. Halitoxin.

In recent years, several new compounds have been identified in sponges. Initially, Burkholder reported that saline suspension of the sponges *Speciospongia vesparina, Ianthella ardis, Cinachyra cavernosa* and *Verongia fistularis* injected i.p. in mice, inoculated with Walker M and Walker 256 transplantable tumors, inhibited the growth of these tumors.[81] Baslow extracted halitoxin from *Haliclina viridis.* This toxic substance reduced the mortality of BALB/c mice inoculated with Ehrlich ascites tumor cells by 30%.[82] Sigel tested crude extracts from *Chondrilla nucula* in lymphatic leukemia-bearing BDF_1 mice. The extracts prolonged the life of the treated animals by 23 to 40% as compared with the untreated controls.[83]

Purification studies by Hollenbeak and co-workers led to the isolation of [2(3,5 dibromo-4-hydroxyphenyl)-ethyltrimethylammonium chloride] from the sponge *V. fistularis.*[81] This substance gave an ED_{50} value of 20 μg/mℓ in the PS cytotoxicity test and also exhibited adrenergic activity.[84] The structure of the molecule is shown in Figure 6.

Hollenbeak and Schmitz also reported that crude extracts of the marine sponge *V. spongelii* gave a T/C value of 135% in 200 mg/kg dose in the murine PS bioassay. Purification monitored by the KB cytotoxicity test led to the isolation of aplysinopsin, a tryptophan derivative.[85] Figure 7 shows the structure of the aplysinopsin molecule. The ED_{50} values for this agent in the KB, PS, and LE cytotoxicity tests were 0.87, 3.8 and 3.7 μg/mℓ, respectively. Schmitz and co-workers isolated halitoxin, a complex mixture of high molecular weight toxic pyridinium salts from the sponges *Haliclina viridis, rubens* and *erina.*[86] Halitoxin is cytotoxic (ED_{50} of 2.8 μg/mℓ in KB cell culture) and is hemolytic and toxic to fish and mice. The general structure suggested by the investigators for halitoxin is depicted in Figure 8.

Gopichand and Schmitz isolated several dibromotyrosine-based compounds from the sponge *Aplysina fistularis* forma fulva. Figure 9 shows the chemical structure of these compounds. Aerothionin(Ia), oxoaerothionin(IIa), and hydroxyaerothionin(IIIa)

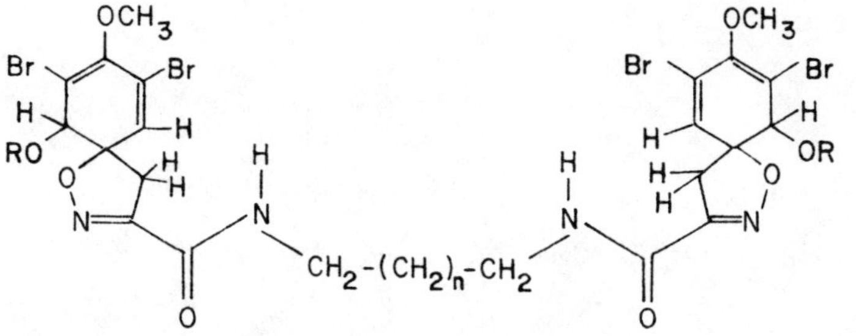

Ia n=2 R=H
b n=2 R=$COCH_3$
IIa R=H
b R=$COCH_3$
IIIa n=1 R=H
b n=1 R=$COCH_3$

FIGURE 9. Aerothionin(Ia), oxoaerothionin(IIa), and hydroxyaerothionin(IIIa).

were all inactive in the NCI standard in vitro test system,[87] but the acetylated derivatives (Ib, IIb, IIIb) displayed marginal cytotoxicity in doses of 2·3 to 33 μg/mℓ.

Gopichand and Schmitz also reported the isolation of two unusual types of brominated metabolites from sponges of the genus *Aplysina*.[88] Figure 10 shows the structure of these compounds. Fistularin-1 (I) gave only marginal cytotoxicity (ED_{50} values of 4.1, 4.1, and 1.3 μg/mℓ in the PS, KB and LE tests, respectively).[89]

Crude extracts from the sponge *Xestospongia muta* showed confirmed in vivo anticancer activity.[90] Schmitz and Gopichand isolated a dibrominated straight-chain C_{16} acetylenic acid from this extract.[91] Figure 11 shows the structure of the unstable oily acid (I) and the esterified, more stable derivative (II). The acid gave an ED_{50} of 24 and 34 μg/mℓ doses in PS and LE cell systems, respectively, and the effective doses for 50% inhibition of cell growth in the same systems for the ester were 29 and 24 μg/mℓ, respectively. However, neither compound was active in the PS in vivo system in mice.

Since sponges provided the first clinically applicable anticancer drug of marine origin, many investigators are actively exploring these animals for anticancer compounds. Further developments can be expected in this research area in the forseeable future.

2. Phylum Cnidaria (Coelenterata) (Corals, Anemones, Jellyfishes)

Early scientists thought that the cnidarians were plants. They are still called the flowers of the sea. All have a centrally located mouth, surrounded by tentacles. Two forms exist, the polyp and the medusa.

Tabrah and co-workers reported that crude extracts of the soft corals *Cephea conifera, Aurelia labiata, Stochiatis sp. (Tahiti) and Nephthea* sp. (Australia) completely inhibit the growth of Ehrlich ascites tumors in mice.[92] Schmitz and co-workers isolated two new cembrane derivatives, nephthenol and epoxynephthenol acetate, from the soft coral *Nephthea* species.[93] Vanderah and co-workers reported the isolation of two new cembrane hydrocarbons,[94] cembrane-A and cembrane-C, from the *Nepthea* species. Cembrane-A gave ED_{50} values in in vitro PS, LE, and KB systems in doses of 0.22, 0.31, and 26 μg/mℓ, respectively. Figure 12 shows the chemical structure of these compounds.

Vanderah and co-workers isolated a new diterpenoid, xenicin, from another soft coral, *Xenia elongata*. The absolute configuration has been determined by single crystal X-ray diffraction but biological test data are not yet available. Figure 13 shows the structure of xenicin.[95]

Weinheimer and co-workers demonstrated that aqueous alcohol extracts of the soft coral, *Sinularis flexibilis*, has confirmed antineoplastic activity in the PS system in mice.[96] Purification of the crude extract, monitored by in vitro PS and KB tests led to the isolation of two new cytotoxic cembranolids,[97] sinularin (I) and dihydrosinularin (II). A third compound, sinulariolide (III) isolated from an alcoholic extract of *S. flexibilis* was reported earlier by Tursch and co-workers.[98] Sinularin shows a structural

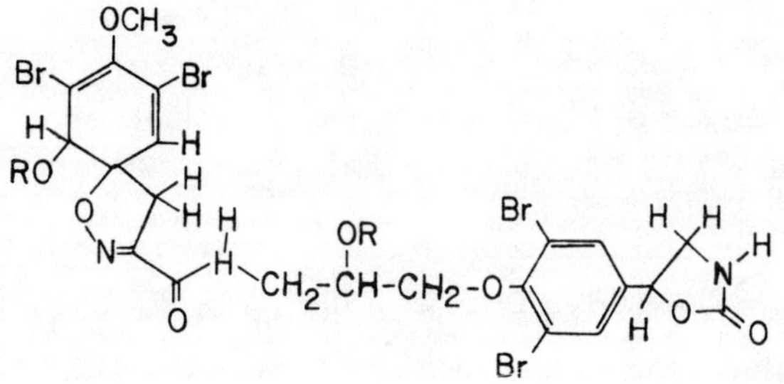

I R = H

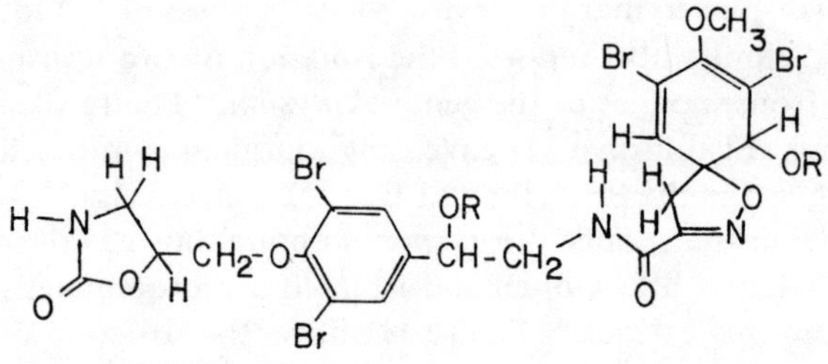

II R = H

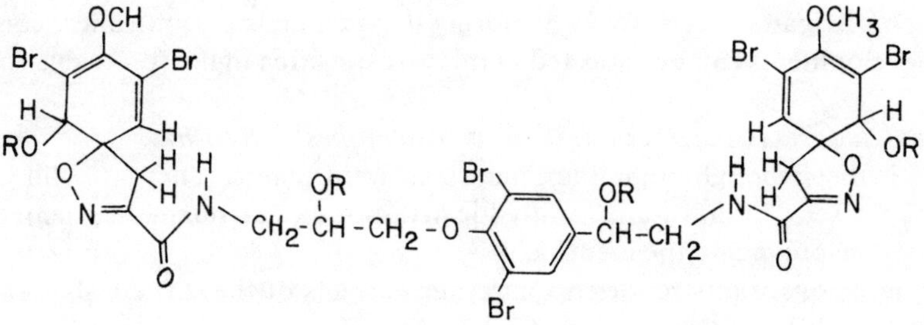

III R = H

FIGURE 10. Fistularin-I, fistularin-II, and fistularin-III.

$$BrCH=CH-CBr=CH-(CH_2)_4-CH=CH-C\equiv C-(CH_2)_3-R$$

I R = CO_2H

II R = CO_2CH_3

FIGURE 11. Dibrominated straight-chain C_{16} acetylenic acid I and its methyl esters II.

and stereochemical similarity to crassin acetate, an antineoplastic cembranolids isolated by Hossain and co-workers from the *Pseudoplexura genus* of *Caribbean gorgonians*,[99] also a Coelenterata. Figure 14 shows the structure of the three cembranolides.

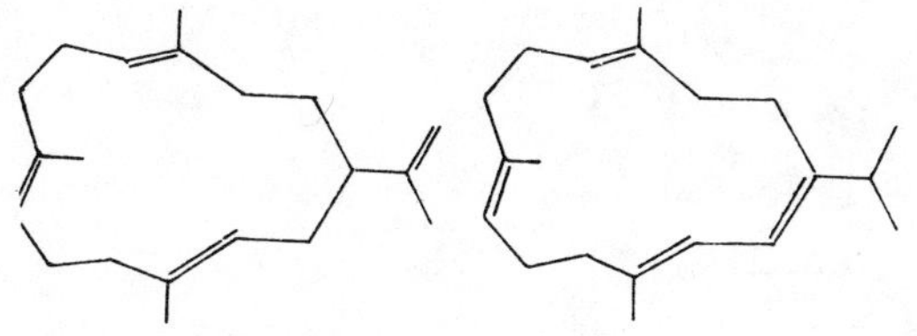

FIGURE 12. Cembrane-A (I) and Cembrane-C (II).

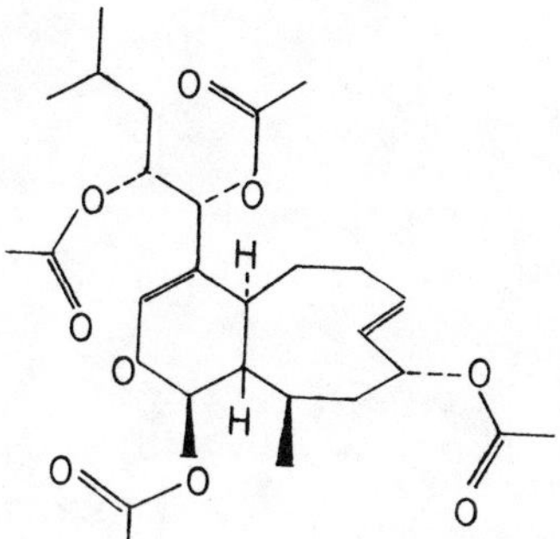

FIGURE 13. Structure of xenicin.

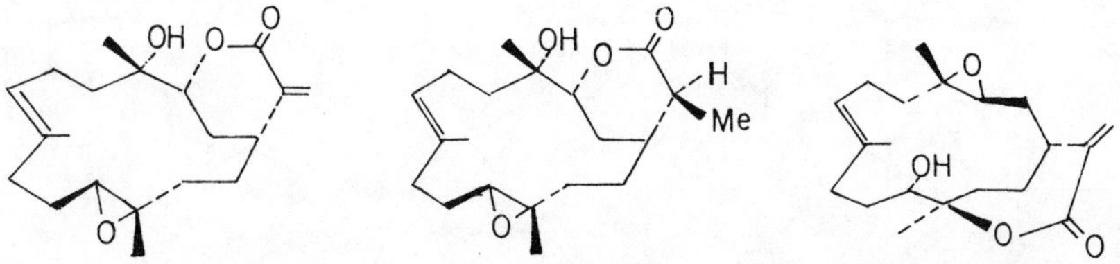

FIGURE 14. Sinularin (I), dihydrosinularin (II), and sinulariolide (III).

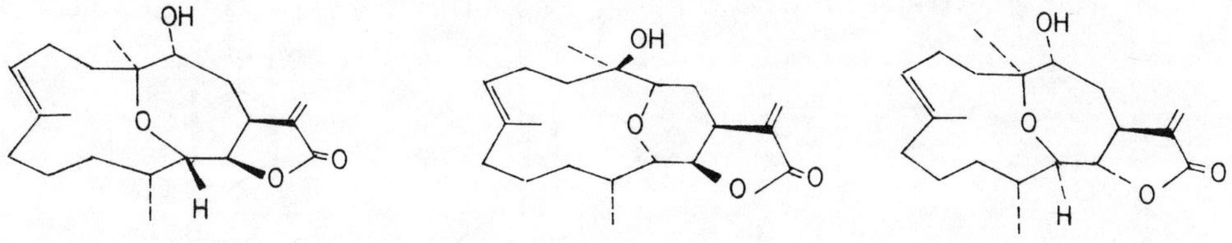

FIGURE 15. Chemical structure of jeunicin (I), eunicin (II) and 13,14-bis-epijeunicin (III).

The effective doses for 50% inhibition by sinularin, dihydrosinularin, and sinulariolide in the KB system are 0.3, 15, and 20 μg/mℓ, and in the PS system, 0.1, 1.1, and 7.0 μg/mℓ, respectively.

Crude extracts from another gorgonian, *Eunicea mammosa,* are also active in the PS system in mice. Fractionation and purification of these extracts have led to the isolation of three cembranolides. Two major components, jeunicin (I) and eunicin (II),[100] were identified by van der Helm and co-workers. A third cembranolide, 13, 14 bis-epijeunicin (III) was subsequently discovered by the same research group.[101] The ED_{50} for compound III for 50% retardation of cell growth in the KB and LE systems was 25 and 7.5 μg/mℓ, respectively.[102] Figure 15 shows the structure of these cembranolides.

Secorgosterol was isolated by Enwall and van der Helm from the gorgonian, *Pseudogorgia americana,* and its molecular structure and absolute configuration has been determined, but biological data have not been provided.[103] Figure 16 shows the structure of the 3-p-iodobenzoate-11-acetate derivative of secorgosterol, a member of this

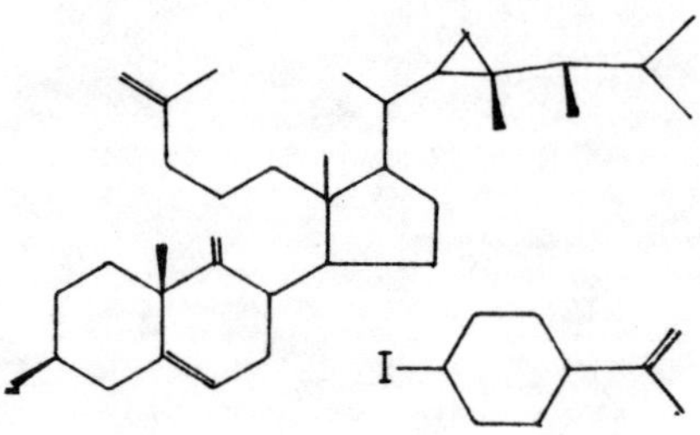

FIGURE 16. Absolute configuration of the 3-p-iodobenzoate-11-acetate of secorgosterol.

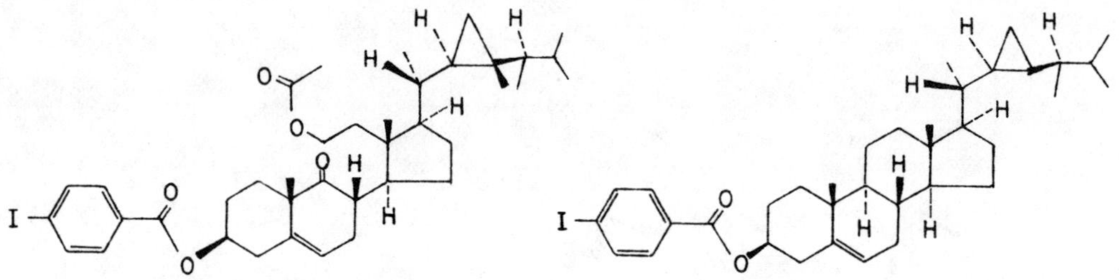

FIGURE 17. 23-dimethylgorgosterol (I) and 9-oxo-9,11-secorgost-5-ene-3β-11-diol 11 acetate (II).

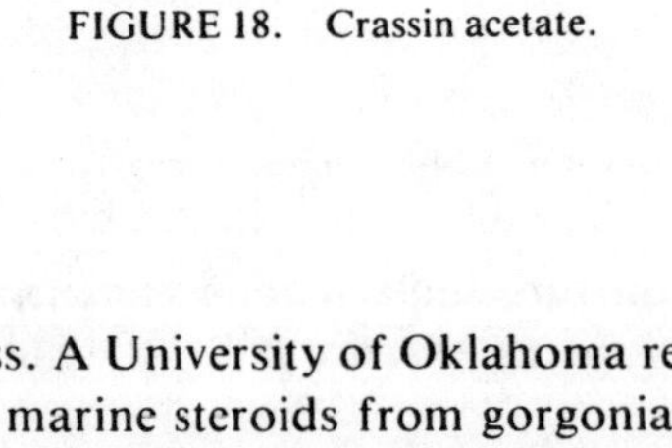

FIGURE 18. Crassin acetate.

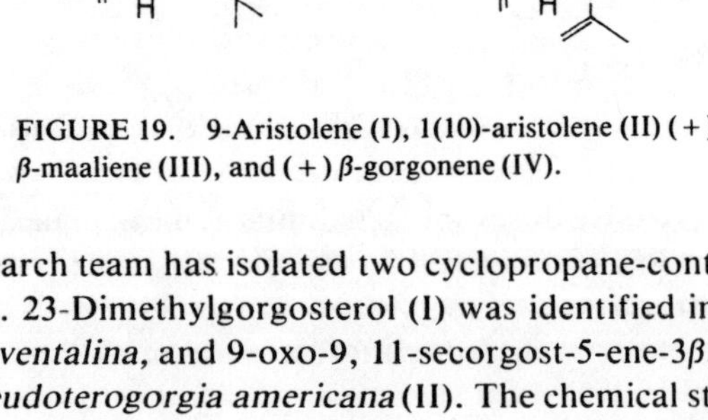

FIGURE 19. 9-Aristolene (I), 1(10)-aristolene (II) (+) β-maaliene (III), and (+) β-gorgonene (IV).

class. A University of Oklahoma research team has isolated two cyclopropane-containing marine steroids from gorgonians. 23-Dimethylgorgosterol (I) was identified in extracts of *Gorgonia flagellum* and *G. ventalina,* and 9-oxo-9, 11-secorgost-5-ene-3β, 11-diol-11-acetate was isolated from *Pseudoterogorgia americana* (II). The chemical structure (Figure 17) and the absolute configuration were determined by single crystal X-ray diffraction studies of their p-iodobenzoate derivatives by Enwall and co-workers.[104]

Weinheimer and Matson[57] reported that crassin acetate, a lactic cembrane diterpene, is the principal antineoplastic agent in *Pseudoplexura prorosa, flagellosa, wagenari* and *crucis.* The isolation was guided by KB and PS in vitro test systems. Crassin acetate gave a T/C of 130% in 50 mg/kg dose and an ED_{50} value of 2.2 μg/mℓ in the in vitro PS test. Figure 18 shows the chemical structure of the crassin acetate molecule.

Weinheimer and Matson also isolated sesquitepene hydrocarbons from *Pseudopterogorgia americana.*[105] The extracts prepared from this gorgonian consisted mainly of 9-aristolene(I), 1(10)-aristolene(II), (+) γ-maaliene(III), and (+)-β-gorgonene(IV).

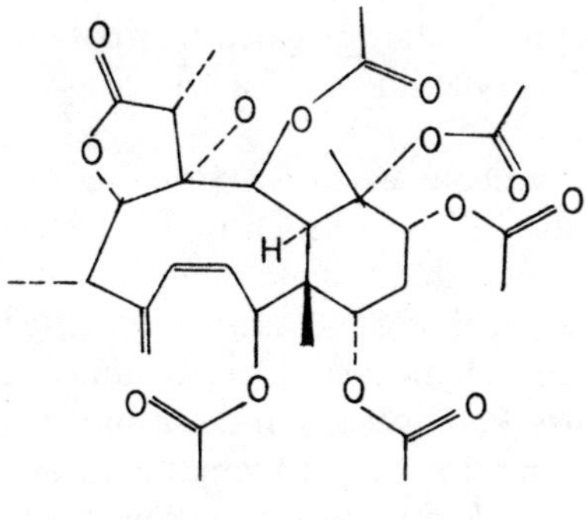

FIGURE 20. Structure of briarein-A acetate.

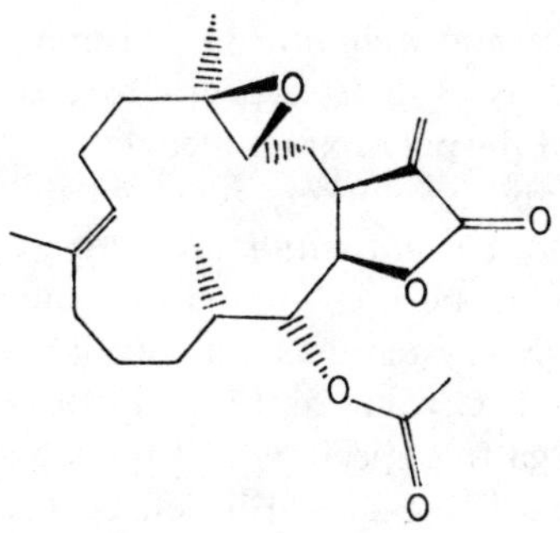

FIGURE 21. Absolute configuration of eupalmerin acetate.

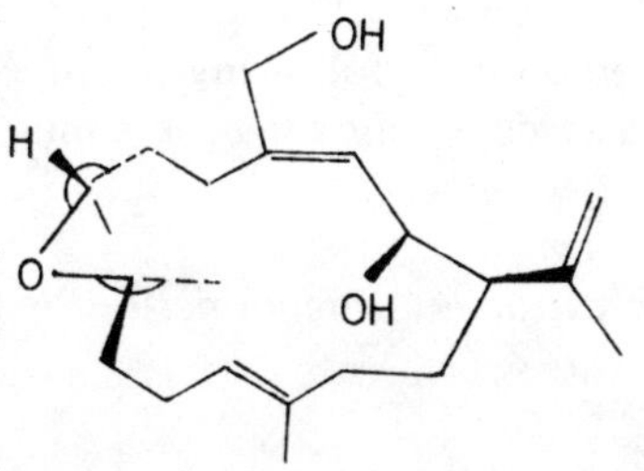

FIGURE 22. Absolute configuration of asperdiol-A.

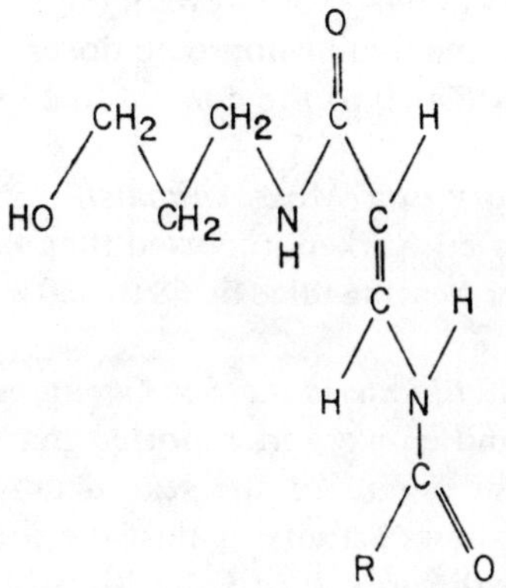

FIGURE 23. Proposed structure for palytoxin

The latter possesses a new isoprenoid skeleton. The graphic representation of the chemical structures of these compounds is shown in Figure 19. Data on the biological activity of these substances are not yet available.

Briarein-A (Figure 20), a chlorine substituted diterpeniod, was isolated from the gorgonian, *Briareum asbestinum,* by Burks and co-workers,[106] but anticancer data were not provided.

Eupalmerin acetate was isolated from the gorgonian, *Eunicea palmeria,* by Rehm.[107] The structure of this compound was determined by van der Helm and co-workers[108] after converting the natural product to the addition product, eupalmerin acetate dibromide, for X-ray studies. Figure 21 shows the structure of the molecule.[109] Biological data were not provided for this agent.

Weinheimer and co-workers isolated Asperdiol-A from the crude aqueous extracts of the gorgonians Eunicea asperula and tourneforti. The crude extract was active in the PS tumor test in mice and the isolation was monitored by in vitro cytotoxicity assay. The pure substance is a nonlactonic cembrane.[110] Figure 22 shows the absolute configuration.[111] The effective doses of asperdiol-A (ED_{50}) in the KB, PS, and LE cell culture systems were 24, 6, and 6 $\mu g/m\ell$, respectively.

Crude alcoholic extracts of the zoanthid, *Palythoa toxica,* exhibited antitumor activity against Ehrlich ascites tumors in mice in doses of 84 mg/kg[112] and showed marginal activity in the PS mice system as well. Moore and co-workers isolated palytoxin, the most toxic substance of marine origin known to date, from the crude extracts of *Pal-*

ythoa toxica and *mammilosa.*[113] Detailed structure studies revealed that palytoxin is a substituted N-(3-hydroxypropyl)-trans-3-amidoacrylamide.[114] Figure 23 shows the structure of the palytoxin molecule.

Norton and co-workers reported that crude aqueous extracts of the sea anemone, *Anthopleura elegantissima,* are active against the PS and Ehrlich tumor systems[115,116] in mice. It has been estimated that the agent responsible for the biological activity in the PS in vivo system has a molecular weight larger than 5000 daltons. Purified material gave a T/C value of 154% in the PS system. Dunn demonstrated that crude extracts of alga free specimens of the sea anemones *Radianthus papillosa* and *A. elegantissima* are active against Ehrlich ascites tumors in mice.[117] The highest level of activity obtained was a T/C of 139% in the PS lymphatic leukemia system in BDF_1 mice.

Tabrah and co-workers[92] found that crude extracts of the jellyfishes *Cephea conifera* and *Aurelia labiata* and of the sea anemone *Stochiatis* sp. Tahiti inhibited the growth of Ehrlich mouse ascites tumor in 100% of the test animals. Purified extracts protected 50% of the animals from developing ascites in doses of 0.22 mg per mouse. Animals treated with maximum nontoxic doses survived 30 days following the inoculation of the tumor while all of the controls died within 21 days following the tumor challenge.

3. Phylum Bryozoa (Moss Animals)

Pettit and co-workers reported that a crude extract of *Bugula neritina* extended the life of tumor-bearing mice by 68 to 100%.[21]

4. Phylum Echinodermata (Sea Cucumbers, Sea Urchins, Sea Stars)

Nigrelli and co-workers reported that holothurin, a steroid saponin extracted from the Cuverian organs of the sea cucumber, *Actinopyga agassizi,* is highly toxic and hemolytic, shows activity against the S-180 and Krebs-2 ascites tumors in Swiss mice, and inhibits the growth of KB cells in culture.[118-123] Pettit and co-workers reported the isolation of lanostane-type saponins from the sea cucumbers, *Stichopus chloronotus,* Brandt, *Thelenota ananas,* Jaeger, and *Actinopyga mauritiana.* The major cytotoxic components from the various types of sea cucumbers were designated stichostatin-1 (PS ED_{50} 2.9 $\mu g/m\ell$), thelenostatin-1 (PS ED_{50} 1.5 $\mu g/m\ell$) and actinostatin-1 (KB ED_{50} 2.6 and LE ED_{50} 2.1 $\mu g/m\ell$). The structure of these saponins were partially characterized.[124] It was suggested by the investigators that stichostatin-1 contains a lanostane nucleus (I) and a holotoxin-A type glycoside system (II), and that thelenostatin-1 and actinostatin-1 are half esters of sulfuric acid, similar to the tentative structure[124] of holothurin A (III). Figure 24[125] illustrates the structure of these saponins. It has been suggested by the investigators that the cytotoxic activity may be expressed through inhibition of protein and RNA synthesis. These saponins gave ED_{50} values in the PS, KB and LE systems ranging from 1.5 to 2.6 $\mu g/m\ell$.

Ryoyama reported that coelomic fluid preparations of the sea urchins, *Anthocidaris crassipina, Pseudocentrotus depressus,* and *Hemicentrotus pulcherrimus* contain a substance which lyses erythrocytes.[126] The hemolytic activity is destroyed by trypsin or 2-mercaptoethanol, indicating that the agent responsible for this activity may be a protein or protein-like substance.[127] The coelomic fluid is also active against Yoshida sarcoma cells in vitro.[128] However, the Thio-TEPA-resistant subline of the Yoshida sarcoma (YS-Thio-TEPA) is 100-times less sensitive to the coelomic fluid preparation than the original Yoshida sarcoma cells. Angeles found[129] that crude and boiled crude extracts prepared from starfish, *Protoreaster donosus,* and sea urchin, *Diadema setosum,* are potential inhibitors of the dehydrogenase enzyme system in HeLa cell agar plate cultures.[130]

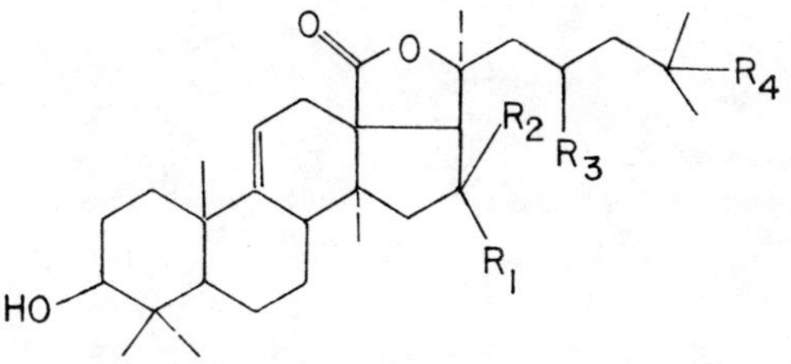

R_1 = OH, R_2 = H, R_3 = $OCOCH_3$, R_4 = H

I

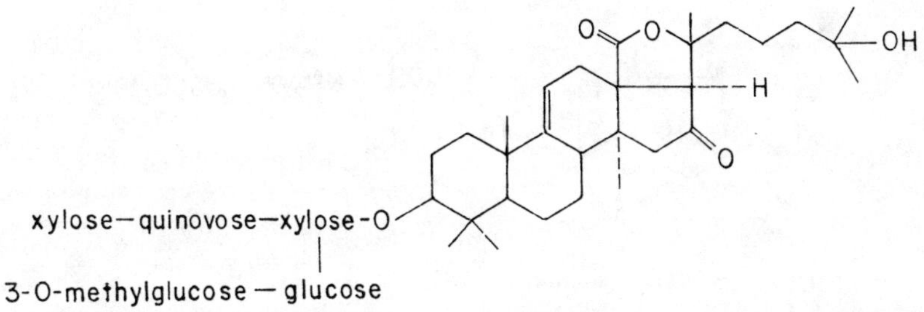

xylose—quinovose—xylose-O
3-O-methylglucose — glucose

II

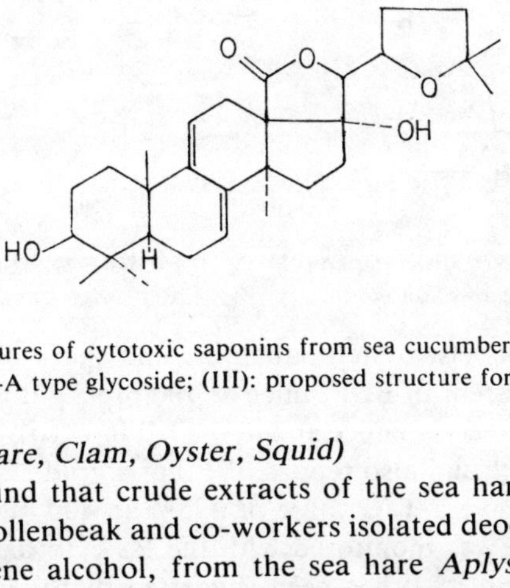

FIGURE 24. Partial structures of cytotoxic saponins from sea cucumbers — (I): lanostane nucleus; (II): holotoxin-A type glycoside; (III): proposed structure for holothurin-A.

5. Phylum Mollusca (Sea Hare, Clam, Oyster, Squid)

Sigel and co-workers found that crude extracts of the sea hare[131] are active in the NCI in vivo test systems. Hollenbeak and co-workers isolated deodactol, a halogenated bisabolene type sesquiterpene alcohol, from the sea hare *Aplysia dactylomela.* This substance shows moderate in vitro toxicity in the LE cell-culture system. The structure and the absolute configuration of the molecule have been determined by X-ray diffraction.[132] Deodactol is isomeric with caespitol and isocaespitol by Gonzales and co-workers,[133-135] suggesting that deodactol might be of algal origin. Deodactol gives an ED_{50} value in the dose of 12 μg/mℓ in the LE system. Figure 25 shows the molecular formula of deodactol.[136] Schmitz and co-workers reported the isolation of 14-bromoobtus-1-en-3, 11-diol from the sea hare *Aplysia dactylamela.* This brominated diterpene diol gave an ED_{50} result in dosages of 4.5 μg/mℓ in KB and 10 μg/mℓ in PS cultures.[137] The chemical structure and absolute configuration were established. Figure 26 shows the structure of this substance.[137]

Pettit and co-workers tested crude extracts of the sea hares *Macrocallista nimbosa* and *Turbo stenogyrus* in the PS system in mice. These extracts were found to be active in 400 mg/kg doses, giving T/C values of 138 and 177% respectively.[138] Purification of this extract resulted in the isolation of taurine as the major cytotoxic component in the crude sea hare extract. Taurine, 2-aminoethanesulfonic acid (Figure 27) is also

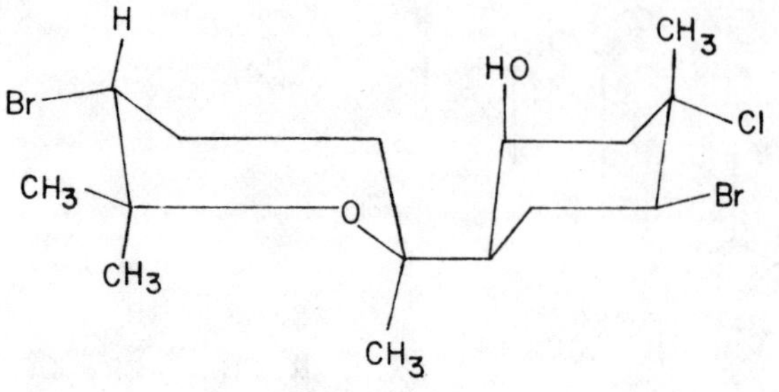

FIGURE 25. Deodactol.

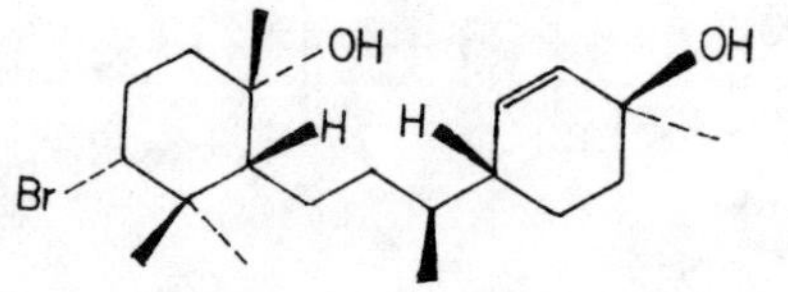

$NH_2CH_2CH_2SO_3H$

FIGURE 27. Taurine.

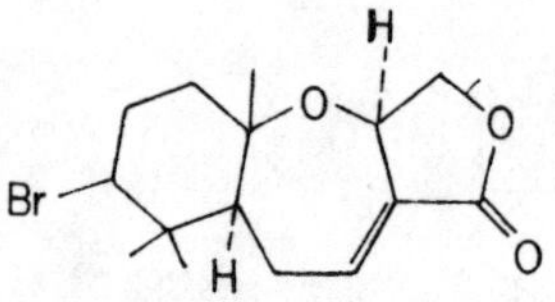

FIGURE 26. 14-Bromoobtus-1-en-3, 11-diol.

FIGURE 28. Structure configuration of aplysistatin.

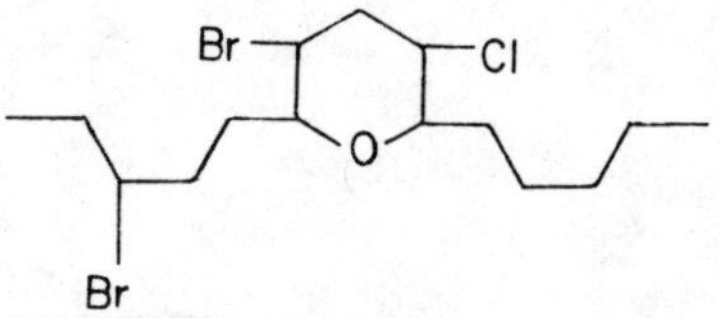

FIGURE 29. Structure of dactylyne.

present in ox bile, mussels, oysters, abalone, and shark blood.[22] It gives a T/C value of 131% in the PS system in BDF_1 mice at 100 mg/kg dose, but is inactive in the LE and the Ehrlich ascites tests in mice and in the KB in vitro cytotoxicity tests.

The same research group also reported[139] that a crude 2-propanol extract of the sea hare *Aplysia angesi* gave a T/C value of 175% in 400 mg/kg dose in the PS in vivo test. The purification was monitored with the PS cytotoxicity method and led to the isolation of aplysistatin.[140] It has been suggested by the investigators that the source of aplysistatin in the sea hare might be the algae, the nutritional source of sea hares. No animal test data have been reported on this substance. Figure 28 shows the structure of the molecule.

Dactylyne, an acetylenic dibromochloro ether was identified by McDonald and co-workers in *Aplysia dactylomela.*[141] Figure 29 illustrates the structure. Biological data were not provided for this compound. Pettit and co-workers also reported the identification of two diterpenes, dolatriol 6-acetate and dolatriol from the extracts of the sea hare *Dolabella auricularia.* The crude extract of this mollusc was active in the PS murine system.[142] Figure 30 illustrates the chemical structure of dolatriol (II) and dolatriol 6-acetate (I) (PS ED_{50} 13 μg/mℓ).

Kato and Scheuer[66,67,143] isolated dibromoaplysiatoxin (I) and aplysiatoxin (II) from the sea hare *Stylocheilus longicauda* and determined the structure of these metabolites (Figures 1 and 31).

Limasset found that extracts prepared from the mussel *Mytilus edulis* inhibit the growth of tobacco mosaic virus.[144] Prescott and co-workers showed that extracts of

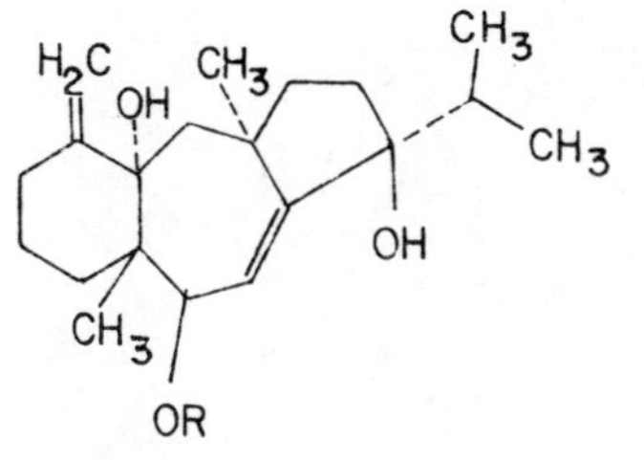

FIGURE 30. Dolatriol-6 acetate(I) and dolatriol (II).

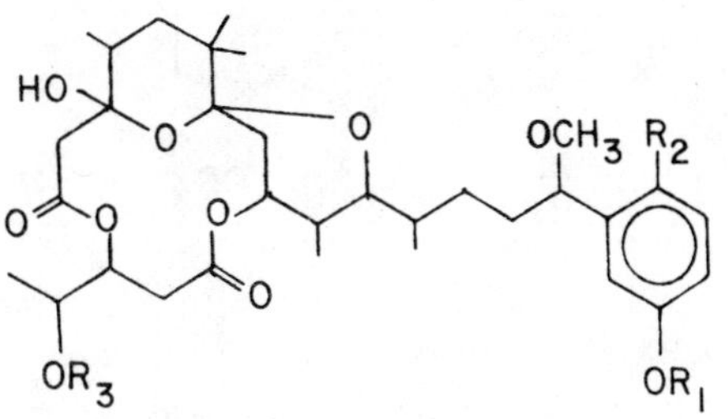

FIGURE 31. Dibromoaplysiatoxin (I) and aplysiatoxin (II).

the abalone, *Halietis rufescens* and the oyster, *Crassostrea virginica*, have antibacterial (Paolin I), as well as antiviral (Paolin II) properties.[145] Schmeer reported that mercenene, extracted from the common clam, *Mercenaria mercenaria*, is active against transplanted sarcoma 180 (S-180) and Krebs (K-2) tumors in swiss mice.[51] Hegyeli found that the antitumor principle in the clam is temperature dependent and the concentration in the molluscs is highest during summer months.[52] He also reported that the concentration of this principle can be enhanced in clams collected in the winter if they are kept for several weeks in warm sea water. On the other hand, extracts of clams collected from polluted sea water failed to show any activity in K-2 and S-180 transplanted tumor systems and against spontaneous mammary tumors in C3H mice. Li and co-workers reported that clam extracts have antiviral activity and inhibit tumor formation in hamsters inoculated with adenovirus type 12.[146] Liu and Cipola found[147] that the liver (digestive diverticulum) of the clam contains most of the antiviral activity, and Li and co-workers demonstrated the therapeutic effect of clam liver extracts against L1210 leukemia in mice.[148] Lavelle reported that clam extracts are general growth inhibitors and counteract chemical carcinogenesis.[149] Prescott and co-workers provided tentative chemical data for Paolin I, Paolin II, and the antineoplastic principle extracted from clam liver.[145] Their studies indicated that all three factors contain carbohydrates, amino acids, and an unknown material, but pyrimidines and purines were not present in these extracts. The same investigators reported that the antiviral component is a thermostable glycoprotein with an approximate mol wt of 10,000 daltons. Column chromatographic separation of the partially purified clam liver extract yielded biologically active fractions of differing molecular weights. The fractions with large molecular weight materials were more active per mg dry weight than the ones containing small molecular weight materials.[150] The investigators suggested that the reason for this phenomenon might be the same as that observed by McMaster and co-workers for Type III pneumococcus polysaccharide antigen,[151] which loses its immunogenic activity as the molecule splits into smaller units.

Johnsson-Hegyeli found that extracts of *Mercenaria* inhibit the growth of KB human carcinoma cells.[152,153] Schmeer tested the clam extracts in the HeLa (At_1) human malignant cell culture and in the FL normal human anmion cell culture systems and found that the malignant cells were completely destroyed by the extract in 12 hr while only a negligible toxicity was observed against the normal cell cultures.[153,154] Schmeer and co-workers reported that the antineoplastic principle isolated from the common clam has a mol wt between 1000 and 2000 daltons, and is a glycopeptide.[155] They prepared active extracts against S-180 tumors in mice from a variety of other marine sources as well.

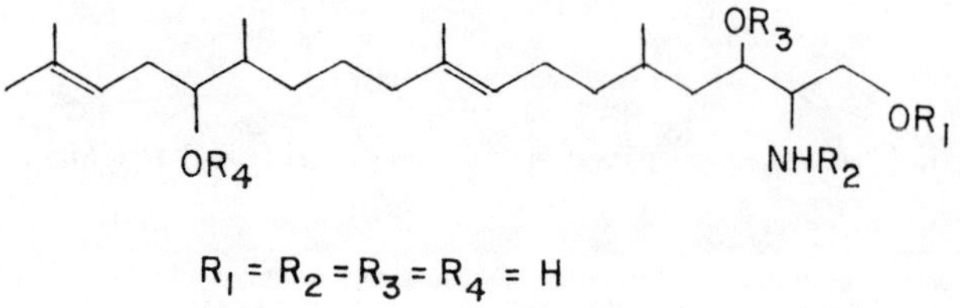

$R_1 = R_2 = R_3 = R_4 = H$

FIGURE 32. Aplidiasphingosine.

These included *Mercenaria campechiensis,*[156] oyster, *Ostrea virginica;* whelk, *Busycon canalicularis,* snail, *Helix* sp.; and squid, *Loligo* sp.[157] They found that the antineoplastic principle prevented the tumor cells in culture to enter the S phase of the cell cycle, the period of DNA synthesis, by stopping the cycle in the G_1 phase.[158]

6. Phylum Annelida (Segmented Worms)

Norton and associates reported that the crude extracts of the sea worm *Lanice conchilega* and the tentacles and body extracts of *Reteterebella queenslandia* prolonged the life of Swiss-Webster female mice inoculated with Ehrich ascites tumor cells, and 60 to 100% of the treated mice were alive 25 days following tumor inoculation while all the control animals died. It was suggested by the investigators that a nonprotein type, large molecular weight component might be responsible for the biological activity.[159,160]

7. Phylum Arthropoda (Horseshoe Crab, Shrimp, Lobster)

Sigel treated BDF_1 mice inoculated with P388 lymphocytic leukemia with the extracts of the horseshoe crab *Limulus polyphemus* and prolonged the life of the treated animals 40 to 172% compared to the untreated controls.[131]

8. Phylum Chordata/Tunicata

Extracts of the tunicate *Aplidium* sp. are cytotoxic to KB and LE tumor cells and inhibit Herpes virus type I. Purification of this extract led to the isolation of aplidiaphingosine, a terpenoid with ED_{50} of 1.9 μg/mℓ in the LE and 8.3 μg/mℓ in the KB cell culture tests. The structure of this agent was identified as 2-amino-5,9,13,17-tetramethyl-8-16-octodecadien-1,3,14-triol[161] by Carter and Rinehart, and the structure is shown in Figure 32.

Cheng and Rinehart indicated that extracts of the marine tunicate *Polyandorcarpa* species are toxic to KB and LE tumors and monkey kidney cells in culture. The extracts exhibited antibacterial activity and slight antiviral activity against herpes virus type I as well.[162] Purification monitored by cytotoxicity tests produced polyandrocarpidine I (Ia) as the major and polyandrocarpidine II (IIb) as the minor bioactive component (Figure 33). A mixture consisting of 90% (Ia) and 10% (IIb) gave ED_{50} of 4.8 μg/mℓ in the LE tests and was also toxic to CV-1 monkey kidney cells.

Pettit and co-workers[21] found that ethanol extracts of the tunicate *Molgula occidentalis* prolonged the life of animals in the PS system (T/C 177%) and extracts of another tunicate *Clavelina picta* gave a T/C of 200% in the same murine test.

9. Phylum Chordata/Vertebrata

Extracts from toads have been used in traditional medicine in China for thousands of years and in Europe for hundreds of years. Modern pharmacology has confirmed that toad extracts contain many medically active components. Some of these have anticancer potential. For instance, marinobufagin (Figure 34) was isolated from the toad

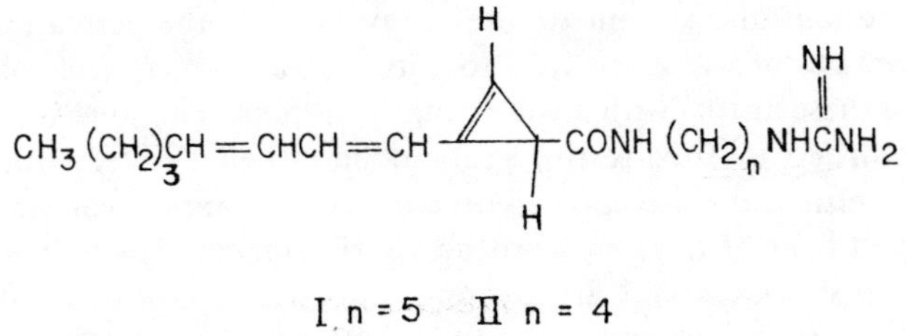

FIGURE 33. Polyandrocarpidine-I (Ia) and II (IIb).

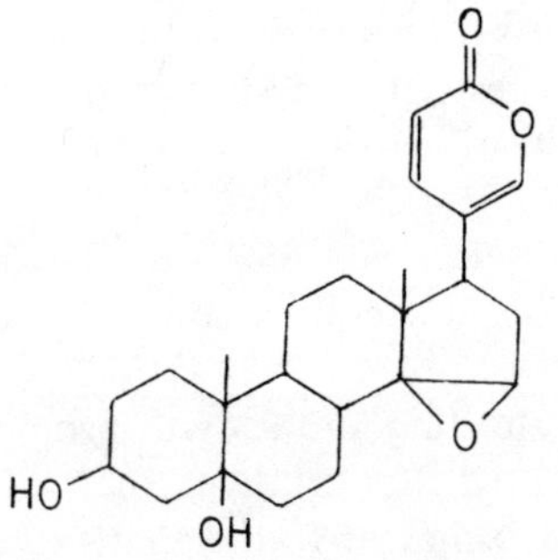

FIGURE 34. Marinobufagin.

Bufo marinus and subsequently synthesized by Pettit and Kamano.[22] Marinobufagin, an antienoplastic bufadienolide, has an ED_{50} of 0.86 μg/mℓ in KB cultures and a T/C of 155% at 1.25 mg/kg dose in the Ehrlich ascites test, but is inactive in the LE and PS systems.

Pettit and associates undertook the first systematic investigation and evaluation of marine vertebrates for antineoplastic substances. They identified 57 species of fish,[22] all of which give rise to extracts confirmed active in the National Cancer Institute PS and KB cancer screens. The purification of one of these extracts from hammer head shark has led to the isolation of pure substances. Further information on these and other shark extracts is presented in the next section, since many of these extracts appear to express their neoplastic action by activating the host's immune system rather than through cytotoxicity.

III. MARINE FACTORS ACTIVE AS RETICULOENDOTHELIAL SYSTEM STIMULANTS, IMMUNE MODIFIERS, AND GROWTH INHIBITORS AND PROMOTERS

Results from recent studies, as well as examples from earlier cancer research, indicate that organisms from bacteria to higher organisms, including man, produce a number of substances that influence the development of malignant tumors in various ways. These substances are not necessarily cytotoxic in nature; or, if they are, they may have selective cytotoxicity for malignant and not for normal cells. Their action may be mediated through the immune system or they may prevent the activation of naturally occurring or manmade procarcinogens into active carcinogens. The following are examples of some of these agents with a wide range of reported actions.

Heller and co-workers isolated restin, a reticuloendothelial system-stimulating agent, from shark liver. Pettit and Ode reported on a number of anticancer components from the blood, fins, and liver of the hammerhead shark *Sphyrna lewini*.[22] Sphyrnastatin 1 and 2 isolated from shark extracts are glycoproteins with a mol wt of at least 40×10^6 daltons.[164] These substances extended the life of mice by 30 to 40% in the PS system. It has been postulated by the investigators that these substances may act by activating the immune system, and not through cytotoxic effects.

A protein isolated from the coelomocytes (macrophages) of the sea star *Asterias forbesi* shows many of the properties of lymphokines produced by mammalian thymus-derived lymphocytes (T-cells) after contact with homologous antigen.[165] The sea star factor (SSF) has delayed skin sensitivity-stimulating property and macrophage-inhibitory action. In in vitro tests, this SSF significantly inhibited the migration of macrophages without the help of sensitized lymphocytes. In C_{57}Bl/6 mice, SSF suppressed the generation of cytologically active T-cells and markedly decreased concavalin-A-induced mitogenesis.[166] Purified SSF was shown to be a basic protein with a mol wt of 38,000 daltons. The molecule is composed of a single pair of heavy and light chains.[167]

Nigrelli described growth-retarding and growth-promoting principles of marine origin.[121] Li and associates observed that crude extracts of abalone retarded the growth of mice.[168] On the other hand, Schmeer found that extracts from hard shell clams are not toxic to normal cells in culture.[154] Szent Gyorgyi and co-workers isolated and investigated growth-retarding and growth-promoting factors in a wide range of organisms including marine species.[169-170]

Johnsson-Hegyeli and Hegyeli reported the isolation of directin, a heat sensitive pyrophosphate derivative with a mol wt of less than 700 daltons.[172] This agent is not cytocidal to malignant cells but induces a measure of control over the behavior of KB and HeLa cells in culture. The cell growth in directin-treated cultures changes from a disorderly, haphazard pattern to polarized directional growth. The agent does not induce any microscopically detectable alterations in the already ordered architecture of normal cells maintained under identical cell culture conditions.[173-178] While directin was originally isolated from human urine, a high molecular weight component with directin-like activity was subsequently isolated from whale stomach as well as from calf thymus. These latter factors have at least 40 times higher biological activity, based on the dry weight of the purified fraction, than the low molecular weight directin isolated from urine,[179] indicating that the active agent might be a small molecule, attached to a protein carrier which is partly inactivated during the separation and purification steps, if one starts from urine.

Groupé and his associates reported that xerosin, an extract isolated from a Gram-negative soil bacterium, prolonged the latent period of tumor production in chicks inoculated with Rous sarcoma virus,[180] and pointed out that virus-induced tumor systems might be utilized, in addition to transplanted tumors, in order to detect potential antitumor or antiviral agents.[181] While a number of antiviral agents have been found in marine species, no marine agents with xerosin-like activity have been reported to date. It is likely that further search along these lines will result in the discovery of other tumor inhibitor agents of marine origin.

IV. CONCLUSION

This review of research on marine antitumor agents indicates that considerable national and international attention is being given to this area at the present time. Many new investigators have entered this field in recent years in the hope of finding agents which will be effective in the prevention or cure of cancer. Cancer studies at every level depend upon an exploratory as well as a systematic approach. The exploration and testing of new ideas and techniques demand as much freedom as possible in order not to inhibit inventiveness. As Sir John G. Bennette has stated, "Explorers are not required to be efficient and economical, but those who follow them to survey and develop the new territory must collaborate efficiently, by integrating their work and avoiding duplication of effort. At the level of exploration we cannot afford to be too economical; at the other level of development and exploitation, we cannot afford not to be."[182]

Current research on antitumor agents of marine origin is virtually virgin territory in spite of the relatively large number of investigators now engaged in this field. The field is still in the exploratory phase and there is a marked unevenness of approaches and levels of effort as of this writing. It is only natural at this stage of development that there be wasted efforts and duplication, as well as false starts. As yet, there are no definite patterns emerging. Antitumor agents have been found in a wide range of marine species and new discoveries continue to be reported. The chemical structures which have been found to be associated with antitumor activity are extremely diverse. As recent investigations have shown, marine antitumor agents may exert their antitumor action in a number of ways, many of which remain to be explored. There is a great need for standardization of test systems and further testing of promising compounds.

REFERENCES

1. **Shimkin, M. B.,** *Contrary to Nature, Cancer,* Department of Health, Education and Welfare Publ. No. (NIH) 76-720, 1977.
2. **Brothwell, D.,** *The Evidence for Neoplasms in Diseases of Antiquity,* Brothwell, D., and Sardison, A. T., Eds., Charles C Thomas, Springfield, Ill., 1967, 320.
3. **Wolff, J.,** *Die Lehre von der Krebskrankheit von den altesten Zeiten bis zur Gegenwart,* Gustav Fischer, Jena, 1929.
4. **Janssen, P. A.,** *Paleopathology. Diseases and Injuries of Prehistoric Man,* John Baker, London, 1970.
5. **Johnsson, R. I. and Hegyeli, A. F.,** *Battelle Technical Review,* 10, June 1966.
6. **Li, C. P., Goldin, A., and Hartwell, J. L.,** *Cancer Chemother. Rep. Part 2,* 4, 97, 1974.
7. **Kim, S.,** *New Med. J. Korea,* 10, 56, 1967.
8. **Ehrlich, P.,** *The Harben Lectures for 1907,* H. K. Lewis, London, 1908.
9. **Bergmann, W. and Feeney, R. J.,** *J. Am. Chem. Soc.,* 72, 2809, 1950.
10. **Bergmann, W. and Feeney, R. J.,** *J. Org. Chem.* 16, 981, 1951.
11. **Bergmann, W. and Burke, D. C.,** *J. Org. Chem.,* 20, 1501, 1955.
12. **Bergmann, W. and Burke, D. C.,** *J. Org. Chem.,* 21, 226, 1956.
13. **Bergmann, W. and Stempien, M. J., Jr.,** *J. Org. Chem.,* 28, 1575, 1957.
14. **Cohen, S. S.,** *Perspect. Biol. Med.,* 6, 215, 1963.

15. **Doering, A., Keller, J., and Cohen, S. S.,** *Cancer Res.,* 26, 2444, 1966.
16. **Cohen, S. S.,** in *Progress in Nucleic Acid Research,* Vol. 5, Davidson, J. N., and Cohn, W. E., Eds., Academic Press, New York, 1966, 1-88.
17. **Ch'ien, L. T., Schabel, F. M., and Alford, C. A.,** in *Selective Inhibitors of Viral Function,* Carter, W. A., Ed., CRC Press, Boca Raton, Fla., 1973, 227.
18. **Nigrelli, R. F.,** *Ann. N.Y. Acad. Sci.,* 90, 615, 1960.
19. *Proceedings of the Food-Drugs from the Sea Conference,* Marine Soc., Washington, D.C., 1967, 1969, 1972, and 1974.
20. **Pettit, R. G., Hartwell, J. L., and Wood, H. B.,** *Cancer Res.,* 28, 2168, 1968.
21. **Pettit, R. G., Day, J. F., Hartwell, J. L., and Wood, H. B.,** *Nature,* 227, 962, 1970.
22. **Pettit, R. G.,** *Biosynthetic Products for Cancer Chemotherapy,* Vol. 1, Plenum Press, New York, 1977.
23. **Halstead, B. W.,** Poisonous and Venomous Marine Animals of the World, Vols. 1 and 2, U.S. Government Printing Office, Washington, D.C., 1965-1969.
24. **Russell, F. E.,** in *Advances of Marine Biology,* Vol. 3, Academic Press, New York, 1965, 255.
25. **Scheuer, P. J.,** in *Progress in the Chemistry of Organic Natural Products,* Zechmeister, L., Ed., Springer-Verlag, New York, 1964, 266.
26. **Scheuer, P. J.,** *The Chemistry of Marine Natural Products,* Academic Press, New York, 1973.
27. **Ruggieri, G. D.,** *Science,* 194, 491, 1976.
28. **Andersen, R. J. and Faulkner, D. J.,** in *Food-Drugs from the Sea Proceedings,* Marine Technology Soc., Washington, D.C., 1973.
29. **Baker, J. T. and Murphy, V., Eds.,** *CRC Handbook of Marine Science, Compounds from Marine Organisms,* Vol. 1, CRC Press, Boca Raton, Fla., 1976.
30. **Dawe, C. J. and Harshbarger, J. D., Eds.,** *Neoplasms and Related Disorders of Invertebrates and Lower Vertebrate Animals,* National Cancer Institute Monograph, 31, 1969.
31. **Schlomberger, H. G. and Lucke, B.,** *Cancer Res.,* 8, 657, 1948.
32. **Harshbarger, J. C.,** *Registry of Tumors in Lower Animals, 1965-1973,* Smithsonian Institution, Washington, D.C., 1974.
33. **Harshbarger, J. C. and Dawe, C. J.,** in *Unifying Concepts of Leukemia,* 39, Dutcher, R. M. and Chieco-Bianchi, L., Eds., S. Karger, Basel, 1973, 1-25.
34. **Harshbarger, J. C.,** *Fed. Proc.,* 32, 224, 1973.
35. **Emerson, G. A. and Taft, G. H.,** *Tex. Rep. Biol. Med.,* 31, 302, 1945.
36. **Nigrelli, R. F. and Zahl, P.,** *Proc. Soc. Biol. Med.,* 81, 379, 1952.
37. **Nigrelli, R. F. and Jakowka, S.,** *Ann. N.Y. Acad. Sci.,* 90, 884, 1960.
38. **Chapman, V. J.,** *Seaweeds and Their Uses,* Metheen, London, 1950.
39. **Jackson, D. E.,** *Algae and Man,* Plenum Press, New York, 1964.
40. **Schwimmer, M. and Schwimmer, D.,** *The Role of Algae and Plankton in Medicine,* Grune and Stratton, New York, 1955.
41. **Russell, F. E.,** *Fed. Proc.,* 26, 1206, 1967.
42. **Lewin, R. A.,** *Physiology and Biochemistry of Algae,* Academic Press, New York, 1962.
43. **Kreig, M. B.,** *Green Medicine,* Rand McNally, New York, 1964.
44. **Burkholder, P. R.,** in *Drugs from the Sea,* Freudenthal, H. D., Ed., Marine Technological Soc., Washington, D.C., 1968, 87.
45. **Freudenthal, H. D., Ed.,** *Drugs from the Sea,* Marine Technological Soc., Washington, D.C., 1968.
46. **Gullion, E. A.,** *Use of the Sea,* Marine Technological Soc., Washington, D.C., 1968.
47. **Baslow, M. H.,** *Marine Pharmacology,* Williams and Wilkins, Baltimore, 1969.
48. **Baslow, M. H.,** *Annu. Rev. Pharmacol.,* 11, 447, 1971.
49. **Boolootian, R. A., Ed.,** *Physiology of Echinodermata,* Interscience, New York, 1966.
50. **Li, C. P., Prescott, B., and Jahnes, W. B.,** *Proc. Soc. Exp. Biol. Med.,* 109, 534, 1962.
51. **Schmeer, M. R.,** *Science,* 144, 413, 1964.
52. **Hegyeli, A. F.,** *Science,* 146, 77, 1964.
53. **Norton, T. R. and Kashiwagi, M.,** *J. Pharm. Sci.,* 61, 1914, 1972.
54. **Norton, T. R., Kashiwagi, M., and Quinn, R. J.,** *J. Pharm. Sci.,* 62, 1464, 1973.
55. **Quinn, R. J., Kashiwagi, M., Moore, R. E., and Norton, I. R.,** *J. Pharm. Sci.,* 63, 257, 1974.
56. **Geran, R. I., Greenberg, N. H., McDonald, M. M., Schumacher, A. M., and Abbott, B. J.,** *Cancer Chemother. Rep. Part 3,* 3, 1, 1972.
57. **Weinheimer, A. J. and Matson, J. A.,** *Lloydia,* 38, 378, 1975.
58. **Moore, R. E.,** *Bioscience,* 27, 797, 1977.
59. **Shiomi, K., Kamiya, H., and Shimizu, Y.,** *Biochim. Biophys. Acta,* 576(1), 118, 1979.
60. **Shibata, S., Nishikawa, Y., Takeda, T., Tanaka, M., Fukuoka, F., and Nakanishi, M.,** *Chem. Pharmacol. Bull.,* 16, 1639, 1968.

61. Shibata, S., Nishikawa, Y., Takeda, T., and Tanaka, M., *Chem. Pharmacol. Bull.*, 16, 2362, 1968.
62. Nishikawa, Y., Takeda, T., Shibata, S., and Fukuoka, *Chem. Pharmacol. Bull.*, 17, 1910, 1968.
63. Heckendorf, A. H. and Shimizu, Y., *Phytochemistry*, 13, 2181, 1974.
64. Myderse, J. S., Moore, R. E., Kashiwagi, M., and Norton, T. R., *Science*, 196, 538, 1977.
65. Myderse, J. S. and Moore, R. E., *J. Org. Chem.*, 43, 2301, 1978.
66. Kato, Y. and Scheuer, P. J., *Pure Appl. Chem.*, 41, 1, 1975.
67. Kato, Y. and Scheuer, P. J., *Pure Appl. Chem.*, 48, 29, 1976.
68. Carter, G. T., Rinehart, K. L., Jr., Li, L. H., Kuentzel, S. L., and Connor, J. L., *Tetrahedron Lett.*, 46, 4479, 1978.
69. Stavric, B. and Gorham, P. R., in *Annu. Meet. Can. Soc. Plant Physiol.*, University of British Columbia, 1966, 20.
70. Devlin, J. P., Edwards, O. E., Gorham, P. R., Hunter, N. R., Pike, R. K., and Stavric, B., *Can. J. Chem.*, 55, 1367, 1977.
71. Sawyer, P. J., Gentile, J. H., and Sasner, J. J., Jr., *Can. J. Microbiol.*, 14, 1199, 1968.
72. Schantz, E. J., Mold, J. D., Stanger, D. W., Shavel, J., Riel, F. J., Bowden, J. P., Lynch, J. M. Wyler, R. S., Reigel, B., and Sommer, H., *J. Am. Chem. Soc.*, 79, 5230, 1957.
73. Sun, H. H. and Erickson, K. L., *J. Org. Chem.*, 43, 1613, 1978.
74. Brennan, M. R. and Erickson, K. L., *Tetrahedron Lett.*, 19, 1637, 1978.
75. Waraszkiewicz, S. M., Sun, H. H., Erickson, K. L., Finer, J., and Clardy, J., *J. Org. Chem.*, 43, 3195, 1978.
76. Endicott, K. M., *J. Natl. Cancer Inst.*, 19, 275, 1957.
77. Lee, W. W., Benitez, A., Goodman, L., and Baker, B. R., *J. Am. Chem. Soc.*, 82, 2648, 1960.
78. Reist., E. J., Benitez, A., Goodman, L., Baker, B. R., and Lee, W. W., *J. Org. Chem.*, 27, 3274, 1962.
79. Shannon, W. M., in *Adenide: An Antiviral Agent*, Pavan-Langston, D., Buchanan, R. A., and Alford, C. A., Jr., Eds., Raven Press, New York, 1975.
80. Shannon, W. M., *Ann. N.Y. Acad. Sci.*, 284, 3, 1977.
81. Burkholder, P. R. and Sharma, G. M., *Lloydia*, 32, 466, 1969.
82. Baslow, M. H. and Turlapaty, P., *Proc. West Pharmacol. Soc.*, 12, 6, 1969.
83. Sigel, M. M., Walham, L. L., Lichter, W., Dudek, L. E., Gargus, J. L., and Lucas, A. H., in *Food-Drugs from the Sea Proceedings*, Youngken, H. W., Jr., Ed., Marine Technology Soc., Washington, D.C., 1969, 281.
84. Kaul, P. N. and Sindermann, C. J., Eds., *Drugs and Foods from the Sea, Myth or Reality?* The University of Oklahoma Press, Norman, 1978, 81.
85. Hollenbeak, K. H. and Schmitz, F., *Lloydia*, 40, 479, 1977.
86. Schmitz, F. J., Hollenbeak, K. H., and Campbell, D. C., *J. Org. Chem.*, 43.
87. Gopichand, Y. and Schmitz, F. J., *Tetrahedron Lett.*, 41, 3921, 1979.
88. Gopichard, Y. and Schmitz, F. J., *Tetrahedron Lett.*, in press.
89. Gopichand, Y. and Schmitz, F. J., *Tetrahedron Lett.*, in press.
90. Kaul, P. N., Kulkarni, S. K., Weinheimer, A. J., Schmitz, F. J., and Karns, T. K. B., *Lloydia*, 40, 253, 1977.
91. Schmitz, F. J. and Gopichand, Y., *Tetrahedron Lett.*, 39, 3637, 1978.
92. Tabrah, F. L., Kashiwagi, M., and Norton, T. R., *Int. J. Pharmacol. Therap. Toxicol.*, 5, 420, 1972.
93. Schmitz, F. J., Vanderah, D. J., and Ciereszko, L. S., *J. Chem. Soc. Chem. Commun.*, 407, 1974.
94. Vanderah, D. J., Rytledge, N., Schmitz, F. J., and Ciereszko, L. S., *J. Org. Chem.*, 43, 1614, 1978.
95. Vanderah, D. J., Staudler, P. A., Ciereszo, L. S. Schmitz, F. J., Ekstrand, J. D., and van der Helm, D., *J. Am. Chem. Soc.*, 99, 5780, 1977.
96. Weinheimer, A. J., and Karns, T. K. B., *Proc. Fourth Food-Drugs from the Sea Symposium*, Mayaguez, Puerto Rico, Marine Technology Soc., Washington, D.C., 491, 1974.
97. Weinheimer, A. J., Matson, J. A., Hossain, M. B., and van der Helm, D., *Tetrahedron Lett.*, 34, 2923, 1977.
98. Tursch, B., Braekman, J. C., Daloze, D., Herin, M., Darlsson, R., and Losman, D., *Tetrahedron*, 31, 129, 1975.
99. Hossain, M. B. and van der Helm, D., *Recl. Trav. Chim. Pay-Bas Belg.*, 88, 1413 1969.
100. van der Helm, D., Enwall, E. L., Weinheimer, A. J., Karns, T. K. B., and Ciereszko, L. S., *Acta Crystallogr.*, B32, 1558, 1976.
101. Hossain, M. B., Nicholas, A. F., and van der Helm, D., *J. Chem. Soc., Chem. Commun.*, 385, 1968.
102. Weinheimer, A. J., Matson, J. A., van der Helm, D., and Poling, M., *Tetrahedron Lett.*, in press.
103. Enwall, E. L. and van der Helm, D., *Recl. Trav. Chim. Pay-Bas*, 93, 53, 1973.
104. Enwall, E. L., van der Helm, D., Hsu, I. N., Pattabhiraman, T., Schmitz, F. J., Spaggins, R. L., and Weinheimer, A. J., *J. Chem. Soc. Chem. Commun.*, 215, 1972.

105. **Weinheimer, A. J., Washecheck, P. H., van der Helm, D., and Hossain, M. B.,** *J. Chem. . Chem. Commun.*, 1070, 1968.
106. **Burks, J. E., van der Helm, D., Chang, C. Y., and Ciereszko, L. S.,** *Acta Crystallogr.*, B33, 704, 1977.
107. **Rehm, S. J.,** Ph.D. thesis, University of Oklahoma, 1971.
108. **van der Helm, D., Ealick, S. E., and Weinheimer, A. J.,** *Cryst. Struct. Commun.*, 3, 167, 1974.
109. **Ealick, S. E., van der Helm, D., and Weinheimer, A. J.,** *Acta Crystallogr.*, B31, 1618, 1975.
110. **Weinheimer, A. J., Matson, J. A., van der Helm, D., and Poling, M.,** *Tetrahedron Lett.*, 15, 1295, 1977.
111. **Johnson, C. K.,** *Oak Ridge National Laboratory Report,* ORNL-3794, 1965.
112. **Quinn, R. J., Kashiwagi, M., Moore, R. E., and Norton, T. R.,** *J. Pharm. Sci.*, 63, 257, 1974.
113. **Moore, R. E. and Scheuer, P. J.,** *Science,* 172, 495, 1971.
114. **Moore, R. E., Dietrich, R. F., Higa, B., and Scheuer, P. J.,** *J. Org. Chem.*, 40, 540, 1975.
115. **Norton, T. R., Moore, R. E., Quinn, R. J., Christiansen, P., and Kashiwagi, M. C.,** *Clin. Pharmacol. Ther.*, 15, 216, 1974, Abs.
116. **Quinn, R. J. Kashiwagi, M., Norton, T. R., Shibata, S., Kuchii, M., and Moore, R. E.,** *J. Pharmacol. Sci.*, 63, 1798, 1974.
117. **Dunn, D. F., Kashiwagi, M., and Norton, T. R.,** *Comp. Biochem. Physiol.*, 50C, 133, 1975.
118. **Nigrelli, R. F.,** *Zoologica,* 37, 89, 1952.
119. **Sullivan, T. D., Ladue, K. T., and Nigrelli, R. F.,** *Zoologica,* 40, 49, 1955.
120. **Sullivan, T. D. and Nigrelli, R. F.,** *Proc. Am. Assoc. Cancer Res.*, 2, 151, 1956.
121. **Nigrelli, R. F.,** *Trans. N.Y. Acad. Sci.*, 20, 248, 1958.
122. **Nigrelli, R. F.,** *Ann. N.Y. Acad. Sci.*, 90, 615, 1960.
123. **Nigrelli, R. F., Strempien, M. F., Jr., Ruggieri, G. D., Liguory, V. R., and Cecil, J. T.,** *Fed. Proc.*, 26, 1197, 1967.
124. **Friess, S. L., Durant, R. C., Fink, W. L., and Chanley, J. D.,** *Toxicol. Appl. Pharm.*, 22, 115, 1972.
125. **Pettit, C. R., Harold, C. L., and Harold, D. L.,** *J. Pharm. Sci.*, 65, 1558, 1976.
126. **Ryoyama, K.,** *Biochem. Acta,* 320, 157, 1973.
127. **Ryoyama, K.,** *Biol. Bull.*, 146, 404, 1974.
128. **Ryoyama, K.,** *Jpn. J. Exp. Med.*, 47, 327, 1977.
129. **Angeles, L. T. and De Vera, F.,** *Antitumor Principles from Marine Life, I. Molluscs and Echinoderms,* PICC, Manila, Phillipines, 1978.
130. **Miyamua, S. and Niwayama, S.,** *Antibiot. Chemother.*, 9, 497, 1959.
131. **Sigel, M. M.,** *Conf. on Food-Drugs from the Sea,* Kingston, Rhode Island, 1969.
132. **Hollenbeak, K. H., Schmitz, F. J., Hossain, M. B., and van der Helm, D.,** *Tetrahedron,* 35, 541, 1979.
133. **Gonzales, A. G., Darias, J., and Martin, J. D.,** *Tetrahedron Lett.*, 26, 2381, 1973.
134. **Gonzales, A. G., Darias, J., Martin, J. D., and Perez, C.,** *Tetrahedron Lett.*, 14, 1249.
135. **Gonzales, A. G. Darias, J., Martin, J. D., Perez, C., Sims, J. J., Lin, G. H. Y., and Wing, R. M.,** *Tetrahedron,* 31, 2449, 1975.
136. **Hollenbeak, K. H., Schmitz, F. J., Hossain, M. B., and van der Helm, D.,** *Tetrahedron,* 2338, 1979.
137. **Schmitz, F. J., Hollenbeak, K. H., Carter, D. C., Hossain, M. B., and van der Helm, D.,** *J. Org. Chem.*, 44(14), 2495, 1979.
138. **Pettit, G. R., Ode, R. H., and Harvey, T. B., III,** *Lloydia,* 36, 204, 1973.
139. **Pettit, G. R., Ode, R. H., Herald, C. L., von Dreele, R. B., and Michel, C.,** *J. Am. Chem. Soc.*, 98, 4677, 1976.
140. **Pettit, G., R., Herald, C. L., Allen, M. S., Von Breele, R. B., Vanell, L. D., Kao, J. P. Y., and Blake, W.,** *J. Org. Chem.*, 40, 665, 1975.
141. **McDonald, F. J., Campbell, D. C., Vanderah, F. J., Schmitz, F. J., Washecheck, D. M., Burks, J. E., and van der Helm, D.,** *J. Org. Chem.*, 40, 665, 1975.
142. **Pettit, G. P., Ode, R. H., Herald, C. L., Von Dreele, R. B., and Michel, C.,** *J. Am. Chem. Soc.*, 98, 4677, 1976.
143. **Kato, Y.,** Ph.D. thesis, University of Hawaii, 1973.
144. **Limasset, P.,** *C. R. Acad. Sci.*, 252, 3154, 1961.
145. **Prescott, B., King, M. L., Caldes, G., Li, C. P., and Young, A. M.,** *Int. J. Clin. Pharmacol.*, 9, 1, 1974.
146. **Li, C. P., Prescott, B., Eddy, B., Caldes, G., Green, W. R., Martino, E. C., and Young, A. M.,** *Ann. N.Y. Acad. Sci.*, 130, 374, 1965.
147. **Liu, O. C. and Cipola, R. J.,** *Bacteriol. Proc.*, 164, 162, 1967.
148. **Li, C. P., Prescott, B., Liu, O. C., and Martino, E. C.,** *Nature,* 219, 1163, 1968.
149. **Lavelle, S. M.,** *10th Int. Cancer Congr.*, Houston, 789 (Abstr.), 1298, 1970.

150. Prescott, B., Li, C. P., Caldes, G., and Martino, E. G., *Proc. Soc. Exp. Biol. Med.*, 123, 460, 1966.
151. McMaster, P. R. B., Schade, A. L., Finnerty, J. F., Caldwell, M. B., and Prescott, B., *Fed. Proc.*, 29, 812, 1970.
152. Johnsson-Hegyeli, R. I. E., unpublished data, 1964.
153. Schmeer, A. C., *Cancer Chemother. Rep.*, 50, 9, 655, 1966.
154. Schmeer, A. C., *Natl. Cancer Inst. Monogr.*, 31, 581, 1969.
155. Schmeer, M. R. (A. C.), Horton, D., and Tanimura, A., *Life Sciences*, 5, 1169, 1966.
156. Schmeer, M. R. (A. C.) and Huala, C. V., *Ann. N.Y. Acad. Sci.*, 118, 603, 1965.
157. Schmeer, M. R. (A. C.) and Beery, G., *Biol. Bull.*, 129, 420, 1965.
158. Schmeer, A. C., *Biol. Bull.*, 135, 434, 1968.
159. Norton, T. R., Kashivagi, M., and Quinn, R. J., *J. Pharm. Sci.*, 62, 1464, 1973.
160. Tabrah, F. L., Kashiwaga, M., and Norton, T. R., *Science*, 170, 181, 1970.
161. Carter, G. T., Rinehart, K. L., Jr., *J. Am. Chem. Soc.*, 100, 7441, 1978.
162. Cheng, M. T. and Rinehart, K. L., Jr., *J. Am. Chem. Soc.*, 100, 7409, 1978.
163. Heller, J. H., Pasternak, V. Z., and Ransom, J. P., *Nature*, 199, 905, 1963.
164. Pettit, G. R. and Ode, R. H., *J. Pharm. Sci.*, 66, 757, 1977.
165. Prendergast, R. A., and Suzuki, M., *Nature*, 324, 277, 1970.
166. Prendergast, R. A., Cole, G. A., and Henney, C. S., *Ann. N.Y. Acad. Sci.*, 234, 7, 1974.
167. Prendergast, R. A. and Liu, S. H., *Scand. J. Immunol.*, 5, 873, 1976.
168. Li, C. P., Prescott, B. E. Eddy, B., Chu, E. W., and Martino, E. C., *J. Natl. Cancer Inst.*, 41, 1249, 1968.
169. Szent Gyorgyi, A., Hegyeli, A., and McLaughlin, J. A., *Proc. Soc. Natl. Acad. Sci. U.S.A.*, 48, 1439, 1962.
170. Hegyeli, A., McLaughlin, J. A., and Szent Gyorgyi, A., *Proc. Natl. Acad. Sci., U.S.*, 49, 230, 1963.
171. Hegyeli, A. F., McLaughlin, J. A., and Szent Gyorgyi, A., *Science*, 142, 1571, 1963.
172. Johnsson, R. and Hegyeli, A., *Proc. Natl. Acad. Sci., U.S.*, 54, 1375, 1965.
173. Johnsson, R. and Hegyeli, A., *Fed. Proc.*, 25, 419, 1966.
174. Hegyeli, A. F., *Biol. Bull.*, 131, 381, 1966.
175. Johnsson-Hegyeli, R. I. and Hegyeli, A., *J. Cell Biol.*, 31, 52 Abs., 1966.
176. Johnsson-Hegyeli, R. I. and Hegyeli, A. F., *Fed. Proc.*, 26, 691, 2427, Abs. 1967.
177. Johnsson-Hegyeli, R. I. and Hegyeli, A., *Proc. Int. Conf. of Tissue Culture in Cancer Res.*, Tokyo, 141, 1968.
178. Hegyeli, A. F. and Johnsson-Hegyeli, R. I., *Fed. Proc.*, 27, 5456, 1854, Abs., 1968.
179. Hegyeli, A. F. and Hegyeli, R. J., unpublished data, 1969.
180. Groupé, V. and Rauscher, F. J., *Cancer Chemother. Rep.*, 44, 1, 1965.
181. Pienta, R. J., Bernstein, E. G., and Groupé, V., *Cancer Chemother. Rep.*, 31, 25, 1963.
182. Bennette, J. G., Cooperation and organization in cancer research and treatment, in *What We Know About Cancer*, Harris, R. J. C., Ed., St. Martin's Press, New York, 1970.

Chapter 2

QUINONES AND OTHER CARBOCYCLIC ANTITUMOR ANTIBIOTICS

Klaus Eckardt

TABLE OF CONTENTS

I. INTRODUCTION

This review covers the field of carbocyclic antitumor antibiotics of microbial origin. The most important antibiotics of this chemical group are the anthracyclines and ansamycins. Accordingly, these compounds are discussed separately in this book and in other review articles and therefore have not been included in this chapter.

The present chapter is concerned mainly with recent developments in the chemical investigations and biochemical aspects. Nuclear magnetic resonance (NMR) and mass spectral data have not been reproduced although, in modern chemistry, structures are often based on these methods. References about these data are given.

In the past few years, progress has been made especially in the chemistry of several quinone antibiotics such as pluramycin antibiotics, isocoumarinquinone antibiotics, and aquayamycin and its structural analogues. Some of these groups are discussed in more detail because no reviews have been published on them. Early-described compounds are considered only briefly. It will become obvious from this review that a high percentage of the carbocyclic compounds exhibiting antitumor or cytotoxic activity is of quinone nature.

II. ANTITUMOR AND CYTOTOXIC QUINONE ANTIBIOTICS*

A. 4H-Anthra(1,2-*b*)pyran Antibiotics

Pluramycin-type antibiotics — kidamycin, iyomycins, hedamycin, pluramycin, neopluramycin, indomycins A, B, and C, rubiflavin and griseorubins — are anthraquinone compounds which have closely related biological properties and very similar chemical characteristics. From this group of antibiotics, the chemical structures of kidamycin, hedamycin, pluramycin, and neopluramycin have been published in detail. They all can be derived from a common structure as shown in Figure 1.

The structures of the four antibiotics vary in the nature of the side chain in position two and in the ring substitution at C-3″ (hydroxyl or carbomethoxy). The structures of iyomycin, rubiflavin, tumimycin, griseophagins, and griseorubins grouped together with pluramycin were not yet described in the literature .[3] Indomycins have been studied in some detail.

Pluramycin-type antibiotics have much in common with regard to their biological properties and their mechanisms of action have been compared in reviews by Tanaka[1] and Gale, and Cundliffe, Reynolds, Richmond, and Waring.[2] The designation 4H-anthra(1,2-*b*)pyran antibiotics is according to Hauser and Rhee.

1. Kidamycin

Kidamycin was the first compound of this family of which the structure has been fully elucidated. The antibiotic could be isolated from a variant strain of *Streptomyces phaeoverticillatus* when the strain was fermented in a culture medium containing an-

* The chemistry of natural quinones has been discussed in an excellent book, *Naturally Occurring Quinones*, R. H. Thomson, Academic Press, New York, 1971.

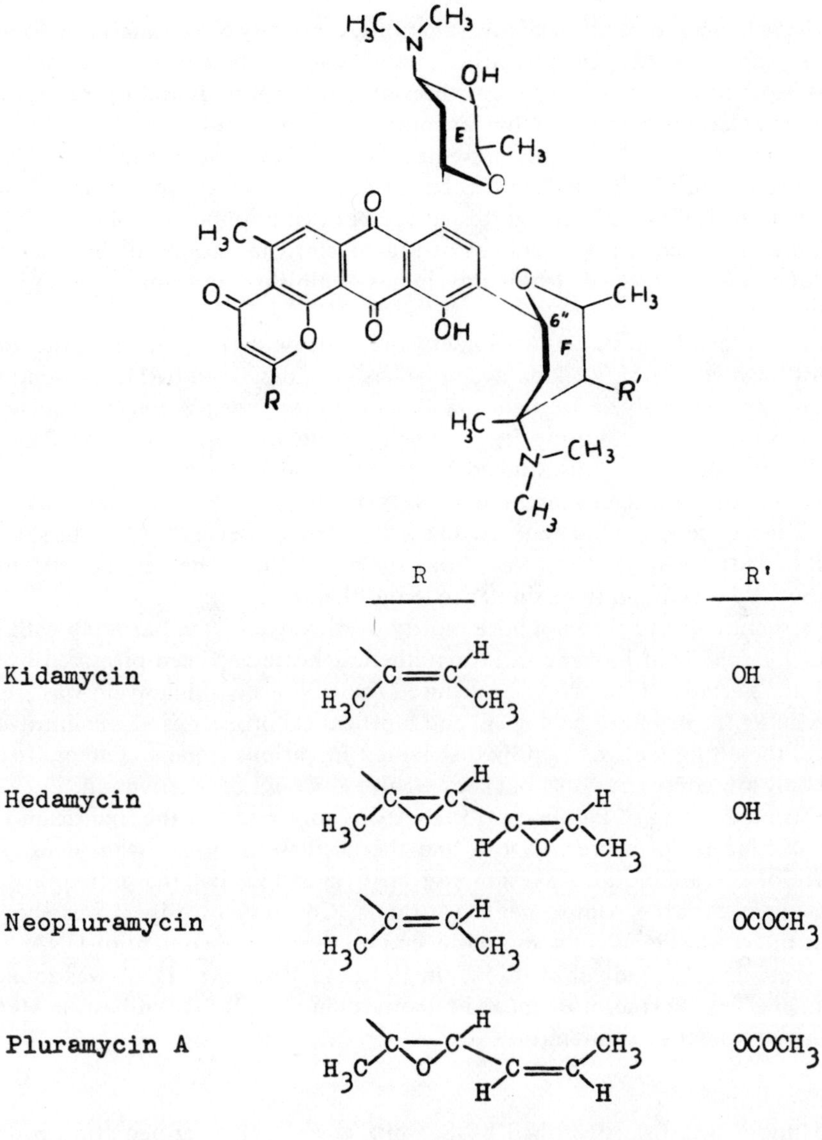

	R	R'
Kidamycin		OH
Hedamycin		OH
Neopluramycin		$OCOCH_3$
Pluramycin A		$OCOCH_3$

FIGURE 1. Structure of several 4H-anthra(1,2-*b*) pyran antibiotics.

thraquinone-2,7-disulfonic acid.[7] Physicochemical characteristics are orange-red needle crystals. Melting point (M.P.): 214 to 217°C (dec.), $[\alpha]_D^{20}$ = +456.7° (with approximately 1. chloroform), λ_{max}, (CH_3OH) 244 nm ($E^{1\%}_{1cm}$ 807), 270 (sh) nm, 434 nm ($E^{1\%}_{1cm}$ 210), I.R. (KBr): 1620, 1610, 1585 cm^{-1}. $C_{39}H_{48}N_2O_9$: proved by mass spectrometry and X-ray analysis.

The structure has been determined by Furukawa et al.[4] From the CO absorption at 1640 cm^{-1} and reductive acetylation of the triacetate, the presence of a quinoid system was suspected. Since three of the nine oxygen functions could not be characterized chemically, they were assumed to be ether oxygens. Attempted acid hydrolysis under various conditions to prove the O-glycosidic linkages failed. The presence of two tetrahydropyran rings each linked through a C-C bond to the anthraquinone ring system and an anthraquinone-fused pyrone ring has been deduced on the basis of NMR and X-ray analysis of bis[des(dimethyl-amino)] kidamycin and isokidamycin. The results of all chemical and physical investigations, especially X-ray analysis of additional kidamycin and isokidamycin derivatives,[5] established the constitution and stereochemistry as depicted in Figure 1. The structures of the C-sugars (E and F) correspond with those of angolosamine and dimethylvancosamine, respectively.

Séquin and Furukawa[6] recently presented a detailed analysis of the ^{13}C-NMR spectra of hedamycin and kidamycin. On this basis, chair conformations for ring E and twist conformations for ring F were suggested for both compounds in solution. In case of isokidamycin, which is the 6″ epimer of kidamycin, chair form for both tetrahydropyran rings was established. Isokidamycin has diminished antitumor and antibacterial activity.[4]

At present, synthetic studies are underway in this field. Very recently, the total synthesis of the methyl ether of kidamycinone has been described by Hauser amd Rhee.[87]

Kidamycin is strongly active against a variety of gram-positive microorganisms.[7] The LD_{50} (mice) is 12.5 to 20.0 mg/kg, i.v. or i.p. The antitumor activity of kidamycin and its acetyl derivative was studied by Kanda et al.[8] and Kanda.[9] Kidamycin was effective in Ehrlich ascites carcinoma, NF-sarcoma, and Yoshida sarcoma and less effective in leukemia L-1210 systems. The acetyl derivative is reputed to be slightly less toxic than kidamycin. Its LD_{50} was about 200 mg/kg, i.v. The antibacterial and antitumor activity was found to be similar to that of kidamycin.[9,10]

The structure of the chromophore moiety of kidamycin is in harmony with the polyketide hypothesis of biogenesis. A hypothetic scheme has been proposed by Furakawa et al.[5] Because of its proposed reduced toxicity, acetyl-kidamycin was studied in greater detail for its pharmacological and biochemical properties. Given intravenously to mice, the antibiotic was rapidly distributed in various organs. Tumor tissue contained only low concentrations, but showed also low rates of inactivation.[11]

The strong binding of kidamycin to DNA was confirmed by the significant increase of the melting temperature of DNA and the decrease of its buoyant density in the presence of the antibiotic. Takeshima et al.[12] suggested that the antibiotic stablizes residual links between complementary strands. Contrary to this, it was shown that binding of acetyl-kidamycin does not depend on the G + C content of DNA. In alkaline sucrose-density gradient solution, single-strand scission of DNA was found as an additional effect. Furthermore, acetyl-kidamycin inhibits RNA synthesis in HeLa cells in a manner similar to α-amanitin.[13]

2. Hedamycin

Hedamycin was first described by Schmitz et al.[14] as a fermentation product of *Streptomyces griseoruber* strain C 1150 (ATCC 15.422). After purification by chromatography on alumina, or by countercurrent distribution, the antibiotic was obtained as orange-red needles. M.p.: 243 to 245°C (dec.), λ_{max} (CH_3OH), 245 nm, 260 to 265 (sh) nm, 430 nm, I.R. (in $CHCl_3$), 1650 cm^{-1}, 1625 cm^{-1}. Séquin et al.[15] suggested an additional carbonyl absorption around 1650 $cm.^{-1}$ The molecular formula is $C_{41}H_{50}N_2$, derived from the mass spectra of hedamycin and its tris(trimethylsilyl) derivative. The structure has been determined by detailed comparison of hedamycin and kidamycin using different spectral methods. Both compounds have almost identical spectra. This strongly suggests that hedamycin also contains the anthraquinone skeleton with the fused γ-pyrone ring, and ^{1}H-NMR double resonance experiments established that hedamycin contains the same tetrahydropyran units as kidamycin.[16] However, differences were found regarding the side chains of both compounds. After excluding alternative structures, the formula shown in Figure 1 with a side chain containing two epoxide rings instead of the olefinic side chain present in kidamycin was proposed as the most probable structure for hedamycin. As mentioned before, Séquin and Furukawa[6] have confirmed this structure of hedamycin, including the stereochemistry by detailed ^{13}C-NMR studies of both kidamycin and hedamycin and some of their derivatives.

Biological properties of hedamycin have been reported by Bradner et al.[17] The antibiotic is strongly active on gram-positive bacteria, especially gram-positive cocci. It is also active on yeasts and protozoa. The compound was shown to be highly cytotoxic to HeLa cells in tissue culture. Antitumor activity was found when tested against Walker 256 intramuscular tumor of rats and adenocarcinoma of the duodenum in hamsters. No effect on growth of the tumors Ca 755, C-1498, and L-1210 in vivo could be found. Hedamycin, like pluramycin A, induces λ phage production in a lysogenic strain of *Escherichia coli*. The LD_{50} is 0.3 mg/kg mice, i.p.[17]

The same authors initially suggested that hedamycin inhibits nucleic acid syntheses. According to Nagai et al.,[18] White and White,[19] and Joel and Goldberg,[20] hedamycin is in vitro more inhibitory towards RNA synthesis than in intact bacterial cells. Hedamycin, like pluramycin A, rubiflavin, and kidamycin, has a strong irreversible binding activity to DNA, even at higher ionic strength. Covalent bonds could be excluded. Because of the increased viscosity of the hedamycin-DNA complex, it has been concluded that binding of the antibiotic to DNA might involve intercalation.

Jernigan et al.[21] recently presented data on the binding of hedamycin to DNA and chromatin of testis and liver. They found that hedamycin binds more effectively to "linker" DNA than to DNA attached to the core of the nucleosomes.

3. Pluramycins and Neopluramycin

Pluramycins were extracted from the culture broth of *Streptomyces pluricolorescens*[22] and purified by column chromatography on alumina or by countercurrent distribution. Pluramycin A was obtained as orange needles or orange prisms. No sharp melting points were observed (157 to 177°C or 200 to 215°C, depending on the type of crystals). λ_{max} (ethylalcohol) 208, 245, and 265 to 270 (sh) nm. As a minor component, pluramycin B was obtained as reddish-brown powder.

Neopluramycin was isolated from fermentations of *S. pluricolorescens* by solvent extraction of the culture filtrate.[23] The neopluramycin-producing strain MB 760-MG1 was found to exhibit very similar properties to the *Streptomyces* strain which produces pluramycins. Two minor components present in the culture broth were found to be closely related to the iyomycin B group and pluramycin B. Neopluramycin forms orange crystals, melts at 180 to 184° (dec.). λ_{max} (ethylalcohol): 216, 243, 270, 430 nm. Elemental composition: $C_{41}H_{50}N_2O_{10}$ Kondo et al.[24] reported on the chemical structures of pluramycin A and neopluramycin. Treatment of neopluramycin with acetic anhydride in pyridine yielded diacetylneopluramycin which was found to be identical with a monoacetyl derivative of kidamycin. Based on the comparison of ^{1}H-NMR spectra, it was concluded that neopluramycin is identical with 3″-0-acetylkidamycin, as shown in Figure 1.

Structure of pluramycin A was obtained by careful ^{1}H-NMR analysis including NOE experiments, as well as by ^{13}C-NMR studies of the diacetyl derivative. After that study, only the configuration of the side chain remained to be established. For this purpose Sêquin and Ceroni[88] synthesized two diastereoisomeric analogues of the pluramycin A side chain which served to settle the complete structure. As shown in Figure 1, the configuration was proved to be *cis* for the olefin and *trans* for the epoxide.

Takeuchi et al.[25,26] and Nishibori[27] studied the biological properties of pluramycins. Interest has been directed to component A which was found to be the most effective. Pluramycin A is active on Gram-positive bacteria and less active on Gram-negative microorganisms. Antitumor activity was found in Ehrlich ascites carcinoma, ascites form of sarcoma-180, and Yoshida sarcoma. It also inhibited growth of solid tumor

Table 1
PHYSICOCHEMICAL PROPERTIES OF INDOMYCINS[a]

Antibiotic	Elemental composition	λ_{max} ($CHCl_3$/dioxane, 1:1)	$[\alpha]^{20}_{D}$ ($CHCl_3$)
Indomycin A	$C_{40}H_{52}N_2O_{10}$	228(sh), 242, 266(sh), 420 nm	+345
Indomycin B	$C_{38}H_{48}N_2O_{10}$	242, 271(sh), 420 nm	+319
Indomycin C	$C_{41}H_{52}N_2O_{11}$	242, 270(sh), 419 nm	+254

[a] Melting points for all three compounds are between 170 and 190°C (dec.)

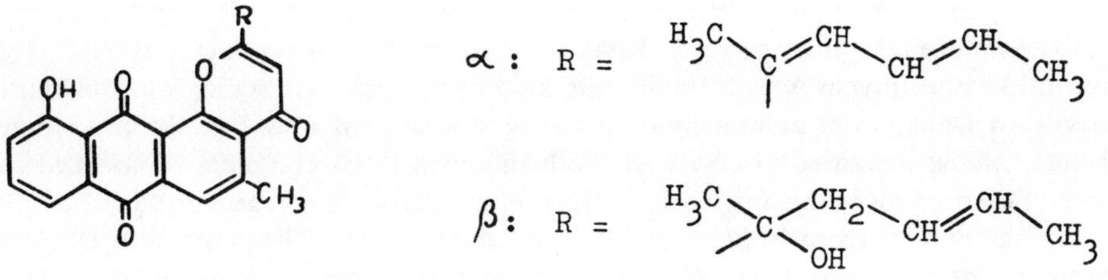

FIGURE 2. Structure of indomycins.

of Ehrlich carcinoma and sarcoma-180. LD_{50} (mice): 1.25 to 2.5 mg/kg i.v., 25 mg/kg, s.c. Pluramycin A had antimitotic effect on HeLa cells in tissue culture.

Pluramycin A is an inhibitor of nucleic acids synthesis.[28] Both RNA and DNA polymerase reactions were significantly inhibited in cell-free systems. In intact bacterial cells, pluramycin A inhibits both protein synthesis and nucleic acids synthesis (the effect on protein synthesis may be a secondary effect). From DNA-melting experiments, Tanaka et al. concluded that the antibiotic complexes with DNA. In comparison to pluramycin, neopluramycin only weakly inhibits Gram-positive microorganisms and is inactive against Gram-negative bacteria. The antibiotic is active against leukemia L-1210 in mice and Yoshida rat sarcoma cells in tissue culture. The toxicity of neopluramycin is lower than that of pluramycin A.[23]

Effects of neopluramycin on ascites tumors implanted in experimental animals have also been described by Hisamatsu and Koeda.[29] Like other antibiotics of the pluramycin family, neopluramycin was shown to be an inhibitor of nucleic acids syntheses.[30]

Thus, the antibiotic exerts its effect on Yoshida rat sarcoma cells primarily by this action. At low concentrations, intercalation of neopluramycin into DNA double strands was suggested. In addition to this, interaction with single-stranded polynucleotide chains (RNA, denatured DNA) was found when high concentrations of neopluramycin were used.

4. Indomycins

Indomycins A, B, and C are yellow-red to brown-red pigments extracted from the mycelium of *Streptomyces* sp. Ind. 927.[31] Some characteristic properties according to Schnell[32] and Brockmann[33] are listed in Table 1. Chemical structures of the indomycins are not yet known. Recently, attention has been directed to the constitution of indomycinons. On the basis of chemical degradation studies, Dahm[34] formulated structures for α-indomycinon and α-indomycinon as shown in Figure 2.

Additional information was obtained on indomycins by Fricke[35] who prepared some indomycinon derivatives. Indomycins showed antibacterial activity against Gram-positive microorganisms and exerted inhibitory activity on Ehrlich ascites carcinoma.[33,35]

5. *Rubiflavin*

Rubiflavin is a dark red substance which showed some physicochemical properties similar to those of pluramycin-type antibiotics. In addition, rubiflavin, hedamycin, and pluramycin were found to share a number of common features in respect to their mode of action.[3]

The antibiotic was isolated from fermentations of a *Streptomyces* species (SC 3.728) and purified by countercurrent separation.[36] Elemental composition: $C_{23}H_{29-31}NO_5$, λ_{max} (ethylalcohol): 224, 265 (sh), 395 (sh), 428, 446 (sh) nm. The antibiotic is active in vitro against Gram-positive and Gram-negative bacteria and molds. LD_{50} 15 mg/kg, i.p., in mice. Rubiflavin was found to be active against sarcoma-180 and carcinoma 755, in vivo.

6. *Iyomycins*

Iyomycin was produced by a strain of *Streptomyces phaeoverticillatus*.[37] The substance is a complex which was found to consist of a high molecular weight fraction (iyomycin A, mainly to be found in the culture filtrate) and a minor component (iyomycin B). Iyomycin B itself is a mixture of five components.[38,39] Column chromatography afforded the components B_1 to B_5 as pure compounds. Iyomycins B_2 to B_5 showed only weak antitumor activity. The main component, iyomycin B_1, was obtained as orange or dull-orange needles (free base). The substance had no clear melting point. λ_{max} (methanol): 243 to 244, 270, 430 nm. The toxicity (LD_{50}) in mice was 4.4 to 5.5 mg/kg, and 12.5 to 15.0 mg/kg, i.p. The LD_{50} in dogs was about 1.5 mg/kg, i.v.[39]

Antitumor activities of iyomycin complex and iyomycin B_1 on transplantable tumors have been described by Hoshino et al.[40] It was found that both compounds exhibit remarkable activity of Ehrlich carcinoma, sarcoma-180, and mouse leukemia SN-36. Iyomycin B_1 was active against Ehrlich ascites carcinoma in mice at 6 to 3 μg per mouse. Attempts were made to enhance the antitumor activity of iyomycin B_1.[41] However, combinations of iyomycin B_1 with high molecular weight substances or formation of metal complexes resulted in inactive preparations. Several derivatives were studied without success. However, acetylated iyomycin B_1 seemed to have enhanced antitumor efficacy.

Because of the therapeutic indexes of iyomycin B_1 type compounds, it was suggested that there was a possible therapeutic value for iyomycin B in treatment of human cancers.

7. *Griseorubins*

Recently, a new complex of antitumor compounds belonging to the pluramycin group of antibiotics has been described by Dornberger et al.[42] The crude, active material was obtained from fermentations of *Streptomyces griseus* strain IMET 20.978. The separation and purification of the components is complicated. Gel chromatography using Sephadex® LH-20 allowed the separation of four major fractions (I to IV). Silica gel chromatography of fraction I afforded the griseorubins A to H, but each substance still contained mixtures of several pigmented compounds. Finally, griseorubins E_1 to E_8 could be obtained as pure substances by thin layer chromatography (TLC).

Very little is known about the chemistry of these compounds. Pluramycin-type structures have been suggested on the basis of UV and infrared spectra and because the similar properties of the leucoacetates to kidamycin.[43] The main component griseorubin E_1 melts at 220°C (dec.). A tentative formula of $C_{44}H_{50}N_2O_{11}$ has been suggested.[43]

One of the complex substances (griseorubin I) has been the subject of more detailed studies in view to its biological activity. Griseorubin I is active in vitro against Gram-

Table 2
PHYSICAL PROPERTIES OF RUBROMYCINS

Rubromycin	Color	Melting point (dec.)	[a]λ_{max} $(CHCl_3)$[b]
α-Rubromycin	Yellow-red	278—281°	319, 352, 365, 415, 484 nm
β-Rubromycin	Red	225—227°	316, 350, 364, 504 nm
γ-Rubromycin	Red	>235°	317, 349, 364, 484, 513, 551 nm
γ-*sio*-Rubromycin	Yellow-red	>295°	324, 362(sh), 438, 470, 495, 532 nm

[a] Taken from Reference 50.
[b] Taken from References 49 and 50.

positive and Gram-negative bacteria, *Mycoplasma gallisepticum,* and *M. hyorhinis.* It is also active against protozoa. Significant cancerostatic activity of griseorubin I was observed in the leukemia L-1210 system by Gutsche et al.[44] Complexing of griseorubin with DNA did not afford preparations with interesting biological properties.[45]

B. Isocoumarinquinone Antibiotics

Isocoumarinquinones (rubromycins, purpuromycin, and griseorhodins) constitute a relatively new family of antibiotics. The skeleton of these red-colored compounds is composed of a hydroxynaphthoquinone part and an isocoumarin moiety. Both parts of the molecule are connected through a spiroketal ring system. Fundamental chemical work in this field was done by Brockmann and Zeeck[51] who studied the structures of rubromycins.

1. Rubromycins

Brockmann et al.[46-48] isolated two crystalline antibiotics — rubromycin and collinomycin — from the mycelium of *Streptomyces collinus.*[48] Rubromycin can be converted into collinomycin by heating the first one in pyridine.[49,50] These compounds have closely related properties and, therefore, the new names β-rubromycin (for rubromycin) and α-rubromycin (for collinomycin) have been proposed. A minor component, γ-rubromycin, could be obtained from fermentations of *S. collinus,* as well as from β-rubromycin, by way of treatment with acetone containing 20% HCl. In addition, when γ-rubromycin is heated in pyridine, an isomeric compound, β-*iso*-rubromycin is obtained. Some characteristic properties of the rubromycins are listed in Table 2.

The rubromycins are derived from naphthazarin, except β-rubromycin which is a 1,2-naphthaquinone.[49,50] This was proven chiefly by spectral data. Pyrolysis of γ-rubromycin afforded 2,5,8-trihydroxy-7-methoxy-naphthaquinone(1,4). The orthoquinoid structure of β-rubromycin was confirmed by the formation of a quinoxaline with o-phenylendiamine. A single carbonyl absorption at 1733 cm^{-1} was assigned to an ester group which is characteristic for rubromycins and purpuromycin, but not for griseorhodins.

The complete structures were established by Brockmann and Zeeck[51] by the following considerations. Treatment of α-, β-, and γ-rubromycin with concentrated H_2SO_4 gave γ-*iso*-rubromycin. Under mild alkaline conditions, β-rubromycin is converted by way of intramolecular β-elimination to α-rubromycin and, analogously, γ-rubromycin to γ-*iso*-rubromycin[51] as shown in Figure 3. The only difference between the structures of α-rubromycin and γ-*iso*-rubromycin is an OH group instead of an OCH_3 group at C-5.

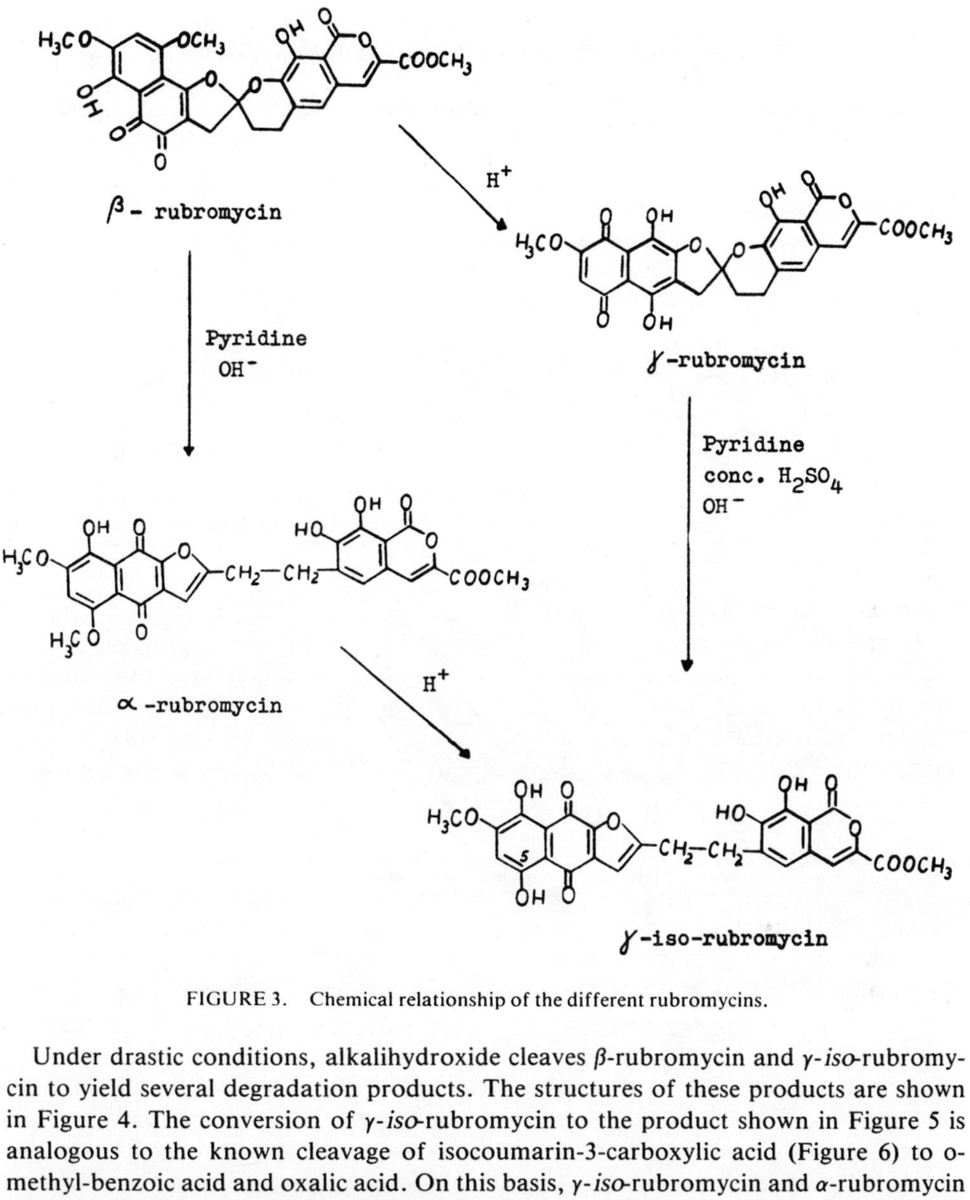

FIGURE 3. Chemical relationship of the different rubromycins.

Under drastic conditions, alkalihydroxide cleaves *β*-rubromycin and *γ*-*iso*-rubromycin to yield several degradation products. The structures of these products are shown in Figure 4. The conversion of *γ*-*iso*-rubromycin to the product shown in Figure 5 is analogous to the known cleavage of isocoumarin-3-carboxylic acid (Figure 6) to o-methyl-benzoic acid and oxalic acid. On this basis, *γ*-*iso*-rubromycin and *α*-rubromycin have been constructed as shown in Figure 3.

The presence of a spiroketal system in *γ*-rubromycin and *β*-rubromycin was concluded from the mechanism of the isomerization of *γ*-rubromycin to *γ*-*iso*-rubromycin and the changes found in the NMR spectra.

2. *Purpuromycin*

Purpuromycin was isolated by Coronelli et al.[52] in the course of a screening for antibiotics from strains of *Actinoplanes.* The red crystalline antibiotic could be obtained from the culture broth of *A. ianthinogenes* and was purified by column chromatography on silica gel buffered with KH_2PO_4. M.p.: > 320°C, dec. at 212°C. Elemental composition: $C_{26}H_{18}O_{13}$ (according to microanalysis and ^1H-NMR spectrum).

Bardone et al.[53] suggested, because of the ultraviolet spectrum that purpuromycin contains a naphthazarin moiety and that it is similar to the rubromycins. The authors

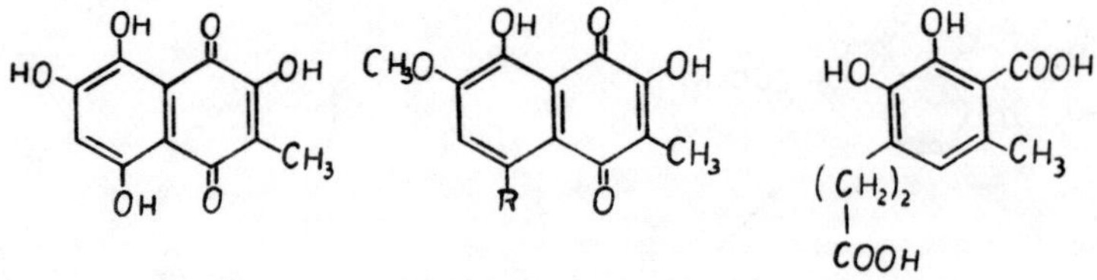

R = OH
R = OCH_3
R = H

FIGURE 4. Alkaline treatment products of rubromycins

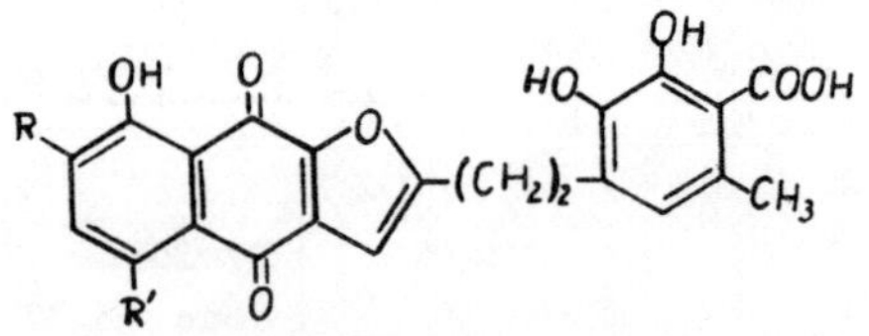

R = OCH_3 R′ = OH

FIGURE 5. Conversion product of γ-*iso*-rubromycin.

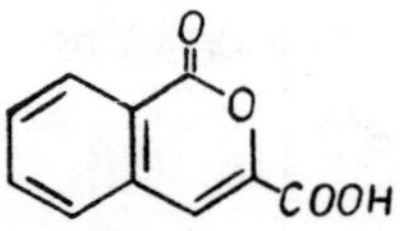

FIGURE 6. Isocumarin -3-carboxylic acid.

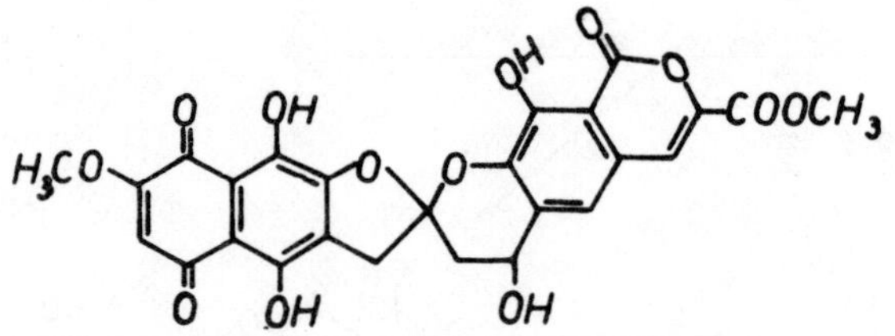

FIGURE 7. Structure of purpuromycin.

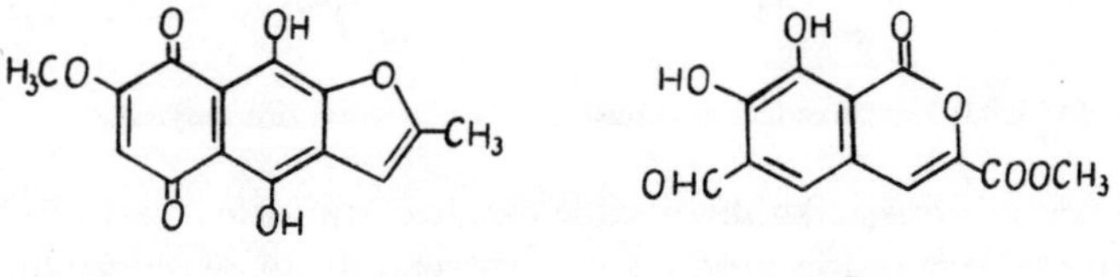

A

B

FIGURE 8. Degradation products of purpuromycin.

carefully compared the mass and NMR spectra with those of γ-rubromycin, and from these and other spectral data, the structure shown in Figure 7 was proposed.

This structure was verified in the following ways: heating of purpuromycin in pyridine at 100° for 12 hr afforded two main products as shown in Figure 8. The naphthazarin chromophore, a, is indicated by its physical properties (redox potential E_o = 0.055 V, UV, and visible absorption spectra, boroacetic anhydride reaction). The presence of a fused furan ring was deduced from the absorptions in the i.r. spectrum at 3130, 1535, 970, 880, and 720 cm^{-1}. Compound a was also obtained by alkali fusion of purpuromycin.

For the second degradation product, structure b has been postulated. The infrared spectrum indicates three CO functions. The ^1H-NMR spectrum shows the presence of only two uncoupled aromatic hydrogens, two chelated OH groups, one aldehyde pro-

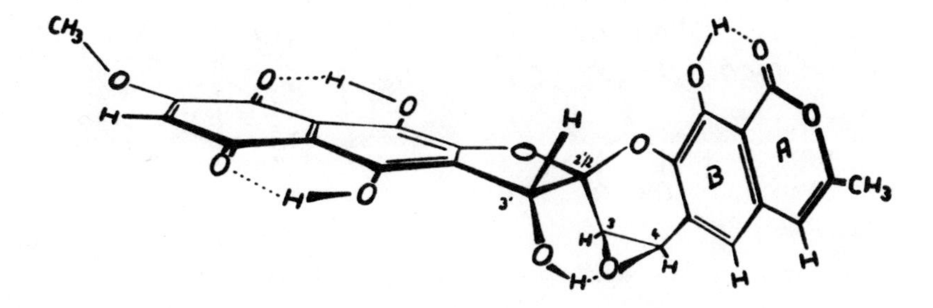

FIGURE 9. Structure of griseorhodin A.

Table 3
PHYSICAL PROPERTIES OF GRISEORHODINS

Griseorhodin	Melting point (°C, dec.)	λ_{max} (nm)	Molecular formula
A	288—290	232, (255), 315, 364, 508, (550) (EtOH)	$C_{25}H_{16}O_{12}$
B	260—262[a]	233(260), 313, 360, 509, (550) (EtOH)	$C_{25}H_{18}O_{12}$
C_2	241—243	234, (260), 314, 360, 509, (550) (EtOH)	$C_{27}H_{20}O_{12}$
C(= FCRC-U)	252—254	230, 275, 310, 360, 505, 540 (MeOH)	$C_{25}H_{18}O_{13}$
G	270—274	233, 315, 355, 505, 545(MeOH)	$C_{25}H_{18}O_{12}$

[a] Taken from Reference 58.

ton, and one OMe group. The *ortho* position of the two hydroxyls was proved by reaction with dichlorodiphenylmethane. All findings could best be brought to harmony by considering the isocoumarin structure.

Finally, the structure of purpuromycin was secured by elimination of the hydroxyl at C-4 which afforded γ-rubromycin.

3. Griseorhodins

Griseorhodins are the coloring matters of certain strains of *Stretomyces griseus*. Eckardt et al.[54] extracted from the mycelium of the strain JA 2640, a red-colored antibiotic complex which consisted of a number of different active compounds. Column chromatography afforded the main component griseorhodin A and the minor components A_2, B, C_2, and C. Some additional polar pigments (griseorhodins K and L) have not been purified. The strain *S. californicus* JA 2640 also produced viomycin.[55]

In the course of screening for antitumor antibiotics, Stroshane et al.[56,57] detected the new compound griseorhodin G, which was produced together with griseorhodins A and C from *S. griseus* FCRC-57 (Figure 9 and Table 3). The griseorhodins are closely related, structurally, to rubromycins and purpuromycin. As a main difference, griseorhodins contain CH_3 instead of $OCOCH_3$ at C-7 of the isocoumarin ring. Accordingly, the infrared spectra of griseorhodins do not show any absorption in the region of 1700 to 1800 cm^{-1}. Furthermore, griseorhodins contain one hydroxyl group at C-3′ which is not present in the rubromycin group. The naphthopurpurin chromophore was proved on the basis of chemical evidences and the preparation of leucoacetates.[59-61]

Degradation of griseorhodin A by heating in pyridine allowed the isolation of two compounds which afforded information about the two parts of the parent molecule (Figure 10). One of them was the quinone derivative, a, previously found as a degra-

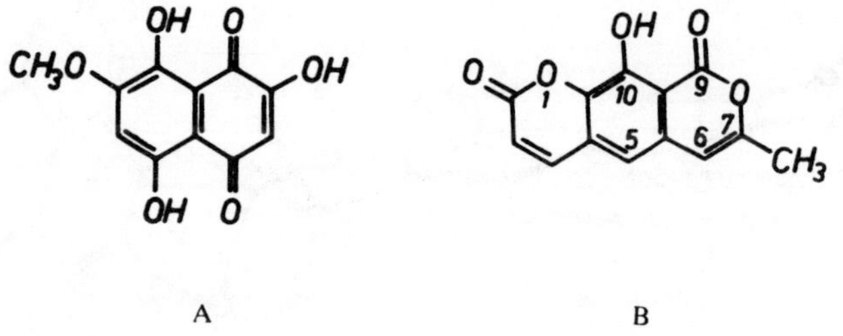

FIGURE 10. Degradation products of griseorhodins.

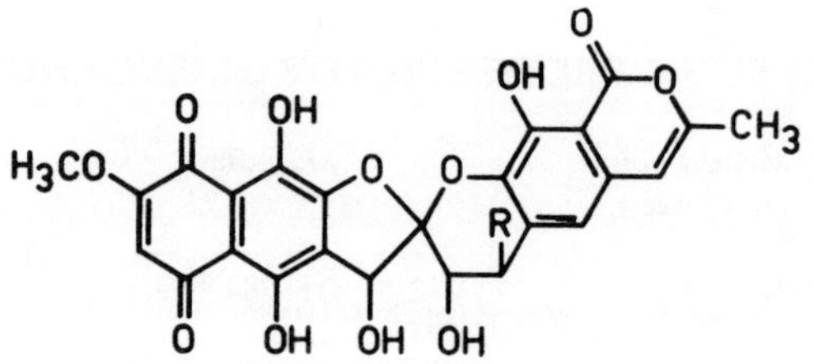

FIGURE 11. Stuctures of FCRC-57-U (griseorhodin C) and FCRC-57-G (griseorhodin G).

dation product of rubromycins. The second compound was obtained in very small quantities as yellow needles. Structure b was deduced for this degradation product mainly on the basis of mass and ^{1}H-NMR spectrometry. This interesting structure contains the complete isocoumarin moiety of the griseorhodin molecule.

The epoxide ring was derived from the NMR chemical shift and coupling constants of the protons at C-3 and C-4 and by conversion to the chlorhydrine and analogous derivatives. A decision between the two enantiomeric forms, C-3′ (R), c-2/2′ (S), C-3 (R), C-4 (R), and its antipode has not been reached.[62]

The isolation and structure elucidation of two griseorhodins (FCRC-57-U and FCRC-57-G) have been described by the Aszalos group.[56,57] The authors used ^{1}H-NMR ^{13}C-NMR, and mass spectrometry for the determination of the structures and, on the basis of these data, they concuded that FCRC-57-U and FCRC-57-G had structures a and b, respectively, as shown in Figure 11.

Both structures were confirmed by synthesis from griseorhodin A. In case of FCRC-57-U, the epoxide ring was cleaved by treatment with trifluoroacetic acid and water. FCRC-57-G was obtained from griseorhodin A using sodium borohydride in isopropanol-dioxane. The authors have proposed the name griseorhodin G for FCRC-57G.

The structure of griseorhodin C has been published independently by Eckardt et al.[63] Accordingly, FCRC-57-U and griseorhodic C appear to be identical.

Treatment of griseorhodin C with concentrated H_2SO_4 in acetone afforded the cyclic acetonide shown in Figure 12 from which the stereochemistry of griseorhodin C has been deduced to be 3′ (R), 2/2′ (S), 3 (R), 4 (S) or the enantiomeric form 3′ (S), 2/2′ (R), 3 (S), 4 (R).[62] Synthetic griseorhodin C, obtained by treatment of griseorhodin A with dil. H_2SO_4, has the same stereochemical arrangement as the natural compound, confirmed by the completely identical CD an NMR spectra.

Mass spectral fragmentation patterns were assigned by Stroshane et al.[57] and Tresselt et al.[60] ^{13}C-NMR structure assignments for all carbon atoms of griseorhodins A, C, and G have been published.[57]

4. Prunacetin A and FCRC-53

Similar to plurallin and iyomycin, two high molecular weight antitumor pigments have been described as isolation products from *S. griseus.* The first, prunacetin A, was isolated from a red variant of *S. griseus,* strain CD-270.[64] The lyophylized active substance was obtained as a deep purple amorphous powder. The compound had no definite melting point: λ_{max}226,280,560,590 nm (0.1 *N*NaOH). It is insoluble in common organic solvents and water.

Treatment of prunacetin A in buffer solution with trichloroacetic acid, followed by extraction with chloroform, afforded the pigment AE which was chemically related to griseorhodins. The apoprotein was supposed to be a glycoprotein with a mol wt of >70.000.

Recently, Chan et al.[76] have described another high molecular weight substance which has been found to be related to prunacetin A. The compound, named FCRC-53, was isolated from the deep purple-colored culture filtrate of a *S. griseus* strain.

Purified FCRC-53 was obtained as a brownish-purple acidic compound with a melting point above 280°C. λ_{max}: 260, 515, 540, 580 nm (0.1 *N*NaOH). The molecular weight of the peptide was calculated on the basis of amino acid analysis and was found to be 5000. The chemical structures of both prunacetin A and FCRC-53 are not yet known.

Prunacetin A is active against gram-positive bacteria and HeLa cells. The LD_{50} (mice) was 93.4 mg/kg, i.p., and 220 mg/kg, i.v. The antibiotic exhibited antitumor activity against Ehrlich and S-180 solid tumors, but was devoid of activity against ascites types of tumors.[64]

The antibiotic FCRC-53 was found to inhibit Gram-positive microorganisms and had cytotoxic properties against KB cells. Contrary to prunacetin A, FCRC-53 was active against P 388 ascites tumor.[76] The rubromycins, purpuromycin, and griseorhodins showed virtually similar activity against Gram-positive bacteria. Gram-negative bacteria were only weakly inhibited.[47,50,52,57,65,66]

Coronelli et al.[52] reported that purpuromycin was active on both Gram-positive and Gram-negative bacteria. This also includes activity against strains which are resistant to other antibiotics used in chemotherapeutic practice. Purpuromycin showed practically no toxicity up to 500 mg/kg mice, i.p. and 1000 mg/kg, orally. When administered orally, purpuromycin is active against lymphoid leukemia L-1210 in mice at a dose of 400 mg/kg.[67] Griseorhodins A, C, and G were found to be significantly active in vitro against cells derived from human carcinoma of the nasopharynx (KB).[57]

C. Granaticin and Structural Analogues

1. Granaticin (Granaticin A)

Another quinone antibiotic — granaticin — has become of interest in recent years. This compound was first detected by Corbaz et al.[68] as an antibacterial agent. In 1975, Chang et al.[69] reported on the isolation of an antitumor antibiotic litmomycin. This antibiotic was found to be identical with granaticin as shown by spectroscopy and by direct comparison of authentic materials.

Granaticin forms red crystals, m.p.:204 to 206°C (dec.)[68] or 223 to 225°C,[69] λ_{max}:223, 286, 496(sh), 532, 576 nm (EtOH). The molecule of granaticin contains two nonaromatic cyclic parts fused to a naphthazarin chromophore[70] as shown in Figure 13. Degradation of granaticin itself, and its acetyl derivative with ozone followed by esterification with diazomethane, afforded fragments as shown in Figure 14 furnishing information about the saturated parts of the granaticin molecule. The study of the ^{1}H-NMR spectra of both degradation products and parent compound has led to two alternative structures for granaticin. The final structure as shown in Figure 13, including the absolute configurations, could be determined by X-ray analysis.[71] Granaticin is active against Gram-positive bacteria. At concentrations between 0.1 and 0.8 μg/mℓ,

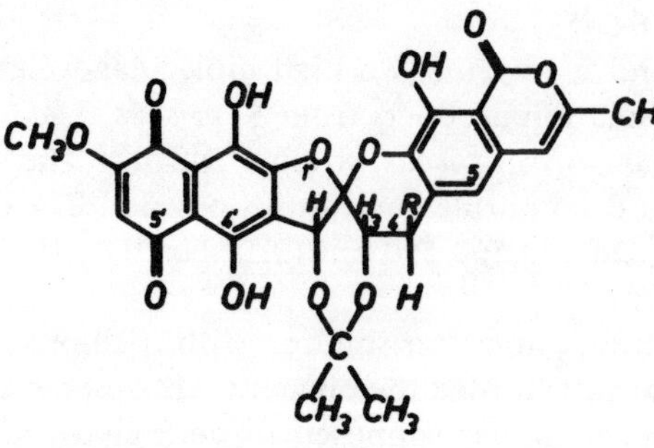

FIGURE 12. Acetonide of griseorhodin C.

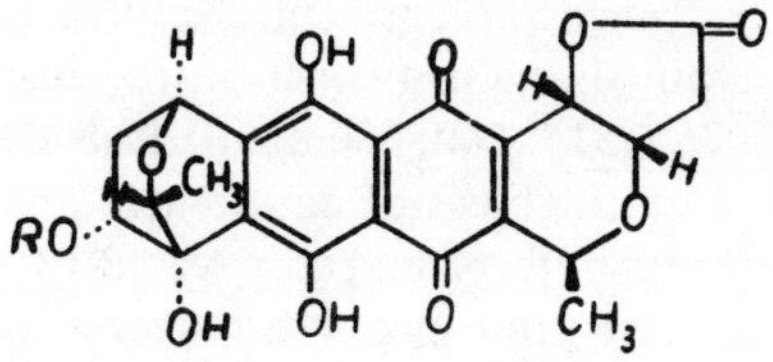

Granaticin, R = H

Granaticin B, R =

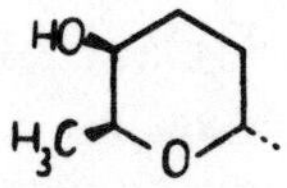

FIGURE 13. Structure of granaticins.

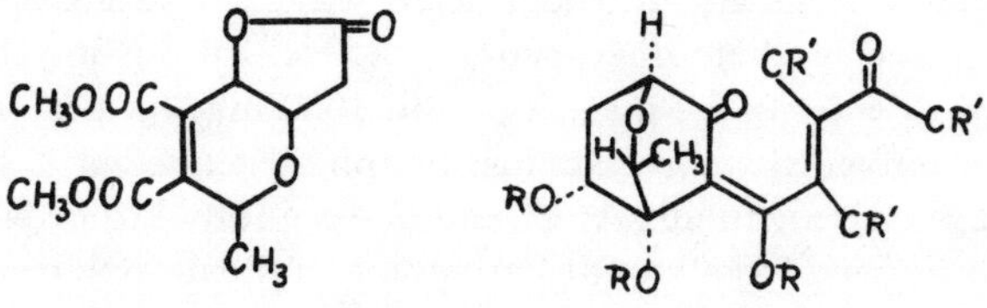

FIGURE 14. Degradation products of granaticin.

the effect of granaticin is bacteriostatic, and at higher concentrations is bactericidic. The mode of antibacterial action has been extensively reviewed by Kersten.[74,77] The antibiotic inhibits, preferentially, RNA syntheses. Ogilvie et al.[90] have shown that the mode of action relates to the interference with the aminoacylation of tRNA.^{LEU}

Antitumor activity was found against P-388 lymphocytic leukemia in mice (T/C 166% at 1.5 mg/kg). The substance was cytotoxic against KB cells as stated by Chang et al.[69] Biosynthesis of gramaticin was studied by Snipes et al.[91]

2. Granaticin B, Dihydrogranaticin, Zg, Zgg, Granaticinic Acid

In fermentations of certain *Streptomyces* strains, granaticin was accompanied by some related pigments. Thus, granaticin B was separated from crude mixtures by countercurrent distribution as a red amorphous powder. λ_{max}:223, 285, 498 (sh), 527, 566 nm (EtOH). The substance was active against some Gram-positive bacteria.[72]

Granaticin B inhibits DNA-dependent RNA synthesis.[89] The structure of granaticin B was determined by the same authors.[70-72] Granaticin B is the L-rhodinoside of gran-

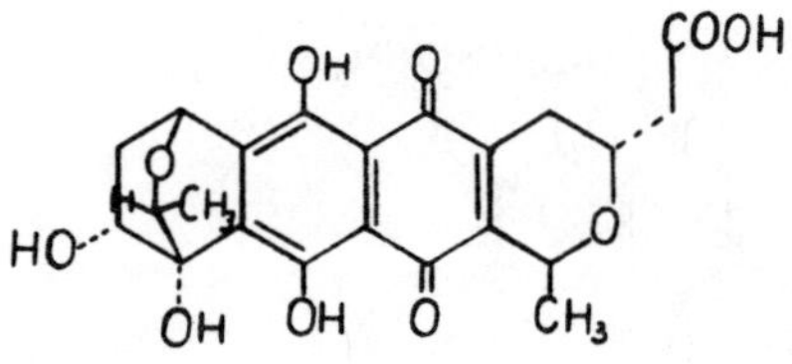

FIGURE 15. Stucture of dihydrogranaticin.

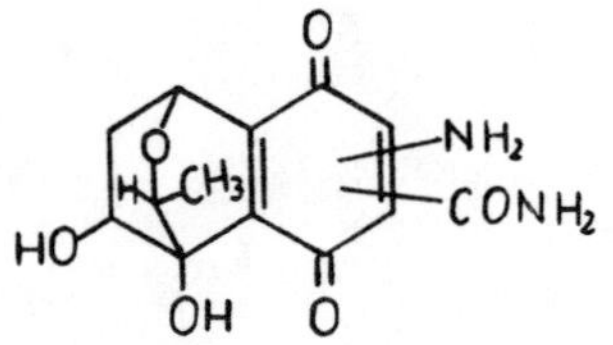

FIGURE 16. Structure of sarcinamycin A.

aticin. The sugar is attached to the hydroxyl at C-12. Rhodinose is a sugar moiety found in anthracyclines.

Dihydrogranaticin (Figure 15) was later isolated from fermentations of *S. thermoviolacens* subsp. pigens var. WR-141, together with granaticin and two related quinizarin derivatives Zgg and Zg.[73] Antitumor properties of these compounds have not been reported. Recently, a new representative of the granaticin family has been isolated from the culture of a thermophilic streptomycete sp. XT-11989.[182] Direct comparisons of ^{13}C- and ^{1}H-NMR spectral data of granaticin and the new compound, termed granaticinic acid, established the chemical structure to be a hydroxyderivative of dihydrogranaticin.

3. Sarcinomycin A

A new compound which was found to have interesting structural relationships to granaticins has recently been described by Reinhardt et al.[75] The antibiotic was extracted from the culture filtrate of *Streptomyces* JA 2861. The red active material could be obtained as pure compound-by-column chromatography, and was given the name sarcinamycin A because of its specific activity against *Sarcina lutea*. The substance forms red crystals. M.p.:194 to 195°C. λ_{max}:261 and 474 nm (chloroform).

The UV spectrum indicated that sarcinamycin A is a benzoquinone derivative. The molecular formula of $C_{13}H_{14}N_2O_6$ was established by mass spectrometry and elemental analysis. Elucidation of the structure was based on ^{1}H-NMP and ^{13}C-NMR investigation of sarcinamycin and its monoacetate, and has led to the structure shown in Figure 16. The precise location of -NH_2 and -$CONH_2$ remains to be determined.

Sarcinamycin A and granaticin are interesting examples of rather different compounds which may have common biogenetic routes leading to identical parts of their structures.

D. Benz[a]anthraquinone-Derived Antibiotics

The detection of the first compound of this group — tetrangomycin — dates back to 1965. The antibiotic was found in cultures of a variant strain of *Streptomyces rimosus.*[78] Other compounds of this family are aquayamycin, SS-228 y, rabelomycin, and yoronomycin. Out of these compounds, aquayamycin and SS-228 y have been shown to exhibit antitumor effects.

1. Aquayamycin

Aquayamycin was isolated from *S. misawanensis* nov. spec. by Hamada et Okami.[79] It was obtained as orange-yellow crystals by extraction of the culture filtrate followed by column chromatography of the antibacterial active material.

According to its spectral data, aquayamycin was clearly a derivative of 5-hydroxy-1,4-naphthaquinone.[80] A double bond conjugated to the naphthaquinone chromophore was suggested from catalytic reduction and the NMR signals of two vicinal olefinic protons. Aquayamycin easily formed a triacetyl derivative. The existence of the

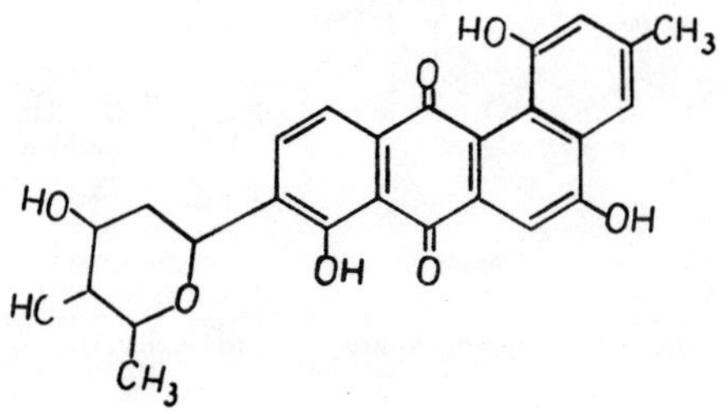

FIGURE 17. Benz[*a*]anthraquinone derivative of aquayamycin.

Table 4
PHYSICAL PROPERTIES OF BENZ [*a*] ANTHRAQUINONE-DERIVED ANTIBIOTICS

Antibiotic	Melting point (°C)	λ_{max} (nm), MeOH	$[\alpha]_D^{20}$	Molecular formula
Aquayamycin	189—190 (dec)	220, 320, 430	+ 160° (1.0, dioxane)	$C_{25}H_{26}O_{10} \cdot C_6H_{12}O_2$
SS-228Y	256—266 (dec)	218, 228(sh), 415(sh), 440-460	−85° (1.0, acetone)	$C_{19}H_{14}O_6$
Tetrangomycin	182—184	206, 267, 330(sh), 400	$[\alpha]_D^{20}$, + 42° (0.861, $CHCl_3$)	$C_{19}H_{14}O_5$
Rabelomycin	193 (dec)	228, 267, 433	−102° ± 10 (1.0 $CHCl_3$)	$C_{19}H_{14}O_6$
Yoronomycin	217—219 (dec)	214, 245(sh), 310, 420	−73.8° (0.8 dioxane)	$C_{19}H_{16}O_8$

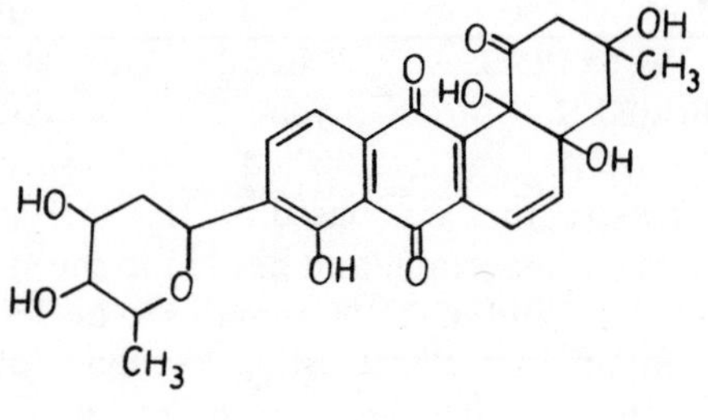

FIGURE 18. Structure of aquayamycin.

two vicinal secondary OH groups were deduced from the reaction of the acetate with periodic acid and the NMR spectra. From consideration of the oxygen functions and the number of double bond equivalents, a pentacyclic structure was suggested.

Various dehydration products were investigated to prove the carbon skeleton: zinc dust distillation afforded a naphthacene derivative. Heating of aquayamycin with barium hydroxide gave a tetracenequinone derivative by way of rearrangement. Heating of aquayamycin with 0.2 N HCl caused aromatization to yield a benz[*a*]anthraquinone (Figure 17 and Table 4) structurally similar to tetrangulol. The structure of the side chain moiety was proved to be 6-methyl-4,5-dihydroxypyran by NMR double resonance experiments. Further chemical reactions and ^{1}H-NMR studies, including NOE experiments, have led to the structure of aquayamycin, as shown in Figure 18).

Aquayamycin inhibits Gram-positive bacteria, except mycobacteria. When aquayamycin was administered to mice inoculated with Ehrlich ascites tumor for 10 days, prolongation of survival period was found at doses >12.5 μg/day. The antibiotic is also active on Yoshida rat sarcoma cells in tissue culture. LD_{50}, in mice, is 12.5 to 25 mg/kg, i.v.[79]

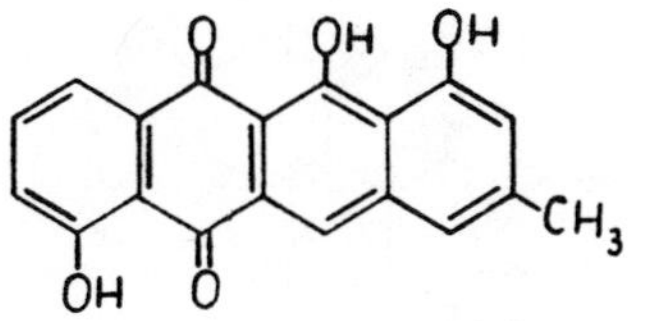

FIGURE 19. Conversion product of SS-228Y.

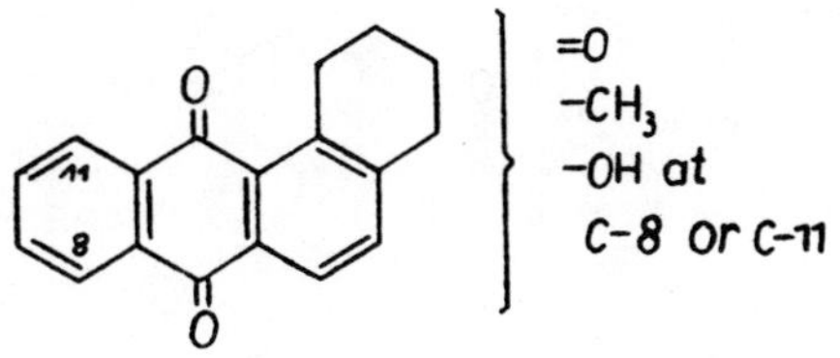

FIGURE 20. Degradation product of SS-228Y.

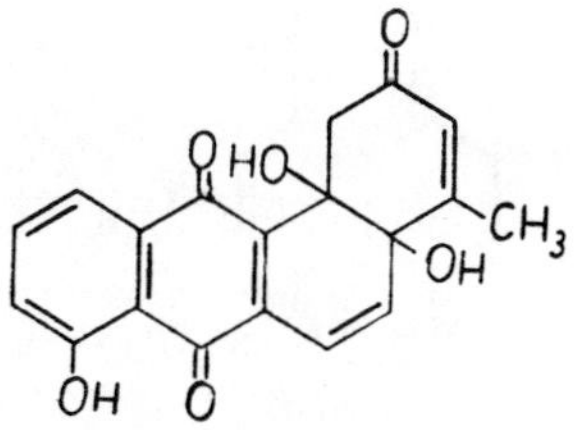

FIGURE 21. Stucture of SS-228Y.

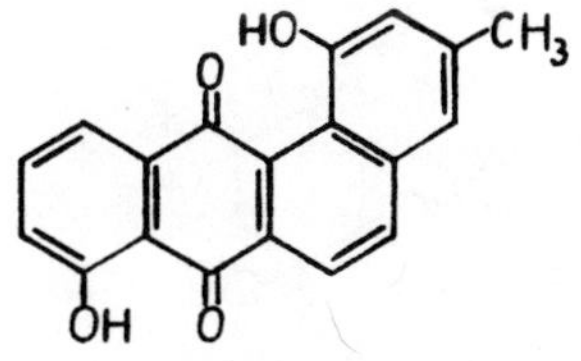

FIGURE 22. Stucture of tetrangulol.

2. *Antibiotic SS-228Y*

This antibiotic was isolated from a *Chainia* species designated *Chainia purpurogena* SS-228 (Okazaki et al.)[81] SS-228Y was obtained as orange to reddish-orange solid material which decomposed at 256 to 266°C.

The antibiotic is not stable. Light exposure or thermal treatment rapidly converts it to SS-228R (Figure 19). At first, this derivative has been studied to determine the structure of SS-228Y.[82] Color tests and zinc dust distillation easily established the hydroxynaphthacenequinone nature. The complete structure of SS-228R was then deduced on the basis of detailed NMR studies especially NOE and spin decoupling experiments.

The NMR pattern observed for the aromatic protons of the parent antibiotic was related to patterns in rabelomycin and tetrangomycin. Hydrogenation in methanol using platinium oxide gave a tetrahydro derivative containing a saturated six-membered ring ketone. Another informative derivative (Figure 20) was obtained when the tetrahydrocompound was heated in xylene with a trace of p-toluenesulfonic acid.

From these results and spectral data Kitahara et al.[82] concluded that SS-228Y had structure as shown in Figure 21.

Like aquayamycin, SS-228Y inhibits in vitro Gram-positive bacteria. It was found to be inactive against Gram-negative bacteria and mycobacteria. The antibiotic was proved to be active on Ehrlich ascites tumor at doses of more than 1.56 μg per mouse per day. LD_{50} in mice is 1.56 to 6.25 mg/kg, i.p.

3. *Tetrangomycin, Rabelomycin, and Yoronomycin*

Tetrangomycin was the first benz[*a*]anthraquinone derivative which has been detected as a metabolite of a living system. The structure elucidation therefore represents pioneer work in this field. The antibiotic (formerly known as LLAE 705 Y) was isolated from fermentations of a variant strain of *Streptomyces rimosus* as a yellow crystalline solid.[78] Treatment of tetrangomycin with mild base resulted in the formation of the new aromatic compound, tetrangulol[83] (Figure 22) which was also isolated as a minor component from the crude fermentation material. This aromatization reaction proved to be characteristic for this type of antibiotics and was analogously used by other authors to establish the structures of related compounds, for example aquayamycin and SS-228Y.

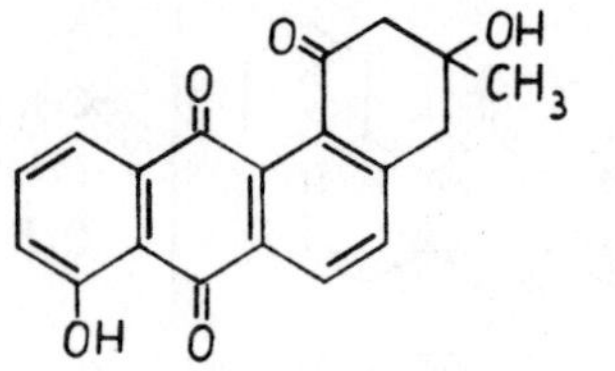

FIGURE 23. Stucture of tetrangomycin.

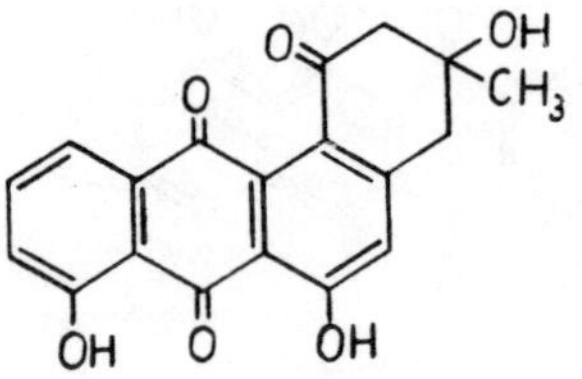

FIGURE 24. Structure of rabelomycin.

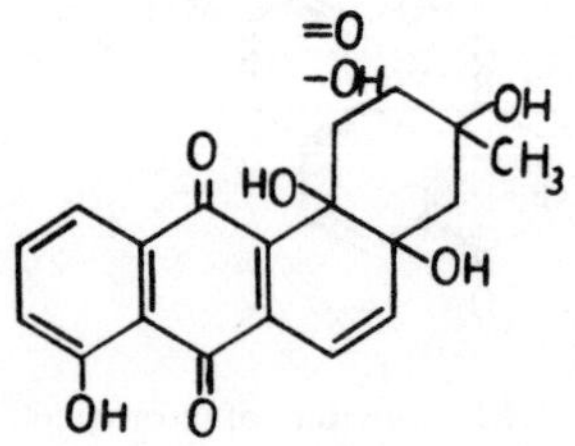

FIGURE 25. Stucture of yoromomycin.

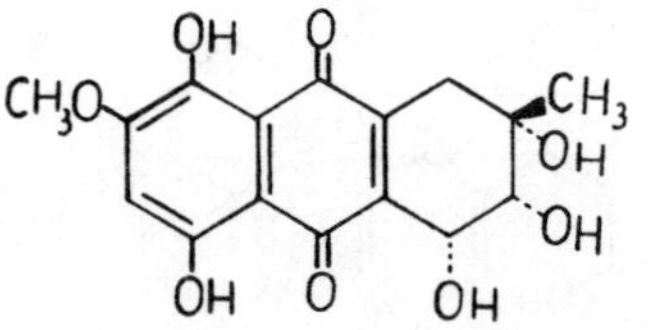

FIGURE 26. Stucture of bostrycin.

Various derivatives of tetrangulol have been prepared and compared with benz[*a*]anthracenes. From these and other data, Kunstmann and Mitscher,[83] and Dann et al.[78] concluded that tetrangulol and tetrangomycin have the structures as shown in Figure 23. Rabelomycin was extracted from the mycelium of *Streptomyces olivaceus* ATCC 21.549, and purified by countercurrent distribution followed by column chromatography on DEAE cellulose and preparative TLC.[84,85] The chemical structures of rabelomycin and tetrangomycin are closely related. The only difference is an additional OH which is present in the molecule of rabelomycin (Figure 24).

Recently, yoronomycin has been reported as a benz[*a*]anthraquinone type antibiotic found in cultures of *S. phaeochromogenes* var. *yoronensis.*[86] The structure of the antibiotic was formulated as shown in Figure 25. The arrangement of the substituents of the saturated ring remained to be determined. Interestingly, little is known about the stereochemistry of these compounds.

Tetrangomyin, rabelomycin, and yoronomycin are active against Gram-positive microorganisms. Yoronomycin is also active against Gram-negative bacteria. The mode of action and other biochemical problems of interest have not been studied so far.

E. Bostrycin

The fungus *Bostrichonema alpestre* produces the antibiotic bostrycin, a naphthaquinone derivative. The compound was extracted from the cells and formed red crystals. M.p.:222 to 224°C (dec). λ_{max}:228, 303, 472, 505, 542 nm (EtOH).[92] The chemical structure was studied by Noda et al.[93,94] who formulated bostrycin as shown in Figure 26 on the basis of spectral data and chemical work. Confirmation of this structure was obtained by X-ray analysis.[95] Bostrycin is effective against *Staphylococcus aureus, Candida albicans,* sarcoma-180 and leukemia virus. LD^{50} is about 109 mg/kg mice, i.p. Stereoselective sythesis of 6-demethoxybostrycin has been described by Roesner et al.[183]

F. Julimycins

Julimycins were found in the course of a screening for antiviral antibiotics in cultures of *Streptomyces shiodaensis.*[96] The crude extraction product designated as julimycin B complex consisted of the components B-O, B-I, B-II, B-III, B-IV, and S.V. Julimy-

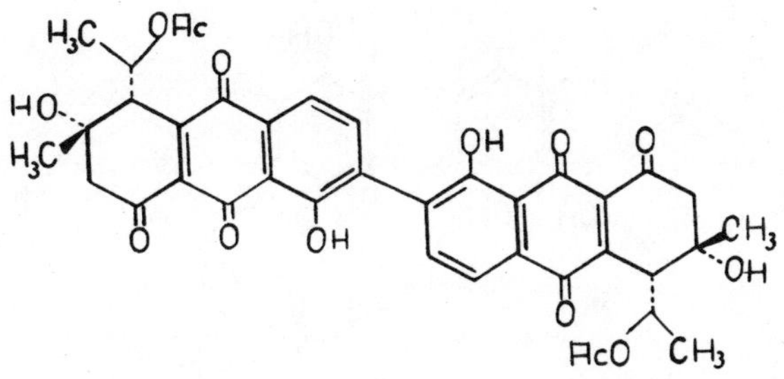

FIGURE 27. Stucture of julimycin B-II.

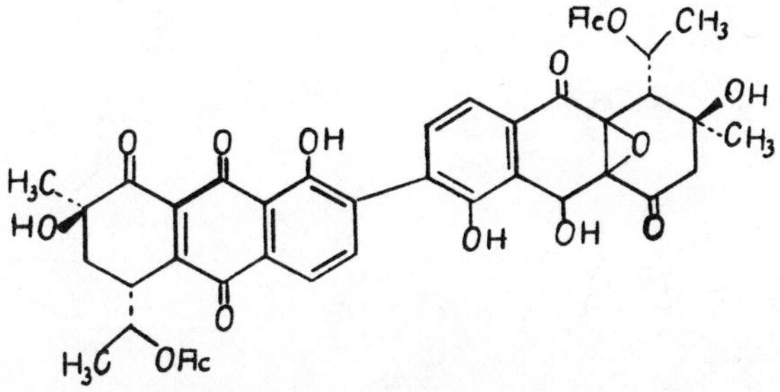

FIGURE 28. Structure of julimycin C.

cin B-II formed reddish-orange plates which decomposed at 215 to 220°, λ_{max}:234, 256 nm (CH_3OH). The compound was proved to be a β-β'-bianthraquinonyl derivative (Figure 27).[97-99] From the other fractions, more than 30 colored pigments (julichromes) were separated by TLC. Some of these pigments could be obtained as pure substances and have been the subject of further structural studies.[100] Julimycin C (Figure 28) was isolated as orange needles from cultures of the same *Streptomyces* strain under modified fermentation conditions.[101]

Julimycin B-II is active against gram-positive microorganisms[96] and is cytotoxic against HeLa cells. Antitumor testing indicated that julimycin B-II was inhibitory to Ehrlich ascites carcinoma in mice and ascites hepatoma AH-130 in rats when injected i.p. The antibiotic demonstrated chemoprophylactic effect against Friend virus leukemia, as well as poliovirus infections in mice.[102] Similar antiviral and cytotoxic effects were found with julimycin C.[101]

III. NONQUINOID CARBOCYCLIC COMPOUNDS

A. Pillaromycin A

Pillaromycin A is a product of *S. flavovirens* No. 65786. This yellow antibiotic is accompanied by the minor components B_I, B_{II}, and C. M.p.:208°C, λ_{max}:220, 275, 298 (sh), 309, 431 nm (EtOH).[103] Pillaromycin A was reported to exhibit a colchicine-like antitumor activity against certain tumor cells. Structure elucidation by Asai[104] including X-ray analysis led to a complete formula for pillaromycin A which was found to be a glycoside of a new sugar, pillarose. However, more recent work by Pezzanite et al.[105] and Walker and Fraser-Reid[106] has shown that the structure assigned to the sugar component was incorrect, and the new formula is shown in Figure 29. Efforts to synthesize pillarose were also made by Paulsen et al.[107] and Brimacombe et al.[108]

B. Pactamycin

Pactamycin was detected in 1960 as a fermentation product of *S. pactum* var. *pac-*

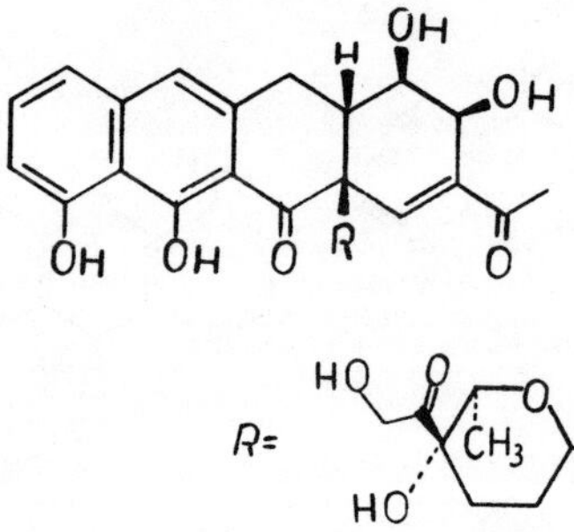

FIGURE 29. Stucture of pillaromycin A.

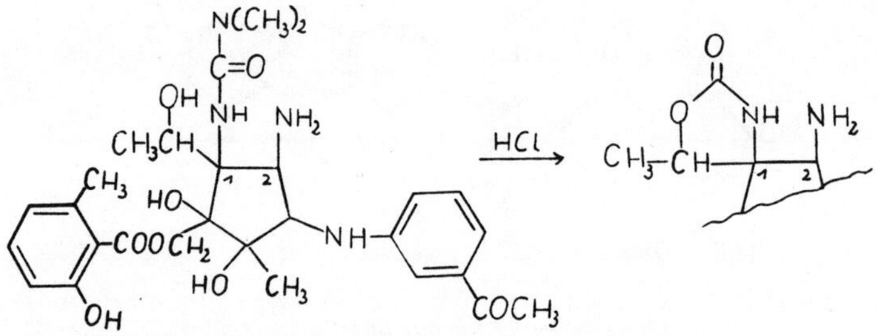

FIGURE 30. Pactamycin and its HCl treatment product.

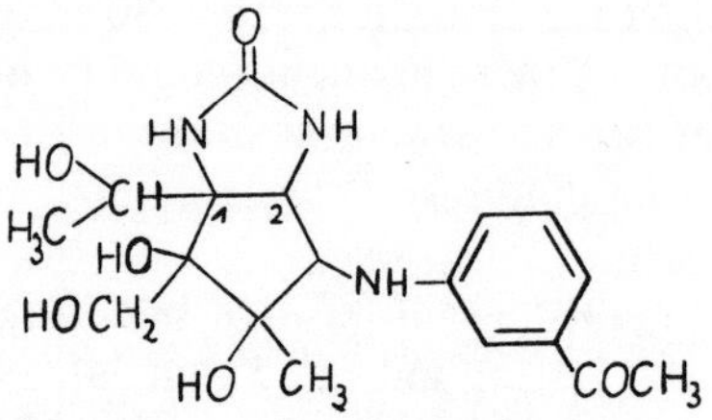

FIGURE 31. Alkaline treatment product of pactamycin.

tum.[109,110] The substance is an amphoteric white powder, λ_{max}:239, 264 (sh), 313, 356 nm (EtOH). Pactamycin is unstable in solution and loses antibiotic activity very rapidly.

Pactamycin is a highly substituted cyclopentan. As a result of mass spectral studies, the previously reported molecular formula[110] had to be revised to $C_{28}H_{38}N_4O_8$.[111] Heating of pactamycin in 2 N HCl gave dimethylamine and a fragment shown in Figure 30. Mild basic hydrolysis of pactamycin afforded the cyclic urea derivative shown in Figure 31 which lacks the methylsalicylic acid moiety. The presence of the *m*-acetoanilino moiety was found by oxydation of the alkaline treatment product with performic acid. This reaction gave *m,m'*-diacetoazoxybenzene and *m*-nitroacetophenone. Some further chemical reactions and examination of IR, mass, and NMR spectra have led to the assignment of the structure of pactamycin, as shown in Figure 30.

Pactamycin is active against Gram-positive and, to a lesser degree, Gram-negative bacteria. It exhibits a strong in vitro activity against KB human epidermoid carcinoma cells. In vivo, pactamycin was active against tumors in mice and hamsters (Ca-755, L-1210 and eight other animal tumors). The LD_{50} in mice was 10.7 mg/kg, orally, and

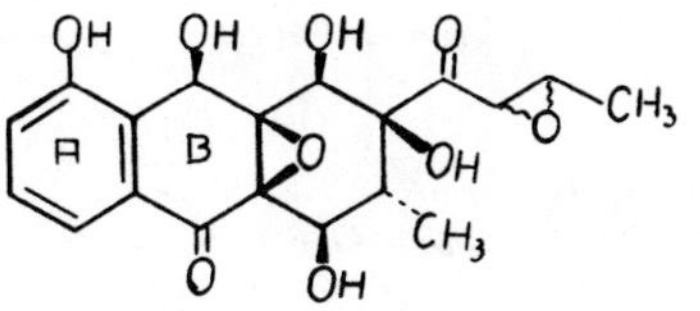

FIGURE 32. Stucture of cervicarcin.

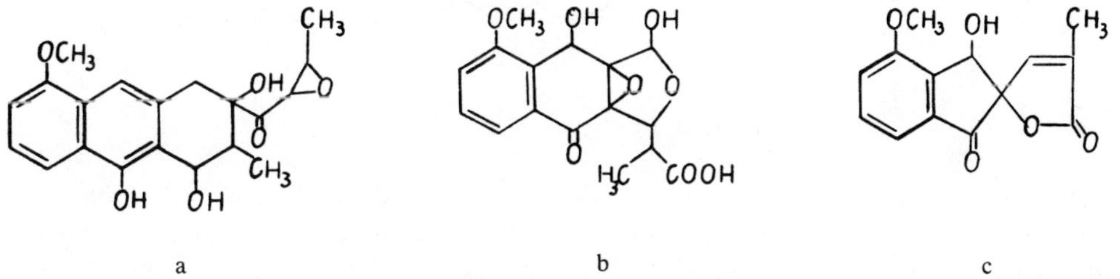

a

b

c

FIGURE 33. Reaction products of cervicarcin.

15.6 mg/kg, i.v. Dogs were much more sensitive. A single dose of 0.75 mg/kg proved to be lethal to dogs.

The mechanism of action of pactamycin was investigated by many workers and has been reviewed extensively by Gale et al.,[112] Goldberg,[113] and, recently, by Vázquez.[114] In general, pactamycin was found to be an inhibitor of protein synthesis in intact cells and in extracts of bacteria and eukaryotes. At low concentrations, the antibiotic binds rapidly to the 30 S ribosomal subunits of bacteria and to the 40 S subunits of mammalian cells.

It was inferred from various results that pactamycin, within a certain range of low concentrations, is a selective inhibitor of polypeptide chain initiation. Main support for this primary action comes from experiments using intact cells and extracts from mammalian sources. Thus, low concentrations of pactamycin were shown to inhibit the initiation of globin chains in vitro without inhibition of protein chain completion or release. Under changed conditions, especially at higher concentrations, chain elongation is also effected.

More recent work has shown that this inhibition of polypeptide chain initiation takes place after the binding of the initiator Met-tRNA$_F$ to the 40 S subunit, and is associated with the ribosomal "joining reaction" by preventing the release and reuse of the joining factor.[115-117] In bacterial systems, inhibition of peptide chain initiation is not as pronounced as that found in reticulocyte systems.

C. Cervicarcin

Ohkuma et al.[118] isolated cervicarcin from the culture filtrate of *S. ogaensis.*[119] The antibiotic formed colorless crystals, m. p.:203 to 205°C (dec), λ_{max}:227, 264, and 323 nm (EtOH). The chemical structure (Figure 32) of this interesting compound was mainly established by careful study of chemical degradation reactions.[120,121] The authors prepared a methyl-derivative which was used as starting material for the following reactions. Periodate oxidation of methyl-cervicarcin glycol, obtained by treatment of methylcervicarcin with acid, liberated acetaldehyde. The same oxidation reaction afforded crotonaldehyde when deoxymethylcervicarcin was used. Both reactions support the α, β-epoxyketone structure of the side chain.

Potassium permanganate oxidation gave 3-methoxyphthalic anhydride. Catalytic hydrogenation of methylcervicarcin with platinium oxide formed a dihydroderivative.

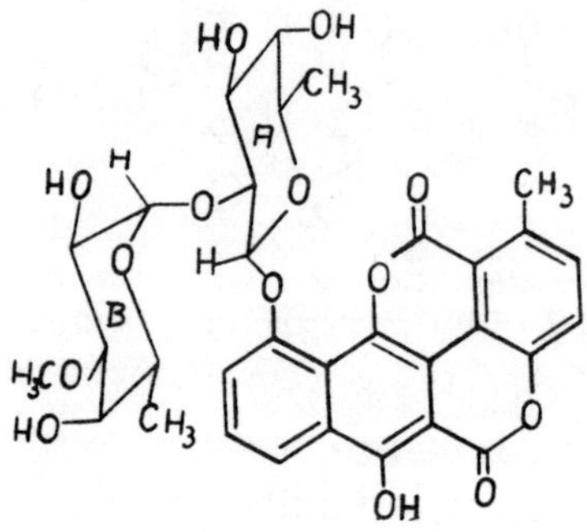

FIGURE 34. Structure of chartreusin.

A naphthalene compound (Figure 33) was formed under reductive conditions using 30% palladium on carbon, and treatment of methylated cervicarcin with sodium periodate gave cervic acid (Figure 33a). This reaction established the structure of ring C. Lactonization of b followed by periodate oxidation led to structure C via several steps. The structure investigation was completed by the isolation of 3-methyl-5,9-dimethoxyanthraquinone-(1,4) as a degradation product, and by the determination of the absolute configurations on spectral grounds.[176]

Cervicarcin is devoid of any activity against bacteria, yeast, and fungi. At dosages not less than 7.5 mg/kg/day the antibiotic prolonged the survival period of mice bearing Ehrlich ascites carcinoma. The compound was also active on solid tumors when tested against Ehrlich carcinoma, sarcoma-180 and NF sarcoma, but had only slight effect on sarcoma-37 and Friend virus leukemia.[122] Out of the derivatives of cervicarcin, only the methyl ether had a marked inhibitory effect on Ehrlich ascites carcinoma which was found to be superior to that of the parent compound.

The LD_{50} of cervicarcin is 48.5 mg/kg, i.p., and of the methyl ether is 13.5 mg/kg mice.

D. Chartreusin

Recently, antitumor activity of the well-known antimicrobial antibiotic, chartreusin, has been detected. This compound (identical with X-465 A,[124,125] 747, [126,127] and lambdamycin[128]) was isolated from several species of *Streptomyces*, especially *S. chartreusis*.[123] Chartreusin forms yellow or yellow-green crystals. M.p.:180°C,[123] or 184 to 186°C,[124] λ_{max}:236, 266, 334, 380, 401, 424 (95% EtOH). The antibiotic is a disaccharide of a substituted 2-phenylnaphthalene. The disaccharide chain is composed of D-fucose (A) and D-digitalose (B). Initial work to derive the chemical structure was done by Sternbach et al.[125] In further investigations, Simonitsch et al.[129] and Eisenhuth et al.[130] formulated chartreusin as shown in Figure 34.

Mild acid hydrolysis leads to the cleavage of D-digitalose and the D-fucoside. Hydrolysis, under drastic conditions, affords fucose and the chartreusin aglycone. Partial synthesis of chartreusin was attempted by Guyot.[136]

The antibiotic is active against Gram-positive bacteria and was found to exhibit significant activity against ascitic P 388, L-1210 leukemia and B-16 melanoma (mice) when administered i.p.[131,132] The drug binds to DNA and inhibits DNA and RNA polymerase activities of L-1210 cells in culture and in vivo.[133] Activity against leukemia L-1210 has also been described by Fleck et al.[128] Ergotropic properties were found by Heinecke et al.[134] It was shown that the chartreusin aglycone is biosynthesized by condensation and subsequent scission of a single polyketide chain of 22 carbon atoms.[135]

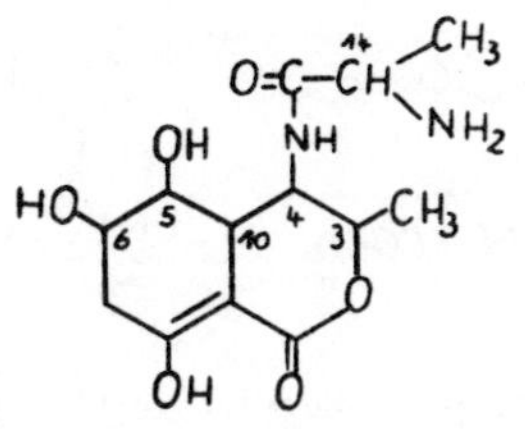

FIGURE 35. Structure of actinobolin.

E. Actinobolin, Bactobolin

Actinobolin is a broad-spectrum antibiotic which has been isolated from fermentations of *S. griseoviridis* var. *atrofaciens* (P-D05000) by means of carbon adsorption and ion exchange techniques.[137,138] The antibiotic was obtained as a white powder with no definite melting point. It is hygroscopic and very soluble in water. λ_{max} (sulfate, in 0.1 *N* HCl):263 nm.

The investigation of actinobolin is an instructive example of the application of computer techniques to structure elucidation. Acid hydrolysis of actinobolin and base hydrolysis of N-acetyl-actinobolin[139-143] afforded a number of degradation products. The structures of these products were determined by further chemical methods, spectral data and use of a computer program which generated all fragment structures consistent with available data. On the basis of these results, the authors formulated actinobolin as shown in Figure 35. The isolation of L-alanyl-L-threonine from mild oxidation of actinobolin and further ^{1}H-NMR data, established the absolute configuration to be 3 (R), 4 (R), 5 (R), 6 (R), 10 (R), 14 (S).[144]

Actinobolin is active against Gram-positive and Gram-negative bacteria[145] and was found to inhibit some tumors in mice and rats. Marked effect ws observed on carcinoma 1025, glioma 26, Walker carcinosarcoma 256, sarcoma-180 ascites tumor, and Ehrlich ascites carcinoma as well as on two transplantable human neoplasms H. S. #1 and H. S. #3.[146,147] LD_{50} (mice) is approximately 800 mg/kg, i.v. There is some evidence that actinobolin interferes with purine metabolism[148] Actinobolin was found to possess interesting properties that make it suitable for topical application in the control of dental caries and periodontal disease.[149]

A new chlorine-containing antibiotic structurally related to actinobolin was isolated from the culture broth of *pseudomonas* BMG 13-A7.[184] The new compound, named bactobolin, was found to inhibit the growth of Gram-positive and Gram-negative bacteria. A marked prolongation effect in the survival period of mice implanted with the mouse leukemia L-1210 cells was observed after treatment with bactobolin. Based on ^{1}H NMR spectral studies of bactobolin and two derivatives as well as chemical degradation, the chemical structure of bactobolin was proposed to be 3-dichloromethyl-actinobolin. But contrary to the structure of actinobolin, the configuration at C-3 in bactobolin was proved to be S.

F. Simple 6-Membered Ring Derivatives

1. Enaminomycins, Epoformin

Enaminomycins A, B, and C (Figure 36) recently have been isolated from the culture broth of *S. baarnensis* No. 13120.[150] The compounds are very unique in their chemical properties (A: amorphous powder, m.p.:105°C (dec), λ_{max}: 245, 293 nm, B: colourless needles, m.p.:160°C (dec), λ_{max}:244, 293 nm, C: colourless needles, m.p.:173°C (dec), λ_{max}:245, 290 nm, (MeOH).[151] Enaminomycins are active against Gram-positive and Gram-negative bacteria and some species of fungi. Enaminomycin A is the only one

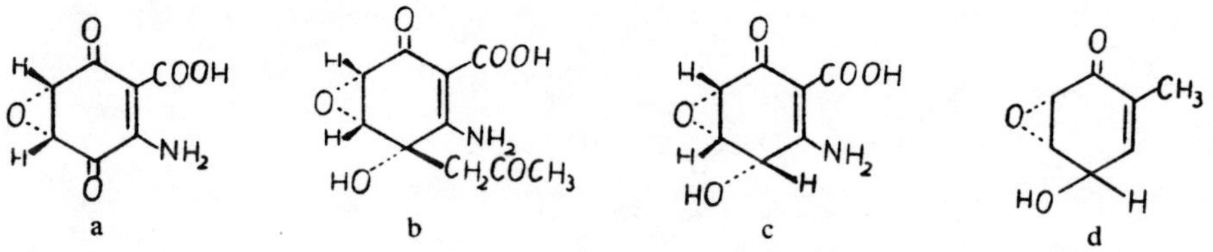

a b c d

FIGURE 36. Structure of enaminomycin A (a). enaminomycin B (b), enaminomycin C (c) and of epoformin (d).

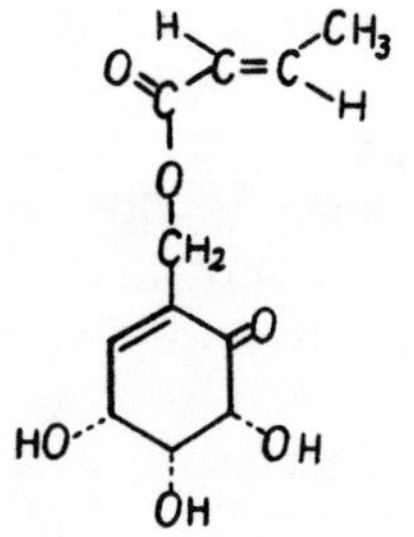

FIGURE 37. Structure of glyoxalase inhibitor.

also exhibiting activity against L-1210 mouse leukemia cells in vitro. LD_{50} for A: 28.5 mg/kg mice, i.v. Structure determinations using physicochemical methods and X-ray analysis have led to structures as shown in Figure 36.[152]

The enaminomycins are structurally related to epoformin,[153] terremutin,[154] panepoxydon,[155] and epoxydon.[156] Epoxydon, panepoxydon, and two other fungal metabolites of the epoxy quinone family (crotepoxide,[178] panepoxydion[155]) were described as antitumor compounds. Recently, an antagonist of chloramphenicol, I-851, was found to have the same constitution formula as reported for enaminomycin A.[179] Epoformin is a product of *Penicillium claviforme* which was found to exhibit activity against bacteria, PS cells, and L-1210 sarcoma.[153] It forms colorless crystals, $[\alpha]_D^{23} = +114.3°$ (with approximately 1.0 EtOH). Recently, syntheses of (±)-epoformin and (±)-epiepoformin was reported by Ichihara et al.[180]

2. 2-Crotonyloxymethyl-4,5,6-trihydroxycyclohex-2-enone

Umezawa et al. have started a screening program for glyoxalase inhibitors as potential anticancer agents. In the course of this screening, a low molecular weight substance was isolated from fermentations of *S. griseosporeus* which was obtained as colorless needles, m.p.:181°C, λ_{max}:211, 312 nm (EtOH). The active agent was found to inhibit glyoxalases prepared from both rat liver and yeast, and was active on HeLa cells an Ehrlich ascites carcinoma. Weak effect was observed on solid type of Ehrlich carcinoma and L-1210 cells inoculated to mice.[157]

The active compound was proved to be 2-crotonyloxymethyl-4(R),5(R),6(R)-trihydroxycyclohex-2-enone (Figure 37) on the basis of NMR and X-ray studies, as well as by conversion to the decrotonyl and the aromatic triacetate derivatives.[158] The glyoxalase inhibition might be caused by the reaction of the compound with enzyme-active sites.[157]

3. Terphenyllin A

A number of structurally interesting p-terphenyl derivatives has been isolated from a toxic strain of *A. candidus*. One of the main components, designated terphenyllin

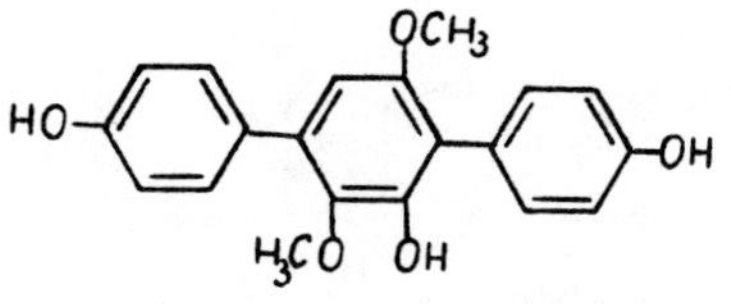

FIGURE 38. Structure of terphenyllin A.

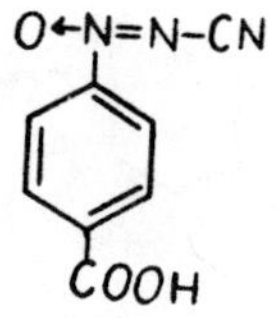

FIGURE 39. Structure of calvatic acid.

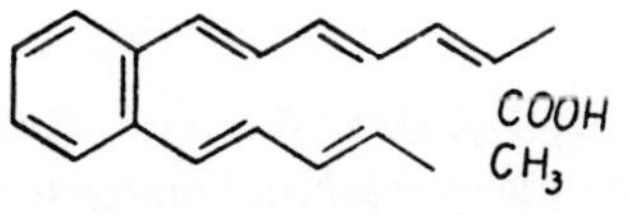

FIGURE 40. Structure of demetric acid.

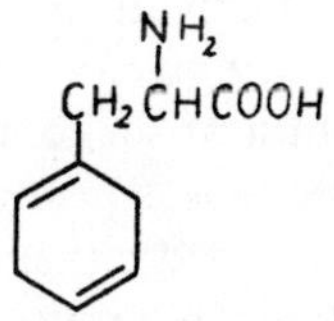

FIGURE 41. Structure of the metabolite of L-phenylalanine.

A, was found to exhibit specific cytotoxicity on HeLa cells. Terphenyllin A is a colorless compound (m.p.:244 to 245°, λ_{max}:222, 275, EtOH) and has structure as shown in Figure 38[159] which was verified by Takahashi et al.[160] Cytotoxicity to cultured HeLa cells was observed with 4″-deoxyterphenyllin, a satellite compound of terphenyllin. p-Terphenyl derivatives are rare in microfungi. A new metabolite of this family, 3-hydroxyterphenyllin, was isolated from cultures of *A. candidus.*[181]

4. Calvatic acid

Calvatic acid was isolated from both culture broth of *Calvatia lilacina* (Berk) Henn. p.[161] and *C. craniformis* (SHW.) FR.[162] The compound was obtained as light yellow powder, m.p.:198 to 199°C (dec),[161] 182 to 183°C (dec), λ_{max}:306 nm (MeOH),[162] A structure as shown in Figure 39 has been established by chemical transformations,[161] as well as by synthesis.[162]

Calvatic acid is active against Gram-positive and some Gram-negative bacteria and fungi. It inhibits the growth of Yoshida sarcoma in cell culture and mouse leukemia L-1210.[162] LD_{50}:125 to 250 mg/kg mice, i.p.

5. Demetric acid

Demetric acid is a cytotoxic product of *Streptomyces umbrosus* var. *suragaoensis* ATCC 19104 obtained as light-yellow needles. λ_{max}:227, 283, 314 nm (EtOH) (Figure 40). The antibiotic inhibits some strains of bacteria, yeasts, fungi, and protozoa and is cytotoxic for HeLa cells, L cells, Ca 755, L-1210 and Ca 1498.[163]

6. L-3-(2,5-Dihydrophenyl)alanine

This antimetabolite of *L*-phenylalanine was isolated from an unclassified *Streptomyces* strain X-13185 as colorless needles, m.p.:206 to 208°C (hydrochloride). The structure as shown in Figure 41 has been confirmed by oxidative conversion to phenylalanine and synthesis.[164] The antibiotic was found to inhibit gram-positive and gram-negative bacteria and *Trichomonas vaginalis.*[164,165] Also, activity was observed against sarcoma-180 (mice) when administered intraperitoneally.[164] The compound was found to be an effective antagonist of phenylalanine in experiments with *Leuconostoc dextranicum* 8086 and young rats.[166]

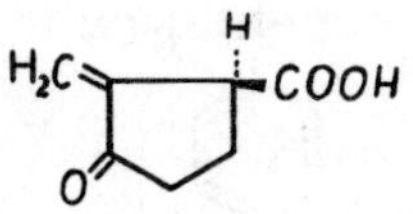

FIGURE 42. Structure of sarkomycin.

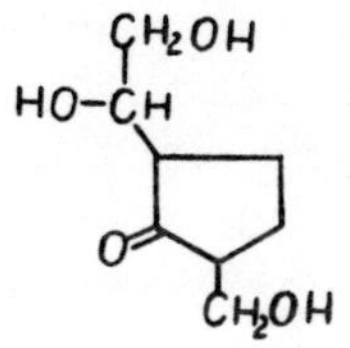

FIGURE 43. Structure of vertimycin.

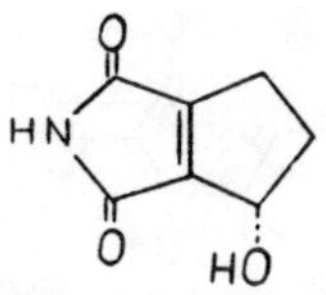

FIGURE 44. Structure of maleimycin.

G. Cyclopentane Derivatives

1. Sarkomycin, Vertimycin

The investigation of sarkomycin (= Sarkomycin A) dates back to the years 1953 to 1958. The antibiotic was produced by a strain of *Streptomyces erythrochomogenes*[167-169] and can be obtained either as free acid (colorless oil), or as salts (Na salt, Ca salt, both white, hygroscopic powders). Hooper et al.[170] established the chemical structure of sarkomycin mainly by ozonolysis which revealed the presence of the exocyclic methylene group and by catalytic reduction furnishing 2-methyl-3-cyclo-pentanone carboxylic acid. Synthesis of sarkomycin has been described by Toli[171] and Yonemitsu et al.[173] which proved the absolute configuration of natural (-)-sarkomycin as shown in Figure 42. Sarkomycin has only weak antibacterial activity, and it was found to be effective against Ehrlich carcinoma of mice. A more detailed review on the biological activity has been published by Sung.[172] A chemical structure similar to that of sarkomycin was established for vertimycin (Figure 43).[174] The light yellow substance is a product of *S. verticillatus* strain JA 4438, m.p.: 163 to 165°C (dec). The antibiotic was active against bacteria and Ehrlich ascites carcinoma both in vitro and in vivo.

2. Maleimycin

Maleimycin is an antibiotic which was isolated from a showdomycin producing strain of *S. showdoensis*. The antibiotic is active against leukemia L-1210 cells and is also active against gram-positive and gram-negative bacteria, including mycobacteria. Crystalline maleimycin melts at 116°, λ_{max}: (H_2O) 225 nm, 230 (sh) nm. By taking into consideration ^1H- and ^{13}C-NMR data, the structure shown in Figure 44 was concluded for the antibiotic. Although both maleimycin and showdomycin are derivatives of maleimide, the ways of biosynthesis seem to be different.[175] The antibiotic was synthetised by Singh and Weinreb[177] starting from cyclopenten-1,2-dicarboxylic acid.

IV. SUMMARY

In this chapter, antitumor antibiotics of carbocyclic structure have been reviewed. Out of these compounds, two belong to the early antibiotics which were found by systematic screening. In 1953, the low molecular weight substance, sarkomycin, was detected in a screening using Ehrlich carcinoma. Sarkomycin seemed to be of interest because of its low chronic toxicity and promising clinical results. At the very same time, several authors described remarkable results with actinomycin.

Pluramycin was found in 1956. The chemical structure of this antibiotic has been determined only recently. Several other antibiotics belonging to the pluramycin group have been described since. These compounds were found to intercalate into DNA and to be inhibitors of RNA and DNA synthesis.

Clinically, carbocyclic antitumor antibiotics were not extensively studied so far. In a few cases, derivatives of the antibiotics showed properties superior to those of the parent compounds. For example, the acetyl derivative of kidamycin was found to be less toxic than kidamycin itself without loss of antitumor activity. Similarly, acetylation of iyomycin B_1 seemed to enhance antitumor efficacy. However, attempts to use pluramycin-type compounds in combinations with carrier molecules failed to improve the therapeutic index in animal experiments.

In general, only a limited number of the carbocyclic antibiotics described above gave favorable preclinical or clinical results offering a chance for practical use. But since only a small number of clinical trials have been reported, definitive conclusions or critical evaluations are not yet justified. At the present time, several new ways for in vitro screening are used. Thus, for example, 2-crotonyloxymethyl-4,5,6-trihydroxycyclohex-2-enone has been detected by Umezawa et al.[162] in the course of a screening for glyoxalase inhibitors. It is hoped that such new screening methods will yield more novel carboxylic compounds, useful to combat cancer.

Significant progress has been made in the chemistry of these antibiotics. In the last few years the chemical structures of many antibiotics have been determined by modern physico-chemical methods. It was found that, except for certain groups of related compounds, the carbocyclic antibiotics are widely varying in their chemical structures. It is therefore understandable that structure-activity relationships among these compounds have not been detected.

Chemical modifications of anthracycline antibiotics have proved to be successful for the development of better anticancer derivatives. It is likely that using the same approach, progress will follow with other types of carbocylic antibiotics. The carbocyclic antibiotics are mostly stable compounds of low molecular weight and may therefore be useful as starting materials for preparation of more effective semisynthetic drugs.

REFERENCES

1. **Tanaka, N.**, in *Antibiotics, I. Mechanism of Action,* Gottlieb, D. and Shaw, P. D., Eds., Springer Verlag, Berlin, 1967, 166.
2. **Gale, E. F., Cundliffe, E., Reynolds, P. E., Richmond, M. H., and Waring, M. J.**, *The Molecular Basis of Antibiotic Action,* John Wiley and Sons, New York, 1972, 242.
3. **Bérdy, J.**, *Adv. Appl. Microbiol.,* 18, 309, 1974.
4. **Furukawa, M., Hayakawa, J., Ohta, G., and Jitaka, Y.**, *Tetrahedron,* 31, 2989, 1975.
5. **Furukawa, M.**, *Tetrahedron Lett.,* 1065, 3287, 1974.
6. **Séquin, U. and Furukawa, M.**, *Tetrahedron,* 34, 3623, 1978.
7. **Kanda, N.**, *J. Antibiot.,* 24, 599, 1971.
8. **Kanda, N., Kono, N., and Asano, K.**, *J. Antibiot.,* 25, 553, 1972.
9. **Kanda, N.**, *J. Antibiot.,* 25, 557, 1972.
10. **Hata, T., Umezawa, I., Komiyama, K., Asano, K., Kanda, N., Fujita, H., and Kono, M., 6th Int. Congr. Chemother., Tokyo, 1969,** *Progress in Antimicrobial and Anticancer Chemotherapy,* Vol. 1, University of Tokyo Press, Tokyo, 1970, 80.
11. **Umezawa, I., Komiyama, K., Takeshima, H., Hata, T., Kono, M., and Kanda, N.**, *J. Antibiot.,* 26, 669, 1973.
12. **Takeshima, H., Okamoto, M., Komiyama, K., and Umezawa, I.**, *Biochim. Biophys. Acta,* 418, 24, 1976.
13. **Okamoto, M., Takeshima, H., Komiyama, K., and Umezawa, I.**, *J. Antibiot.,* 29, 1334, 1976.
14. **Schmitz, H., Crook, K. E. Jr., and Bush, J. A.**, *Antimicrob. Agents Chemother. (1961-70),* 606, 1966.
15. **Séquin, U., Bedford, C. T., Sung, K. Ch., and Scott, A. J.**, *Helv. Chim. Acta,* 60, 896, 1977.
16. **Séquin, U.**, *Tetrahedron,* 34, 761, 1978.
17. **Bradner, W. T., Heinemann, B., and Gourevitch, A.**, *Antimicrob. Agents Chemother.,* 613, 1967.
18. **Nagai, K., Yamaki, T., Tanaka, N., and Umezawa, H.**, *J. Biochem. (Japan),* 62, 321, 1967.
19. **White, H. L. and White, J. -R.**, *Biochemistry,* 8, 1030, 1969.

20. Joel, P. B. and Goldberg, J. H., *Biochim. Biophys. Acta*, 224, 361, 1970.
21. Jernigan, H. M., Irvin, J. L., and White, J. R., *Biochemistry*, 17, 4232, 1978.
22. Maeda, K., Takeuchi, T., Nitta, K., Yagishita, K., Utahara, R., Osato, T., Ueda, M., Kondo, S., Okami, Y., and Umezawa, H., *J. Antibiot. Ser. A*, 9, 75, 1956.
23. Kondo, S., Wakashiro, T., Hamada, M., Maeda, K., Takeuchi, T., and Umezawa, H., *J. Antibiot.*, 23, 354, 1970.
24. Kondo, S., Miyamoto, M., Naganawa, H., Takeuchi, T., and Umezawa, H., *J. Antibiot.*, 30, 1143, 1977.
25. Takeuchi, T., Nitta, K., and Umezawa, H., *J. Antibiot. Ser. A*, 9, 22, 1956.
26. Takeuchi, T., Hikiyi, T., Nitta, K., and Umezawa, H., *J. Antibiot. Ser. A*, 10, 143, 1957.
27. Nishibori, A., *J. Antibiot. Ser. A*, 10, 213, 1957.
28. Tanaka, N., Nagai, K., Yamaguchi, H., and Umezawa, H., *Biochem. Biophys. Res. Commun.*, 21, 328, 1965.
29. Hisamatsu, T. and Koeda, T., *J. Antibiot.*, 24, 200, 1971.
30. Tsukada, J., Hamada, M., and Umezawa, H., *J. Antibiot.*, 24, 189, 1971.
31. Ogawara, H., Maeda, K., Nitta, K., Okami, Y., Takeuchi, T., and Umezawa, H., *J. Antibiot. Ser. A*, 19, 1, 1966.
32. Schnell, J., Dissertation, University of Göttingen, 1963.
33. Brockmann, H., *Angew. Chem.*, 80, 493, 1968.
34. Dahm, K. H., unpublished results.
35. Fricke, J., Dissertation, University of Göttingen, 1973.
36. Aszalos, A., Jelinek, M., and Berk, B., *Antimicrob. Agents Chemother.*, 68, 1964.
37. Matsumae, A. and Hata, T., *J. Antibiot. Ser. A*, 17, 112, 1964.
38. Nomura, S., Yamamoto, H., Matsumae, A., and Hata, T., *J. Antibiot. Ser. A.*, 17, 104, 1964.
39. Sano, Y., Kanda, N., and Hata, T., *J. Antibiot. Ser. A*, 17, 117, 1964.
40. Hoshino, M., Umezawa, J., Mimura, Y., and Hata, T., *J. Antibiot. Ser. A*, 20, 30, 1967.
41. Hata, T., Umezawa, I., and Kanda, N., *Antimicrob. Agents Chemother.*, 540, 1966.
42. Dornberger, K., Berger, U., Eckardt, K., Jungstand, W., Gutsche, W., Wohlrabe, K., and Knöll, H., GDR Patent, 188.597.
43. Dornberger, K., Berger, U., and Knöll, H., *J. Antibiot.*, in preparation.
44. Gutsche, W., Jungstand, W., and Wohlrabe, K., *Zentralbl. Pharm.*, 110, 1013, 1971.
45. Dornberger, K., unpublished results.
46. Brockmann, H. and Renneberg, K. H., *Naturwissenschaften*, 40, 49, 1953.
47. Brockmann, H. and Renneberg, K. H., *Naturwissenschaften*, 40, 166, 1953.
48. Lindenberg, W., *Arch. Mikrobiol.*, 17, 361, 1952.
49. Brockmann, H., Lenk, W., Schwantje, G., and Zeeck, A., *Tetrahedron Lett.*, 30, 3525, 1966.
50. Brockmann, H., Lenk, W., Schwantje, G., and Zeeck, A., *Chem. Ber.*, 102, 126, 1969.
51. Brockmann, H. and Zeeck, A., *Chem. Br.*, 103, 1709, 1970.
52. Coronelli, C., Pagani, H., Bardone, M. R., and Lancini, G. C., *J. Antibiot.*, 27, 161, 1974.
53. Bardone, M. R., Martinelli, E., Zerilli, L. F., and Coronelli, C., *Tetrahedron*, 30, 2747, 1974.
54. Eckardt, K., in *Antibiotics — Adv. Res., Prod., Clin. Use*, Herold, M. and Gabriel, Z., Eds., Butterworths London and Czechoslovak Medical Press, Prague, 1966, 1414.
55. Thrum, H., Eckart, K., Fuegner, R., and Bradler, G., *Z. Allg. Mikrobiol.*, 7, 121, 1967.
56. Stroshane, R. M., Chan, J. A., Aszalos, A. A., Rubalcaba, E. A., and Roller, P., 12th Middle Atlantic Regional Meeting, Am. Chem. Soc., (Abstr. OR 37), 1978.
57. Stroshane, R. M., Chan, J. A., Rubalcaba, E. A., Garretson, A. L., and Aszalos, A. A., *J. Antibiot.*, 32, 197, 1979.
58. Eckardt, K., Tresselt, D., and Ihn, W., Symposium Papers, *11th Int. Symp. Chem. Nat. Products*, JUPAC. Vol. I, 217, 1978.
59. Eckardt, K., Tresselt, D., and Ihn, W., *Tetrahedron*, 34, 399, 1978.
60. Tresselt, D., Eckardt, K., and Ihn, W., *Tetrahedron*, 34, 2693, 1978.
61. Eckardt, K., *Chem. Ber.*, 98, 24, 1965.
62. Eckardt, K., Tresselt, D., and Schönecker, B., *Tetrahedron*, in press.
63. Eckardt, K., Tresselt, D., and Ihn, W., *J. Antibiot.*, 31, 970, 1978.
64. Arai, T., Kushikata, S., Takamiya, K., Yanagisawa, F., Komyama, T., *J. Antibiot. Ser. A*, 20, 334, 1967.
65. Brockmann, H., et al., German Patent 918.162, 1954.
66. Eckardt, K., Thrum, H., Bradler, G. and Fuegner, R., *Antibiotiki*, 7, 603, 1965.
67. Pagani, H., Sesto, S. G., Coronelli, C., Bardone, M. R., Lancini, G., German Offen. 2.412.890.
68. Corbaz, R., Ettlinger, L., Gäumann, E., Kalvoda, J., Keller-Schierlein, W., Kradolfer, F., Manukian, B. K., Neipp, L., Prelog, V., Reusser, P., and Zähner, H., *Helv. Chim. Acta*, 40, 1962, 1957.
69. Chang, Ch.-J., Floss, H. G., Soong, P., and Chang, Ch.-T., *J. Antibiot.*, 28, 156, 1975.

70. Keller-Schierlein, W., Brufani, M., and Barcza, S., *Helv. Chim. Acta*, 51, 1257, 1968.
71. Brufani, M. and Dobler, M., *Helv. Chim. Acta*, 51, 1269, 1968.
72. Barcza, S., Brufani, M., Keller-Schierlein, W., and Zähner, H., *Helv. Chim. Acta*, 49, 1736, 1966.
73. Pyrek, J. St., Achmatowicz, O. Jr., and Zamojski, A., *Tetrahedron*, 33, 673, 1977.
74. Kersten, W., in *Progress in Molecular and Subcellular Biology*, Vol. 2, Hahn, F. E., Ed., Springer-Verlag, Berlin, 1971, 48.
75. Reinhardt, G., Bradler, G., Tresselt, D., Ihn, W., Radics, L., and Eckardt, K., Abstracts Int. Symp. on Antibiotics, Weimar, GDR, 1979, B-31.
76. Chan, J. A., Wei, T. T., Kalita, C. C., Warnick, D. J., Garretson, A. L., and Aszalos, A. A., *J. Antibiot.*, 30, 1140, 1977.
77. Kersten, H. and Kersten, W., *Molecular Biology, Biochemistry and Biophysics, Vol. 18, Inhibitors of Nucleic Acid Synthesis*, Kleinzeller, A., Springer, G. F., and Wittmann, H. G., Eds., Springer-Verlag, Berlin, 1974, 135.
78. Dann, M., Lefemine, D. V., Barbatschi, F. Shu, P., Kunstmann, M. P., Mitscher, L. A., and Bohonos, N., *Antimicrob. Agents Chemother.*, 832, 1966.
79. Sezaki, M., Hata, T., Ayukawa, S., Takeuchi, T., Okami, Y., Hamada, M., Nagâtsu, T., and Umezawa, H., *J. Antibiot.*, 21, 91, 1968.
80. Sezaki, M., Kondo, S., Maeda, K., Umezawa, H., and Ohno, M., *Tetrahedron*, 26, 5171, 1970.
81. Okazaki, T., Kitahara, T., and Okami, Y., *J. Antibiot.*, 28, 176, 1975.
82. Kitahara, T., Naganawa, H., Okazaki, T., Okami, Y., and Umezawa, H., *J. Antibiot.*, 28, 280, 1975.
83. Kunstmann, M. P. and Mitscher, L. A., *J. Org. Chem.*, 31, 2920, 1966.
84. Liu, W.-Ch., Parker, W. L., Slusarchyk, D. S., Greenwood, G. L., Graham, S. F., and Meyers, E., *J. Antibiot.*, 23, 437, 1970.
85. Meyers, E., Slusarchyk, D. S., Liu, S.-Ch., and Parker, W. L., German Patent 2.124.711, 1971.
86. Ozaki, M., Enami, Y., Matsumura, S., and Kumagaya, K., Abstracts of Papers, Annu. Meeting Agric. Chem. Soc., Kyoto, Japan, 255, 1976.
87. Hauser, F. M. and Rhee, R., *J. Am. Chem. Soc.*, 101, 1628, 1979.
88. Séquin, H. and Ceroni, M., *Helv. Chim. Acta*, 61, 2241, 1978.
89. Wêiser, J., Janda, J., Mikulik, K., and Tax, J., *Folia Microbiol. (Prague)*, 22, 329, 1977.
90. Ogilvie, A., Wiehauer, K., and Kersten, W., *Biochem. J.*, 152, 511, 1975.
91. Snipes, C. E., Chang, C.-J., *Planta Med.*, 33, 304, 1978., *J. Am. Chem. Soc.*, 101, 701, 1979.
92. Abe, J. et al., Japanese Patent 7.005.036, 1970.
93. Noda, T., Take, T., Otani, M., Miyauchi, K., Watanabe, T., and Abe, J., *Tetrahedron Lett.*, 58, 6087, 1968.
94. Noda, T., Take, T., Watanabe, T., and Abe, J., *Tetrahedron*, 26, 1339, 1970.
95. Takenaka, A., Furusaki, A. and Watanabe, T., *Tetrahedron Lett.*, 58, 6091, 1968.
96. Shōji, J.-I., Kimura, Y. and Katagiri, K., *J. Antibiot.* Ser. A, 17, 156, 1964.
97. Tsuji, N., *Tetrahedron*, 24, 1765, 1968.
98. Tsuji, N. and Nagashima, K., *Tetrahedron*, 24, 4233, 1968.
99. Nakai, H., Shiro, M. and Koyama, H., *J. Chem. Soc. B*, 498, 1969.
100. Tsuji, N. and Nagashima, K., *Tetrahedron*, 25, 3017, 1969, 26, 5719, 1970.
101. Katagiri, K. and Tsuji, N., Japanese Patent, 6.805.720, 1968.
102. Matsuura, S., Shiratori, O., Harada, Y., and Katagiri, K., *J. Antibiot. Ser. A*, 20, 282, 1967.
103. Asai, M., *Chem. Pharm. Bull.*, 18, 1699, 1970.
104. Asai, M., *Chem. Pharm. Bull.*, 18, 1706, 1713, 1720, 1724, 1970.
105. Pezzanite, J. O., Clardy, J., Lau, P.-Y., Wood, G., Walker, D. L., and Fraser-Reid, B., *J. Am. Chem. Soc.*, 97, 6250, 1975.
106. Walker, D. L. and Fraser-Reid, B., *J. Am. Chem. Soc.*, 97, 6251, 1975.
107. Paulsen, H., Roden, K., Sinwell, V., and Koebernick, W., *Chem. Ber.*, 110, 2146, 1977.
108. Brimacombe, J. S., Mather, A. M., and Hanna, R., *Tetrahedron Lett.*, 13, 1171, 1978.
109. Bhuyan, B. K., Dietz, A. and Smith, C. G., *Antimicrob. Agents Chemother.*, 184, 1961.
110. Argoudelis, A. D., Jahnke, H. K. and Fox, J. A., *Antimicrob. Agents Chemother.*, 191, 1961.
111. Wiley, P. F., Jahnke, H. K., MacKellar, F., Kelly, R. B., and Argoudelis, A. D., *J. Org. Chem.*, 35, 1420, 1970.
112. Gale, E. F., Cundliffe, E., Reynolds, P. E., Richmond, M. H., and Waring, M. J., *The Molecular Basis of Antibiotic Action*, John Wiley and Sons, New York, 310, 1972.
113. Goldberg, J. H., in *Antibiotics, III, Mechanism of Action of Antimicrobial and Antitumor Agents*, Corcoran, J. W. and Hahn, F. E., Eds., Springer-Verlag, Berlin, 498, 1975.
114. Vâzquez, D., *Inhibitors of Protein Biosynthesis*, Springer-Verlag, Berlin, 37, 1979.
115. Kappen, L. S., Suzuki, H. and Goldberg, J. H., *Proc. Natl. Acad. Sci.*, 70, 22, 1973.
116. Kappen, L. S. and Goldberg, J. H., *Biochem. Biophys. Res. Commun.*, 54, 1083, 1973.

117. Kappen, L. S. and Goldberg, J. H., *Biochemistry*, 15, 811, 1976.
118. Ohkuma, K., Suzuki, S., Itakura, C., Sega, T., and Sumiki, Y., *J. Antibiot. Ser. A*, 15, 247, 1962.
119. Nagatsu, J., Ansai, K., Ohkuma, K., and Suzuki, S., *J. Antibiot. Ser. A*, 16, 207, 1963.
120. Marumo, S., Sasaki, K. and Suzuki, S., *J. Am. Chem. Soc.*, 86, 4507, 1964.
121. Marumo, S., Sasaki, K., Ohkuma, K., Anzai, K., and Suzuki, S., *Agric. Biol. Chem.*, 32, 209, 1968.
122. Itakura, C., Sega, T., Suzuki, S., and Sumiki, Y., *J. Antibiot. Ser. A*, 16, 231, 1963.
123. Leach, B. E., Calhoun, K. M., Johnson, L. E., Teeters, C. M., and Jackson, W. G., *J. Am. Chem. Soc.*, 75, 4011, 1953.
124. Berger, J., Sternbach, L. H., Pollock, R. G., LaSala, E. R., Kaiser, S., and Goldberg, M. W., *J. Am. Chem. Soc.*, 80, 1636, 1958.
125. Sternbach, L. H., Kaiser, S. and Goldberg, M. W., *J. Am. Chem. Soc.*, 80, 1639, 1958.
126. Ghione, M. and Zavaglio, V., *Giorn. Microbiol.*, 2, 176, 1955.
127. Arcamone, F., Bizioli, F., and Scotti, T., *Antibiot. Chemother.*, 6, 283, 1955.
128. Fleck, W., Strauss, D., Prauser, H., Jungstand, W., Heinecke, H., Gutsche, W., and Wohlrabe, K., *Z. Allg. Mikrobiol.*, 16, 521, 1976.
129. Simonitsch, E., Eisenhuth, W., Stamm, O. A., and Schmid, H., *Helv. Chim. Acta*, 47, 1459, 1964.
130. Eisenhuth, W., Stamm, O. A., and Schmid, H., *Helv. Chim. Acta*, 47, 1475, 1964.
131. McGovren, J. P., Neil, G. L., Crampton, S. L., Robinson, M. I., and Douros, J. D., *Cancer Res.*, 37, 1666, 1977.
132. McGovren, J. P., Douros, J. D., and Neil, G. L., *Proc. Am. Assoc. Cancer Res.*, 17, 91, 1976.
133. Li, L. H. and Clark, T. D., Proc. Am. Assoc. Cancer Res., 17, 96, 1976; Cancer Res., 38, 3012, 1978.
134. Heincke, H., Strauss, D. and Fleck, W., *Arch. Tiereraehr.*, 27, 333, 1977.
135. Canham, P. and Vining, L. C., *J. Chem. Soc., Chem. Comm.*, 319, 1976; *Can. J. Chem.*, 55, 2450, 1977.
136. Guyot, M., *Bull. Mus. Natl. Hist. Nat. Sci. Phys. Chim.*, 5, 9, 1975.
137. Fusari, S. A. and Machamer, H. E., *Antibiot. Annu.*, 510, 1958.
138. Haskell, T. H. and Bartz, Q. R., *Antibiot. Annu.*, 505, 1958.
139. Munk, M. E., Sodano, C. S., McLean, R. L., and Haskell, T. H., *J. Am. Chem. Soc.*, 89, 4158, 1967.
140. Munk, M. E., Nelson, D. B., Antosz, F. J., Herald, D. L. Jr., and Haskell, T. H., *J. Am. Chem. Soc.*, 90, 1087, 1968.
141. Nelson, D. B., Munk, M. E., Gash, K. B., and Herald, D. L. Jr., *J. Org, Chem.*, 34, 3800, 1969.
142. Nelson, D. B. and Munk, M. E., *J. Org. Chem.*, 35, 3832, 1970.
143. Struck, R. F., Thorpe, M. C., Coburn, W. C. Jr., and Shealy, Y. F., *Tetrahedron Lett.*, 17, 1589, 1967.
144. Antosz, F. C., Nelson, D. B., Herald, D. L. Jr., and Munk, M. E., *J. Am. Chem. Soc.*, 92, 4933, 1970.
145. Pitillo, R. F., Fisher, M. W., McAlpine, R. J., Thompson, P. E., and Ehrlich, J., *Antibiot. Annu.*, 497, 1958.
146. Sugiura, K. and Reilly, H. C., *Antibiot. Annu.*, 522, 1958.
147. Teller, N. M., Merker, P. C., Palm, J. E., and Woolley, G. W., *Antibiot. Annu.*, 518, 1958.
148. Pitillo, R. F., Schabel, F. M. Jr., and Gilbert Quinnelly, B., *Antibiot. Chemother.*, 11, 501, 1961.
149. Hunt, D. E., Armstrong, P. J. Jr., Black, C. III, and Narkates, A. J., *Proc. Soc. Exp. Biol. Med.*, 140, 1429, 1972.
150. Arai, M., Itoh, Y., Enokita, R., Takamatsu, Y., and Manome, T., *J. Antibiot.*, 31, 829, 1978.
151. Itoh, Y., Miura, T., Katayama, T., Haneishi, T., and Arai, M., *J. Antibiot.*, 31, 834, 1978.
152. Itoh, Y., Haneishi, T., Arai, M., Hata, T., Aiba, K., and Tamura, C., *J. Antibiot.*, 31, 838, 1978.
153. Yamamoto et al., *Takeda Kenkyusho, HO*, 32, 532, 1973.
154. Miller, W., *Tetrahedron*, 24, 4839, 1968.
155. Kis, Z., Closse, A., Sigg, H. P., Hruban, L., and Snatzke, G., *Helv. Chim. Acta*, 53, 1577, 1970.
156. Closse, A., Mauli, R., and Sigg, H. P., *Helv. Chim. Acta*, 49, 204, 1966.
157. Takeuchi, T., Chimura, H., Hamada, M., Umezawa, H., Yoshioka, O., Oguchi, N., Takahashi, Y., and Matsuda, A., *J. Antibiot.*, 28, 737, 1975.
158. Chimura, H., Nakamura, H., Takita, T., Takeuchi, T., Umezawa, H., Kato, K., Saito, S., and Tomisawa, T., *Antibiotics*, 28, 743, 1975.
159. Marchelli, R. and Vining, L. C., *J. Chem. Soc. Chem. Commun.*, 555, 1973.
160. Takahashi, C., Yoshihira, K., Natori, S., and Umeda, M., *Chem. Pharm. Bull.*, 24, 613, 1976.
161. Gasco, A., Serafino, A., Mortarini, V., Menziani, E., Bianco, M. A., and Scurti, J. C., *Tetrahedron Lett.*, 38, 3431, 1974.
162. Umezawa, H., Takeuchi, T., Jinuma, H., Ito, M., Ishizuka, M., Kurakata, Y., Umeda, Y., Nakanishi, Y., Nakamura, T., Obayashi, A., and Tanabe, O., *J. Antibiot.*, 28, 87, 1975.

163. **DeVault, R. L., Schmitz, H., and Hooper, J. R.,** *Antimicrob. Agents Chemother.*, 796, 1965.
164. **Scannell, J. P., Pruess, D. L., Demny, T. C., Williams, T., and Stempel, A.,** *J. Antibiot.*, 23, 618, 1970.
165. **Genghof, D. S.,** *Can. J. Microbiol.*, 16, 545, 1970.
166. **Snow, M. L., Lauinger, C., and Ressler, C.,** *J. Org. Chem.*, 33, 1774, 1968.
167. **Umezawa, H., Takeuchi, T., Nitta, K., Okami, Y., Yamamoto, T., and Yamaoka, S.,** *J. Antibiot. Ser. A*, 6, 147, 1953.
168. **Umezawa, H., Takeuchi, T., Nitta, K., Yamamoto, T., and Yamaoka, S.,** *J. Antibiot. Ser. A*, 6, 101, 1953.
169. **Maeda, K. and Kondō, S.,** *J. Antibiot. Ser. A*, 11, 37, 1958.
170. **Hooper, J. R., Cheney, L. C., Cron, M. J., Fardig, O. B., Johnson, D. A., Johnson, D. L., Palermiti, F. M., Schmitz, H., and Wheatley, W. B.,** *Antibiot. Chemother.*, 5, 585, 1955.
171. **Toki, K., Wada, H., Suzuki, Y., and Saito, Ch.,** *J. Antibiot. Ser. A*, 10, 35, 1957.
172. **Sung, S.,** in *Antibiotics I, Mechanism of Action,* Gottlieb, D. and Shaw, P. D., Eds., Springer-Verlag, Berlin, 1967, 156.
173. **Yonemitsu, O., Sato, Y., Nishioka, S., and Ban, Y.,** *Chem. Ind.*, 490, 1963.
174. **Strauss, D.,** in *Antibiotics, Advances Res. Prod. Clin. Use,* Herold, M. and Gabriel, Z., Eds., Butterworths, London and Czechoslovak Medical Press, Prague, 1966, 451.
175. **Elstner, E. F., Carnes, D. M., Suhadolnik, R. J., Kreishman, G. P., Schweizer, M. P., and Robins, R. K.,** *Biochemistry,* 12, 4992, 1973.
176. **Marumo, S., Harada, N., Nakanishi, K., and Nishida, T.,** *J. Chem. Soc. Chem. Commun.*, 1693, 1970.
177. **Singh, P. and Weinreb, S. M.,** *Tetrahedron,* 32, 2379, 1976.
178. **Kupchan, S. M., Hemingway, R. J., and Smith, R. M.,** *J. Org. Chem.*, 34, 3898, 1969.
179. **Imagawa, Y., Shima, S., Hirota, A., and Sakai, H.,** *Agric. Biol. Chem.*, 42, 681, 1978.
180. **Ichihara, A., Moriyasu, K., and Sakamura, S.,** *Agric. Biol. Chem.*, 42, 2421, 1978.
181. **Kurubane, I., Vining, L. C., McInnes, A. G., and Smith, D. G.,** *J. Antibiot.*, 32, 559, 1979.
182. **Maehr, H., V. Cuellar, H., Smallheer, J., Williams, Th. H., Sasso, G. J., and Berger, J.,** *Monatsh. Chem.*, 110, 531, 1979.
183. **Roesner, A., Tolkiehn, K., and Krohn, K.,** *J. Chem. Res.*, 9, 306, 1978.
184. **Kondo, S., Horiuchi, Y., Hamada, M., Takeuchi, T., and Umezawa, H.,** *J. Antibiot.*, 32, 1069, 1979.

Chapter 3

ANSAMYCINS

V. Sagar Sethi

TABLE OF CONTENTS

I. INTRODUCTION

The term *ansamycin* has been coined for chemical structures with an aromatic chromophore which is joined at two nonadjacent positions by an aliphatic *ansa* chain.[1,2] The *ansa* chain (sometimes called bridge) is linked as an amide to the amino group of the chromophore and does not contain a lactone bond. The aromatic chromophore in most of the known ansamycins is capable of forming a quinone-hydroquinone structure. Ansamycins include antibiotics and plant products with interesting biological properties (Table 1). Maytansines are promising antitumor agents. Rifampicin is used clinically as an antitubercular and antibacterial drug. Some of the ansamycins are used as important inhibitors in cellular and molecular biological research. The molecular weights of ansamycins range between 560 and 750 daltons. Structurally, ansamycins can be divided into two broad classes, namely, naphthalenic ansamycins and benzenic ansamycins (Figures 1 and 2). Naphthalenic ansamycins include rifamycins, streptovaricins, tolypomycin, and naphthomycin. Benzenic ansamycins have maytansine, geldanamycin, and colubrinol as important members. The asymmetric chiral centers in the aliphatic *ansa* chains from C6-C13 have similar configurations suggesting a common pathway of their biosynthesis. Different numbering schemes have been used in the literature for ansamycins.[3,9,10] In order to compare the chemical and biological properties of various ansamycins, a common numbering system starting with C-1 of the amide function of the *ansa* ring, as proposed by Rinehart and Shield,[3] is used here (Figure 1). The numbering system for rifamycins according to Prelog and Oppolzer[2] and IUPAC rules is shown in Figure 3. The numbering system in Figure 3A for rifamycins is used frequently in the literature.

The three-dimensional structures of ansamycins as determined by X-ray crystallography suggests that in each case the *ansa* ring lies above the aromatic system. The coiling of the bridge above the aromatic ring also leads to atropisomerism,[1] which has been observed only for streptovaricins. The rigidity of the quaternary carbon (C-16) of rifamycins and tolypomycin presumably hinders the formation of atropisomers of these compounds. The aromatic rings of the benzenic ansamycins may be too small to prevent their passing through the center of the *ansa* ring. Ansamycins appear to be highly lipophilic compounds. In rifamycins, the hydrophilic oxygen containing functional groups seem to be on the same side of the molecule, thus giving a hydrophobic and a hydrophilic face to the molecule. Such molecules tend to aggregate and form miscelles.[10]

During the past few years, several review articles on ansamycins have been written.[3-10] My purpose in this chapter is to review, critically, the recent literature, especially with respect to the antitumor potential of these drugs.

II. BIOGENESIS OF ANSAMYCINS

Although ansamycins are produced by a variety of organisms, there are distinct structural similarities. For example, in the *ansa* ring, rifamycins, streptovaricins, and tolypomycin have the same number of carbon atoms (C17). Geldanamycin, maytansins, and colubrinol also have the same number of carbon atoms (C15) in the *ansa* bridge. These findings suggest a common biosynthetic pathway. The experimental data for biosynthesis of rifamycins, streptovaricins, and geldanamycin, indeed, support this hypothesis.[11-16]

Biosynthesis of ansamycins has been studied by incorporation of ^{14}C- or ^{3}H-labeled precursors into radioactive ansamycins followed by extensive degradation of the labeled products and assignment of the label for various carbon atoms. More recently, however, the use of ^{13}C-magnetic resonance (cmr) spectroscopy, along with pulsed

Table 1
BIOLOGICAL ACTIVITY, MOLECULAR WEIGHT, AND SOURCES OF NATURALLY OCCURRING ANSAMYCINS

Ansamycins	Mol wt	Sources	Biological activity
Naphthalenic			
Rifamycins	695—750	*Nocardia mediterranei* (*Streptomyces mediterranei*)	Antibacterial
		Micromonospora halophytica	Antifungal, Antiviral, Antitumor
Streptovaricins	750—830	*Streptomyces spectabilis*	Antibacterial, Antiviral
Tolypomycin	822	*Streptomyces tolypophorus*	Antibacterial
Naphthomycin	683	*Streptomyces collinus*	Antibacterial
Benzenic			
Geldanamycins	560	*Streptomyces hygroscopicus*	Antibacterial, Antiprotozoal
Maytansins	691—720	*Maytenus* sp. celastraceal (plant)	Antitumor
		Nocardia sp. (fermentation)	Antimitotic
Colubrinol	735	*Colubrina texensis*, rhamnaceae, (plant)	Antitumor

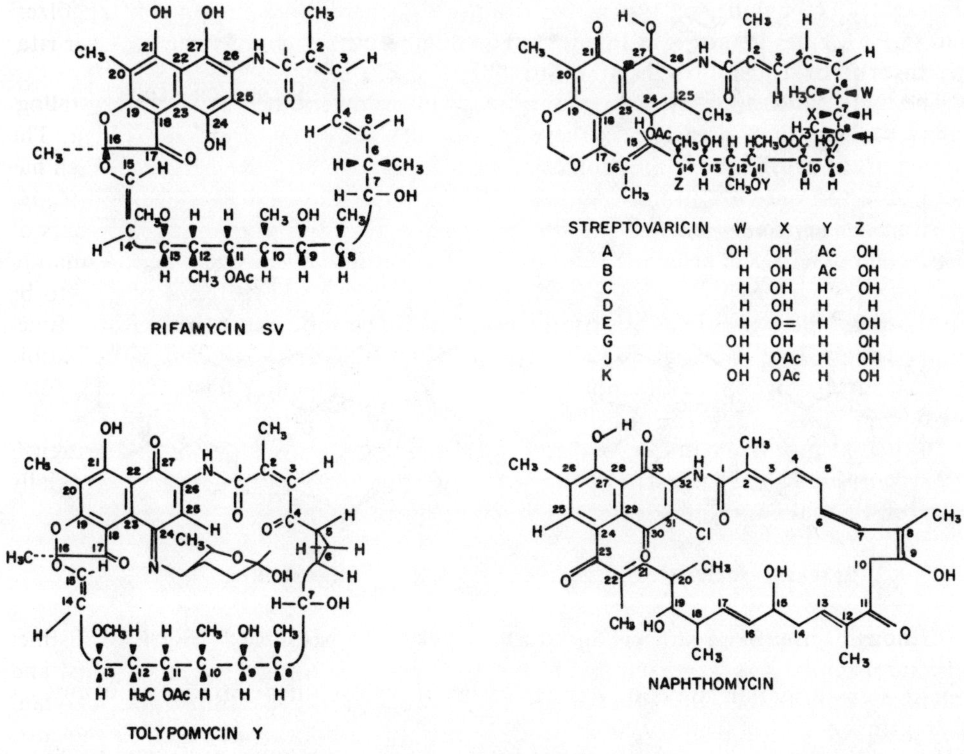

RIFAMYCIN SV

STREPTOVARICIN	W	X	Y	Z
A	OH	OH	Ac	OH
B	H	OH	Ac	OH
C	H	OH	H	OH
D	H	OH	H	H
E	H	O=	H	OH
G	OH	OH	H	OH
J	H	OAc	H	OH
K	OH	OAc	H	OH

TOLYPOMYCIN Y

NAPHTHOMYCIN

FIGURE 1. Structures of naphthalenic ansamycins.

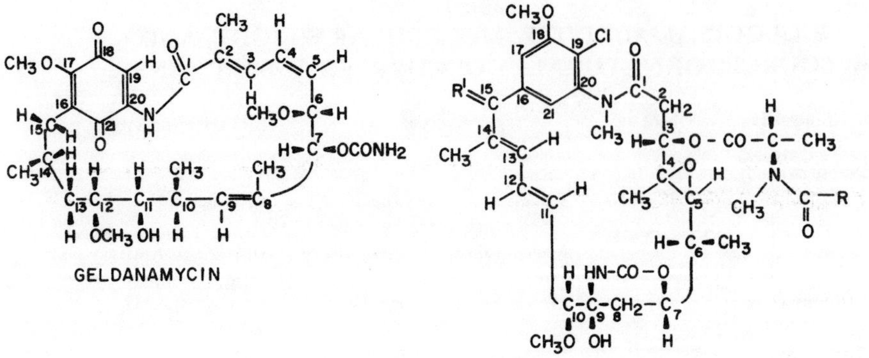

GELDANAMYCIN

Maytansine: R = CH_3 R' = H, H
Maytanprine: R = CH_2CH_3 R' = H, H
Maytanbutine: R = $CH(CH_3)_2$ R' = H, H
Colubrinol: R = $CH(CH_3)_2$ R' = H, OH
Colubrinol Acetate: R = $CH(CH_3)_2$ R' = H, $OCOCH_3$

FIGURE 2. Structures of benzenic ansamycins.

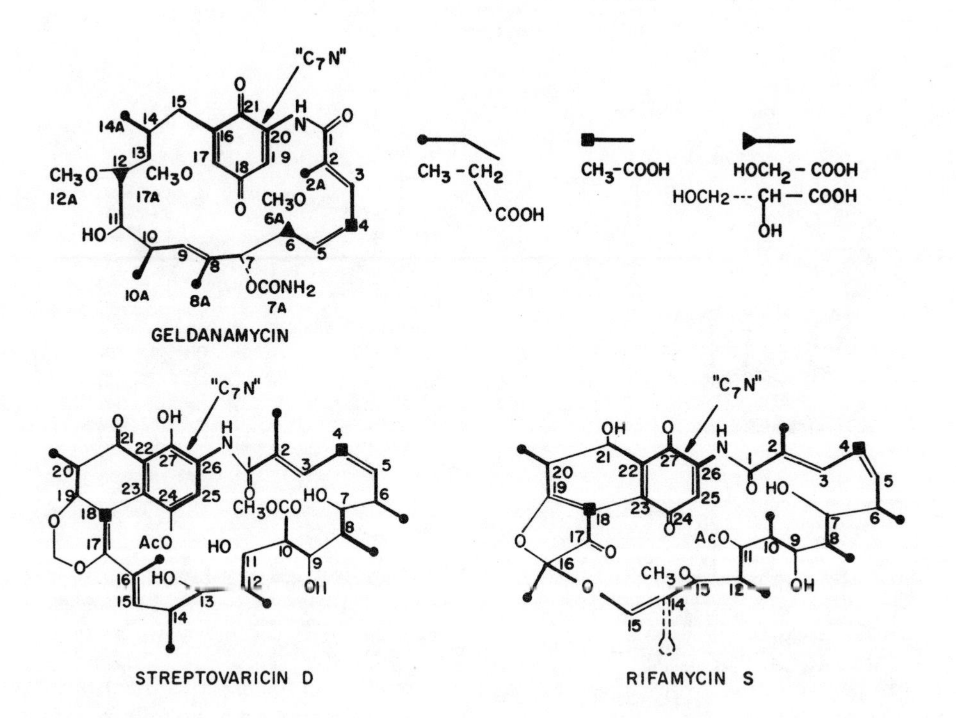

FIGURE 3. Numbering systems of rifamycin B, according to Prelog (A) and IUPAC rules (B).

Fourier transformation method, has facilitated the biosynthetic studies of complex ansamycins.

The origin of various carbon atoms of geldanamycin, streptovaricin D, and rifamycin S from propionic acid and acetic acid or malonic acid is shown in Figure 4. The direction of biosynthesis has been shown to be in the amidehead direction and not in the amidetail direction. The biosynthesis of the "C_7N" moiety of the molecule, which is common to all ansamycins, has been proposed to originate via shikimic acid intermediate (Figure 5). We have no information so far on the biosynthesis of maytansine

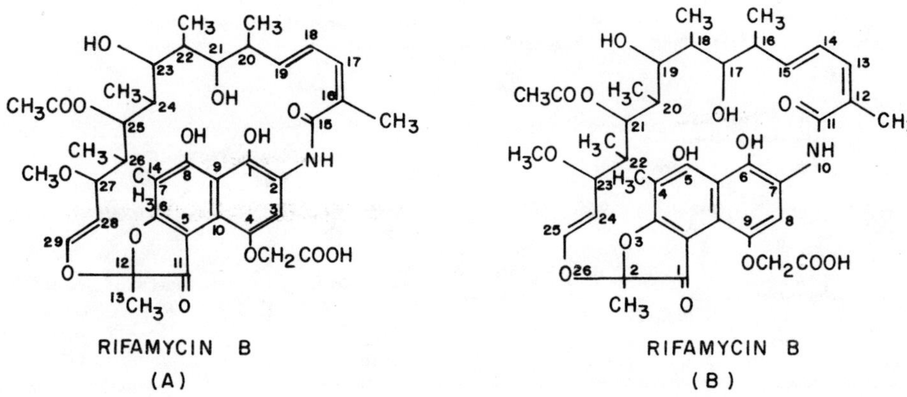

FIGURE 4. Origin of the carbon atoms of ansamycins.[16]

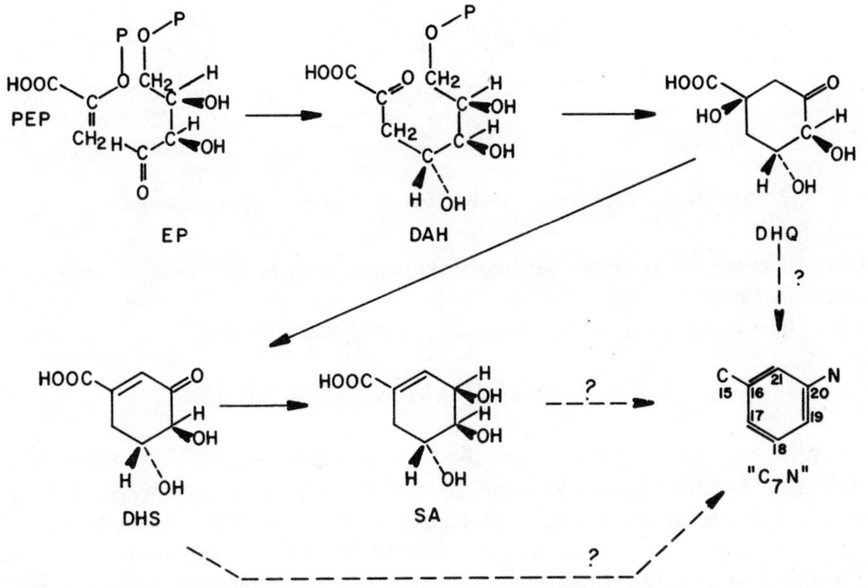

FIGURE 5. Proposed biosynthetic scheme of "C_7N" formation.[16] PEP = phosphoenol pyruvic acid, EP = erythrose phosphate, DAH = 3-deoxy-D-arabinoheptulosonic acid 7-phosphate, DHQ = 3-dehydroquinic acid, DHS = dehydroshikimic acid, SA = shikimic acid.

and tolypomycin, but it has been proposed to follow a similar path as other ansamycins,[14] (Figure 6). Further support of the above schema comes from the following facts:

1. Rifamycin B and tolypomycin Y are cosynthesized by *Streptomyces tolypophorus*.[17]
2. Rifamycin W, a precursor of rifamycin S, has been isolated, and it can be transformed by the parent *Nocardia* strain via rifamycin S to rifamycin B.[14]
3. The precursors of streptovaricins, such as protostreptovaricins and damavaricins have been isolated.[18,19]
4. Protorifamycin I, which is closely related to rifamycin W, and the protostreptovaricins have been isolated from the fermentation broth of a *Nocardia* strain.[20]

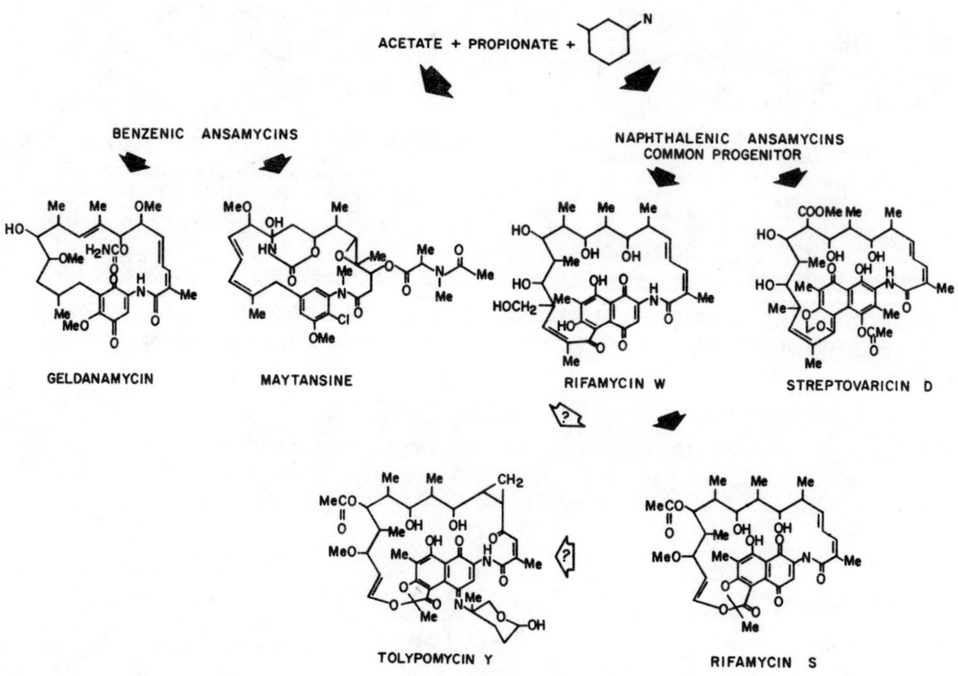

FIGURE 6. Proposed biogenetic interrelationships for the ansamycins.[14]

Thus, the ansamycins are most probably synthesized by microbes from common progenitors. The synthesis starts with the "C_7N" moiety, followed by propionic acid, and acetic acid or malonic acid molecules in the amide-head direction.

III. RIFAMYCINS

Rifamycins have been isolated from the fermentation broths of *Streptomyces mediterranea* and *Nocardia* species as a complex mixture of at least five (A, B, C, D, E) antibiotics.[21] On addition of diethylbarbituric acid to the broth, rifamycin B is produced predominantly.[22] Upon standing at room temperature or on aeration, rifamycin B is converted to a more potent antibiotic rifamycin S via rifamycin O as an intermediate (Figure 7). Reduction of rifamycin S reforms the hydroquinone structure and yields a clinically potent antibiotic, rifamycin SV.[23,24] The structure of rifamycins B, O, S, and SV has been elucidated by chemical[25,26] and X-ray crystallographic methods.[27]

Other rifamycins (rifamycin Y, rifamycin L, 13-O-demethylrifamycin B, 13-O-demethylrifamycin SV, and its deacetyl derivative) have been isolated from the fermentation broths of *S. mediterranei* or its mutants.[9] Besides the natural rifamycins, several hundred derivatives of rifamycins B, S, and SV have been prepared in the laboratories of Lepetit and Ciba-Geigy to optimize the activity of these antibiotics against Gram-positive bacteria, Gram-negative bacteria, and *M. tuberculosis,* and to improve absorption and retention behavior, and to lower toxicity. These extensive studies have yielded three clinically active agents, rifamycin SV, rifampicin, and rifamide.[9,10]

The carboxyl group of rifamycin B has been converted to various amides, hydrazides, and esters.[9,28,29] Among these derivatives, rifamide has proven to be a clinically active compound (Figure 8). The C-24 carbonyl group of rifamycin S (Figure 7) has been condensed with substituted amines under conditions where the hydrogen-bonded C-27 carbonyl group is inert, to form amines, hydrazides, amidrazones, and amino-

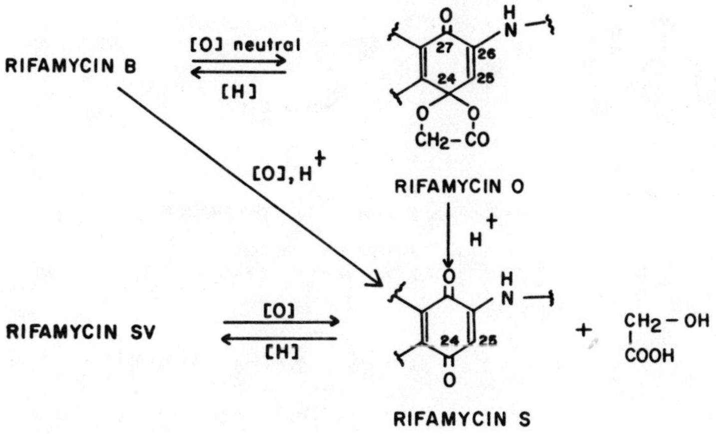

FIGURE 7. Conversion of rifamycin B to rifamycin O, rifamycin S and rifamycin SV.

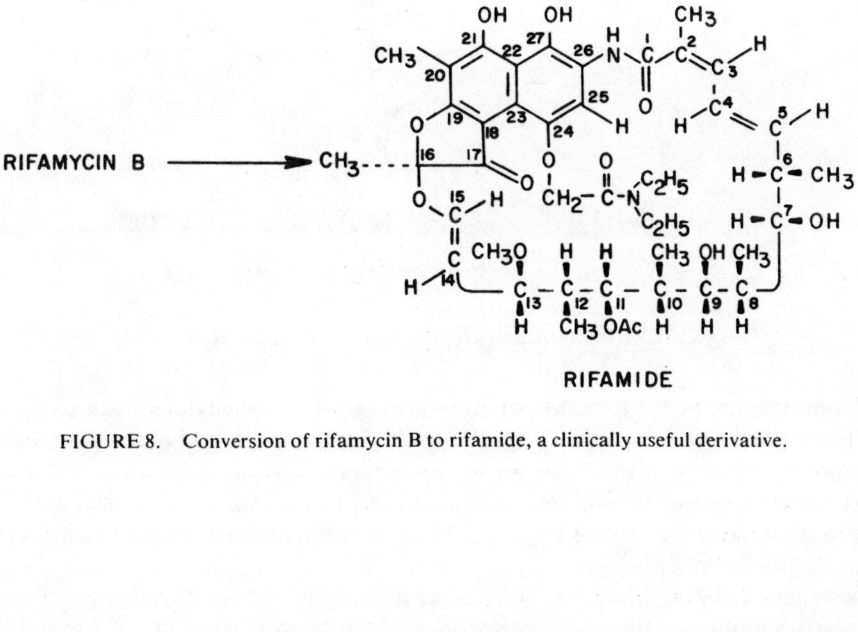

FIGURE 8. Conversion of rifamycin B to rifamide, a clinically useful derivative.

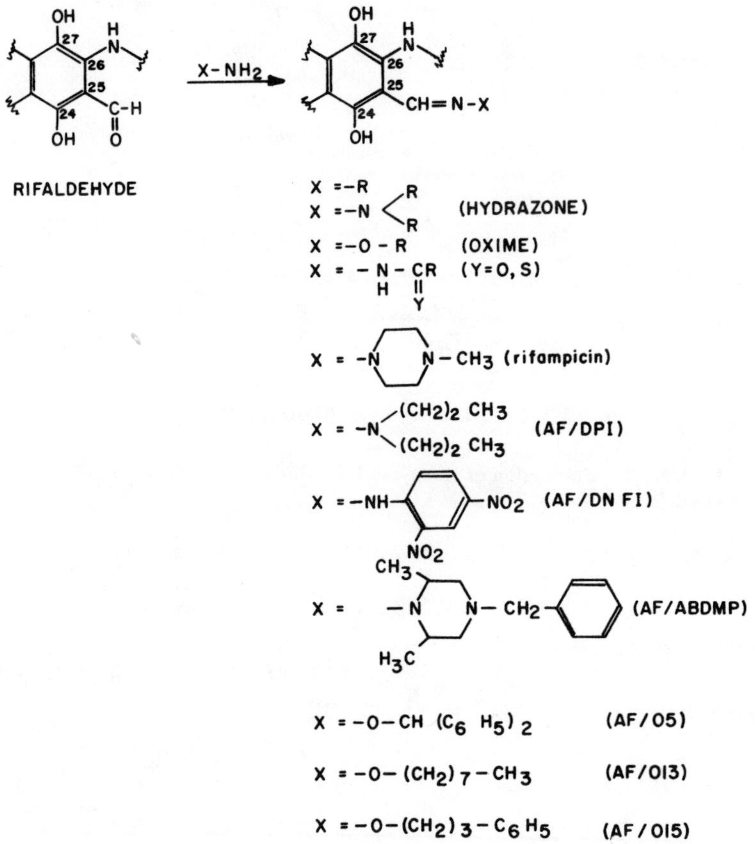

RIFALDEHYDE

$X - NH_2$ → $CH{=}N{-}X$

X = –R
X = –N(R)R (HYDRAZONE)
X = –O – R (OXIME)
X = – N(H) – C(=Y)R (Y = O, S)
X = –N(piperazine)N – CH_3 (rifampicin)
X = –N[$(CH_2)_2CH_3$]$_2$ (AF/DPI)
X = –NH–(2,4-dinitrophenyl) (AF/DN FI)
X = –N(dimethylpiperazine)N – CH_2–C_6H_5 (AF/ABDMP)
X = –O–CH $(C_6H_5)_2$ (AF/05)
X = –O– $(CH_2)_7$ – CH_3 (AF/013)
X = –O–$(CH_2)_3$ – C_6H_5 (AF/015)

FIGURE 9. Derivatives of rifamycin S and rifamycin SV.

guanidines (Figure 9).[9,30] Reaction of rifamycins with 1,2 phenyldiamines give phenazines (X = alkyl, halo, or condensed aromatic ring), and with O-aminophenols, give phenoxazinones (X = alkyl, halo, or nitro). Simple aliphatic primary and secondary amines undergo conjugate addition to the quinone ring of rifamycins instead of direct addition to the quinone carbonyl, giving C-25-substituted derivatives of rifamycin SV (Figure 9).[9,30]

Rifaldehyde (25-formylrifamycin SV) has been condensed with substituted amines and hydroxylamines to produce hydrazones and oximes (Figure 10).[30] Among these derivatives, rifampicine is clinically active. Some other hydrazones and oximes are potent inhibitors of nucleic acid polymerases. Rifaldehyde is produced by the Mannich reaction of rifamycin S with secondary amines in the presence of formaldehyde, from which the aminomethyl derivative gives rise to 25-formylrifamycin SV upon oxidation in acidic medium.

Rifamycins selectively inhibit the synthesis of cellular RNA in sensitive bacteria because they are potent inhibitors of bacterial DNA-dependent RNA polymerase.[31] Rifamycins form a stable complex with highly purified RNA polymerase in a molar ratio of 1:1[33] with K_{eq} of 2.7×10^{-10} *M* at 0°C.[34] The half life (t½) of the complex is 720 min at 0°C, and it is 9 min[34] at 37°C. The enzyme-antibiotic interaction seems to involve hydrophobic bonds as the complex can be dissociated with dimethylsulfoxide[10] and detergents.[35,36] Fifty percent inhibitory concentration (I_{50}) of rifampicine for RNA polymerase activity is in the range of ng/mℓ. In contrast, I_{50} for mammalian nucleic

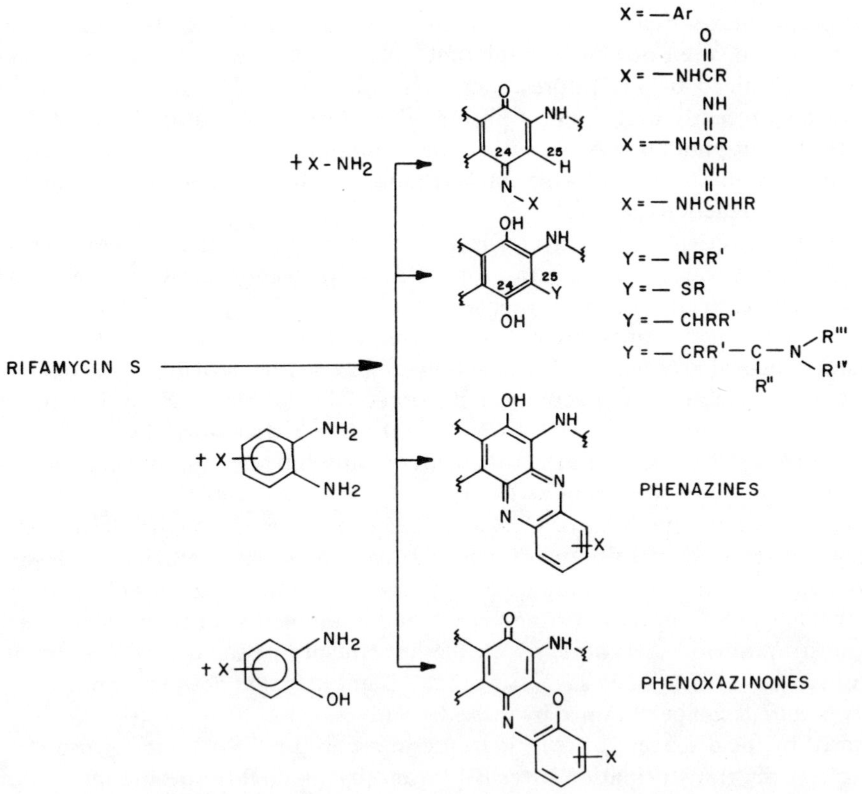

FIGURE 10. Derivatives of 25-formylrifamycin SV (rifaldehyde).

acid polymerases and viral reverse transcriptase are 10^2- to 10^3-fold higher than for the bacterial RNA polymerase.[9,10,37]

The over-simplified sequential reactions common to all nucleic acid polymerases can be written as follows: interaction of the enzyme with template (primer) to form a complex, binding of initiating nucleotide triphosphate to the enzyme-template (primer) complex, formation of the first phosphodiester bond on addition of another nucleotide triphosphate, elongation of nucleic acid chain by synthesis of complementary nucleic acid by movement of the enzyme along template and concomitant release of inorganic pyrophosphate; and, finally, termination of nucleic acid synthesis by dissociation of the enzyme from the complex.[38,39] The mode of action of rifampicin has been studied most extensively with *E. coli*, DNA-dependent RNA polymerase. The DNA-dependent RNA polymerase consists of 5 subunits designated as β', β, α, σ, and ω with mol wt of 165×10^3, 155×10^3, 40×10^3, 80×10^3, and 20×10^3 daltons, respectively.[40-42] The composition of these subunits in the enzyme are $\beta'\beta\alpha_2\sigma\omega$.[40-42] Reconstitution experiments with isolated subunits from rifamycin-sensitive and resistant RNA polymerases have shown that the β subunit of the enzyme contains the drug-resistant site.[43] Streptovaricin and streptolydigin resistance for RNA polymerase activity also resides in the β subunit.[44]

Rifampicin inhibits the initiation reaction of RNA synthesis.[45] However, when RNA polymerase is bound to the promotor (template), or during RNA chain elongation, it becomes refractory to the antibiotic action.[46] It has been suggested that the relative resistance of RNA polymerase-T7 DNA complexes to rifampicin in the presence of nucleoside triphosphates is due to the rapid rate of RNA chain initiation.[47] However, it has been shown by direct binding studies that the on-rate of drug binding to enzyme-promotor complex is reduced by 100- to 500-fold, and to the RNA elongation complexes by five to seven orders of magnitude[34] that of the free enzyme. Similar results have been obtained by the fluorescence-quenching method.[48] However, it has been demonstrated recently that the first phosphodiester bond formation in the RNA synthesis from λPR′phage DNA is only partially inhibited by the antibiotic, while the second phosphodiester bond formation is completely blocked with an accumulation of pppApU in the reaction mixture.[49]

Antibacterial activity of rifamycins is due to the inhibition of RNA polymerase activity. By testing various rifamycins for RNA polymerase inhibition, the following structure-activity relationships have been established:[9,27,34] the naphthoquinone and naphthohydroquinone forms are equally active; acetylation of phenolic OH group in C-21 and hydroxyl groups at C-7 and C-9 gives an inactive derivative; the opening of the *ansa* bridge causes loss of activity; reduction of the double bonds in the *ansa* chain reduces activity; epoxidation at C4-C5 and C14-C15 reduces activity; hydrolysis of acetyl group in C-11 does not affect the activity; and introduction of bulky lipophilic groups in naphthoquinone ring at C-24 and C-25 gives a much higher activity. As noted above, several hundreds of C-24 and C-25 substituted derivatives have been synthesized and some clinically active compounds have been obtained. Sensi[9] has applied the Hansch approach to correlate quantitatively the in vitro activity of oximes of 25-formylrifamycin SV, amides of rifamycin B, and iminomethylpiperazines of rifamycin SV against *Staphylococcus aureus* with the lipophilicity of the derivatives. Parabolic types of curves were obtained suggesting that it is unlikely that new, more potent agents against *S. aureus* can be obtained by these derivatives.

Inspired by the discovery of reverse transcriptase in the RNA tumor viruses[50,51] and their role in the transformation of cells,[52,53] and by the highly specific interaction of bacterial DNA-dependent RNA polymerase with rifamycin, several laboratories tested hundreds of rifampicin derivatives to find a specific inhibitor for viral reverse transcriptase.[9,37,54-56,67,68] On closer analysis of these results, one finds that (1) the amount of rifamycin derivatives needed to cause inhibition of reverse transcriptase activity are in the order of 10^3- to 10^4-fold higher concentrations than needed to inhibit *E. coli* RNA polymerase,[9,10,37] (2) on addition of a nonspecific protein like bovine serum albumin, the concentrations of the agent needed to inhibit enzyme activity is further increased,[57] and (3) at such high concentrations of the rifamycin derivatives, other mammalian nucleic acid polymerases such as RNA polymerases, DNA polymerases are also inhibited[36,37,58,59] (Table 2). Although some specificity for reverse transcriptase is exhibited by AF/DNFI, other active compounds such as AF/05, AF/013, and AF/015 also inhibit DNA polymerase, RNA polymerase, and poly(A) polymerase activities. Active rifamycin derivatives inhibit mammalian nucleic acid polymerases by interacting with the enzyme protein. The nascent polynucleotide chain formation is less affected by the inhibitors,[60] and on addition of increasing amounts of nonionic detergent (Triton X-100) to the antibiotic-inhibited enzyme reaction, the enzyme activity is completely restored.[35,36,53] Since the active rifamycin SV derivatives contain a highly lipophilic side chain, it has been proposed that these derivatives interact with the lipophilic parts of the nucleic acid polymerases.[36] These results are consistent with the hypothesis that all nucleic acid polymerases may have common reaction mechanisms. It has been shown recently that some rifamycin SV C-25-substituted derivatives with a bulk or hydrophobic side chain, such as demethyl dimethyl rifampicin, prevent the formation of enzyme-template complex,[61] while rifazacyclo-16 promotes the dissociation of polymerase-template-primer complexes.[61]

The highest nontoxic dose levels of rifamycins listed in Table 2 on cultured 3T3 mouse fibroblasts are in the range of 3 to 5 μg/mℓ,[37,59] whereas for inhibition of mammalian DNA or RNA polymerase activity in vitro, 20- to 30-fold higher concentrations

Table 2
FIFTY PERCENT INHIBITORY DOSE (I_{50}) IN μg/mℓ OF RIFAMYCIN SV AND ITS C-25-SUBSTITUTED DERIVATIVES FOR NUCLEIC ACID POLYMERASES[36]

Rifamycins	Reverse transcriptase	DNA polymerase α	DNA polymerase β	RNA polymerase I	RNA polymerase II	Poly(A) polymerase
Rifamycin SV_3	>100	>100	>100	>100	>100	>100
C-25′-substituted derivatives[a]						
Rifampicin	>100	>100	>100	>100	>100	>100
AF/DPI	90	>100	>100	>100	>100	>100
AF/ABDMP	95	>100	>100	>100	>100	>100
AF/DNFI	40	>100	>100	100	100	95
$-CH=N-N-(C_6H_5)_2$	77	104	55	65	37	64
AF/05	65	75	48	70	37	66
AF/013	50	65	36	60	30	80
AF/015	>100	108	73	79	66	88

[a] For structures, see Figure 10.

are needed, indicating that the primary target for these antibiotics could be other than nucleic acid polymerases. Recent experiments from Buss et al.[62] suggest that primary action of rifampicin may be on protein synthesis rather than on RNA synthesis; however, the drug concentrations needed to elicit such a change are in the range of 50 to 200 μg/mℓ. In isolated nuclei from L-cells, rifamycin AF/013 at 40 μg/mℓ concentration causes a reduction of total RNA synthesis and the formation of polyphosphorylated 5′termini.[63] These results are in agreement with the effect of these agents on initiaion of RNA synthesis.[60]

Rifamycin derivatives which inhibit reverse transcriptase in vitro should also, theoretically, inhibit transformation of permissive cells by RNA tumor viruses. Early attempts to substantiate this hypothesis have led to the incubation of the viruses with high concentrations of the active rifamycin derivatives and to the measuring of the loss of trasforming capability of the treated viruses.[64,65] When these agents are tested at the highest dose levels nontoxic to the permissive cells, the rifamycin SV oximes and hydrazones show a significant inhibition of transformation foci over the control.[58] However, some rifamycin antibiotics such as rifamycin S, rifamycin O, and rifamycin SV which do not inhibit viral reverse transcriptase in vitro, also show inhibition of transformation foci.[58,66] The metabolic conversion of these rifamycins to inhibitors of reverse transcriptase cell culture cannot be ruled out. Protein synthesis in 3T3 cells is slightly inhibited by rifamycin S at 5 μg/mℓ and the growth of 3T3 cells and their transformants by murine sarcoma virus (MSV) remain unaffected over a 4-day period at 2 to 4 μg/mℓ concentrations.[58] The yield of MSV from MSV-transformed 3T3 cells also remains unaffected, but the MSV yield from freshly infected 3T3 cells is reduced by 1 to 2 log in the early phase of the virus infection.[58] Some C-25-substituted alkylaminomethyl derivatives of rifamycin SV have been tested for inhibition of viral reverse transcriptase and mammalian DNA polymerases, RNA polymerases, and poly(A) polymerase activities. The enzyme activities are not inhibited, but these compounds are capable of inhibiting the MSV foci formation on 3T3 cells at noncytotoxic dose levels.[59] Three compounds, particularly the N-methyl-N-hydroxyethylaminomethyl, the N,N-dimethyl aminomethyl, and the N^4-methylpiperazinomethyl rifamycin SV derivatives, have shown superior inhibition of MSV focus formation as compared to adenine arabinoside and virazole.[59] These compounds inhibit oncornavirus production probably by a further delay in the growth of 3T3 cells that results from infection of the cells with MSV.[59]

With the discovery of inhibition of bacterial DNA-dependent RNA polymerase by rifampicin, two groups[69,70] explored the possibility of inhibition of mammalian viruses which contained their own DNA-dependent RNA polymerase. At relatively high concentrations (60 to 100 μg/mℓ), rifampicin completely inhibits the growth of DNA-containing pox viruses.[7,69,70] The sensitivity of pox viruses to the antibiotic appears to be independent of the virus strain, host cells, or method used to assay virus replication. Other DNA viruses that grow in the cytoplasm, such as frog virus 3, have been claimed also to be inhibited by rifampicin.[7] The replication of DNA viruses which grow in the nucleus, such as SV40, polyoma viruses, and herpes simplex viruses, is not affected by the drug. However, degenerative changes in the cultured cells at 100 μg/mℓ drug concentration cannot be ruled out. There is a 20 to 30% cell growth inhibition[59] at 10 μg/mℓ rifampicin concentration and at 10 to 100 μg/mℓ drug concentration, protein synthesis is inhibited.[62] Nevertheless, rifampicin-resistant vaccinia virus strains have been obtained[7,70] which strongly suggest the antiviral effect of the drug. The assembly of vaccinia virus is blocked at a step in the viral envelope formation.[71] Accumulation of unique virion precursor structures has been observed by electron microscopy.[7,72,73] DNA-dependent RNA polymerase contained in the vaccinia virus cores is not inhibited by rifampicin at 100 μg/mℓ concentration.[7,74] Many poxviral proteins are synthesized in the presence of the drug. Some structural protein precursors of higher molecular weight are also synthesized but fail to be cleaved in the presence of the inhibitor.[7] Whether these events are connected to the primary effect of the drug in the virus replication is not clear.

In clinical treatment, rifampicin is administered orally at 300 to 600 mg/day concentration. Peak serum concentrations of 7 to 28 μg/mℓ have been obtained.[75] Kidney, liver, and lung contain higher rifampicin concentrations. Long-term administration of the drug has been shown to shorten the biological half-life of anticoagulants, glucocorticoids, oral contraceptives, and rifampicin itself, presumably through alterations in the hepatic metabolism. Rifampicin may induce abnormalities in hepatic functions leading to necrosis. The antibiotic inhibits both in vivo and in vitro humoral and cellular immunological responses and thus acts as an immunosuppressive agent.[62]

The antitumor effects of rifamycins and their derivatives are not known. In experimental animal systems with P388 lymphocytic leukemia and L1210 ascites leukemia, promising derivatives of rifamycin SV which inhibit mammalian nucleic acid polymerases and RNA tumor viral reverse transcriptase have been tested with doses up to 100 mg/kg (Table 3). No significant antitumor activity has been noted. Thus, with all the available evidence, it seems unlikely that a selective antiviral or antitumor drug could be developed from rifamycins.

IV. STREPTOVARICINS

Streptovaricins were the first ansamycin antibiotics to be discovered, over 20 yr ago, from the fermentation broth of *Streptomyces spectabilis*.[76] The separation of various streptovaricins from its complex and its structural elucidation studies have been very elegantly accomplished by Rinehart and collaborators.[3,6,77,78] Streptovaricins differ mainly in the degree of oxidation of *ansa* ring substituents (Figure 1). Streptovaricins are chemically related to the rifamycins. The configuration of the eight chiral centers C-6 to C-13 of the *ansa* chain is identical to those in rifamycins and tolypomycin Y. There are, however, some important structural differences between streptovaricins and rifamycins. For example, the *ansa* chain is linked to the chromophore by a C-C double bond, and not via oxygen; the configuration of the conjugated double bonds in the

Table 3
ANTILEUKEMIC ACTIVITY OF RIFAMYCINS IN P388 and L-1210 SYSTEMS[a]

Compound	P388 lymphocytic leukemia		L-1210 ascites cells	
	% T/C	mg/kg	% T/C	mg/kg
Rifamycin SV, 25-Formyl-rifamycin SV derivatives[b]	110	12.50	103	8
AF/ABP	100	50	100	6.25
AF/05	N.A.[c]	N.A.	102	50
AF/013	100	100	106	100
AF/015	N.A.	N.A.	104	80

[a] Data provided by Douros, J. and Acierto, A. M., National Cancer Institute, Bethesda, Md.,
[b] For structures see Figure 10.
[c] N.A. = not available.

ansa ring is different; the hydroxyl group at C-24 is acetylated; the B-ring of the naphthalene chromophore is not a benzene ring, but part of a quinone methide system; and C-17 and C-19 are linked via methylenedioxide bridge (Figure 1).

In contrast to rifamycins, few derivatives of streptovaricin have been synthesized. Primarily, it has been the efforts of Rinehart's laboratory[3] to prepare several derivatives and study their biological properties. Streptovaricins can undergo lactonization to yield products like streptovaricin Fc (Figure 11). The antibiotics can also undergo atropisomerization to yield their respective isomers. On reduction in methanol-ammonia, damavaricin and its lactone, damavaricin Fc are formed. On oxidation in sodium peroxide, the *ansa* chain is cleaved and streptoval C and its lactone, streptoval Fc, are produced (Figure 11). Several acetates of streptovaricins have been synthesized by acetylation. Acetonides of streptovaricin C at position 7, 9 and 9, 11 with $= C(CH_3)_2$ substitutions have been prepared. A series of 19-O-alkyl derivatives of damavaricin C have also been synthesized.[3]

Biological activities of streptovaricins and their derivatives have been extensively reviewed.[3,8] Among the known ansamycins, streptovaricins appear to be the least toxic in whole animals and cultured cells. Streptovaricins are active in vitro against most gram-positive organisms. The high activity against *Mycobacterium* species is particularly interesting. Streptovaricins inhibit RNA and protein synthesis, but not DNA synthesis of microorganisms. Synthesis of all major RNA species are inhibited equally. Streptovaricins inhibit bacterial DNA-dependent RNA polymerase,[79] and this enzyme from streptovaricin-resistant *E. coli* is not inhibited by the agent.[80,83] Analogous to the rifamycin mode of action, streptovaricin inhibits the initiation reaction of RNA synthesis.[79] It is probably the β subunit of the enzyme that is affected by the drug.[44] The dose of streptovaricin complex required to inhibit *E. coli* growth and RNA synthesis is in the range of 10 to 12 $\mu g/m\ell$, while inhibition for *Staphylococcus aureus* is in the range of 1 to 0.5 $\mu g/m\ell$. The latter sensitive strain is used for the antimicrobial assay of the antibiotic. Rinehart et al.[84] have compared the in vitro antimicrobial activity with inhibition of DNA-dependent RNA polymerase activity from *E. coli.* These activities appear to go hand in hand. The order of activity in these assays is streptovaricin A, G $>$ B, C $>$ D, J $>$ E $\gg$ F, while the acetates are totally inactive. A similar order of activity has been obtained in our laboratory: streptovaricin A = B = C $>$ G = J $>$ D $>$ E $>$ Fc; streptovaricin complex = fraction a = b = c = d (Table 4).[37,93] It must be

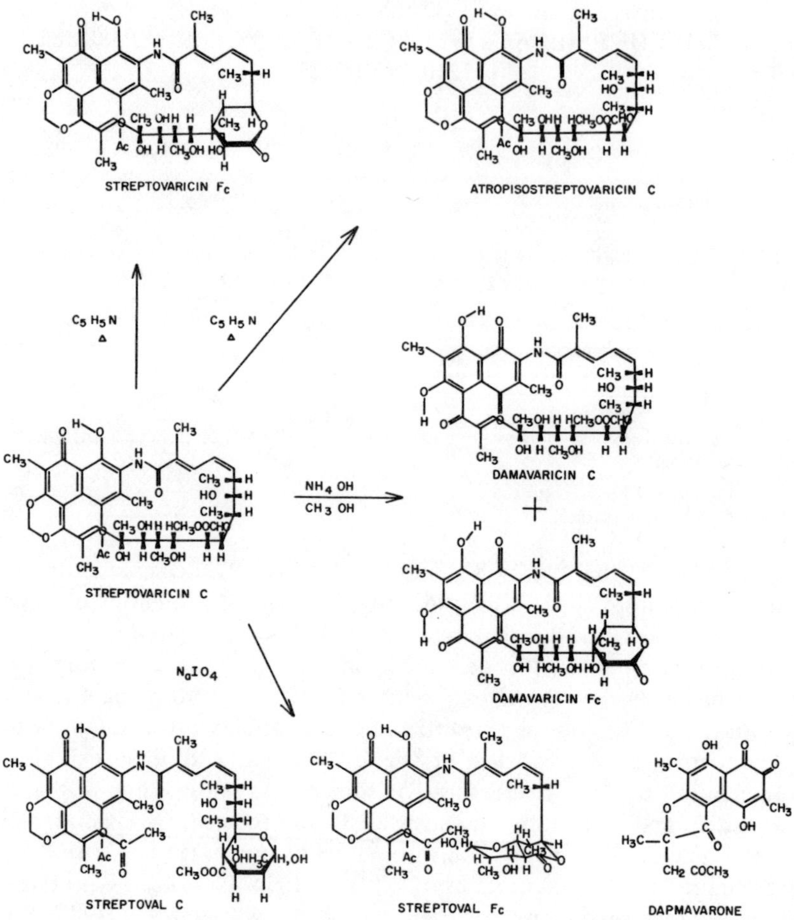

STREPTOVARICIN Fc

ATROPISOSTREPTOVARICIN C

DAMAVARICIN C

STREPTOVARICIN C

DAMAVARICIN Fc

STREPTOVAL C

STREPTOVAL Fc

DAPMAVARONE

FIGURE 11. Structures of streptovaricin products.

pointed out that these comparative studies were carried out under nonlinear inhibitory conditions.

There have been several studies in the literature on inhibition of reverse transcriptase activity of RNA tumor viruses,[8,37,85,86,93] mammalian DNA polymerases,[37,86,87,93] and mammalian DNA-dependent RNA polymerases[37,86,93] by streptovaricins and by their degradation products. Inhibition of reverse transcriptase,[85-87] terminal deoxynucleotidyl transferase,[87] and DNA polymerase α, β, and γ has been obtained at very high concentrations (200 to 600 μg/mℓ) of the drug. In interpreting such studies, one must make the following considerations:

1. Similar to rifamycins, streptovaricins presumably interact directly with the enzyme protein and not with the template or substrate.
2. The purity of the nucleic acid polymerases and their molecular weights are quite varied and hence it is difficult to compare drug inhibition data.
3. Inhibition should be studied in the linear range of reaction kinetics and at saturating conditions of the reactants.
4. Some nonspecific protein like bovine serum albumin should be added in the assay mixtures in order to prevent any nonspecific inactivation of the enzyme activity.

These criteria have not been met in these studies.[85-87] By taking a range of 50 to 500 ×

Table 4
INTERACTION OF STREPTOVARICINS WITH NUCLEIC ACID POLYMERIZING ENZYME ACTIVITIES IN VITRO AT 100 μg/mℓ CONCENTRATION[a]

Compounds	NSC number	Reverse transcriptase	DNA polymerase α	DNA polymerase β	RNA polymerase I	RNA polymerase II	Poly (A) polymerase	*E. coli* RNA polymerase
Streptovaricin A	48810	102	98	76	74	67	74	21
Streptovaricin A diacetate	210761	98	91	92	98	94	89	91
Streptovaricin B	156215	66	89	98	91	87	73	22
Streptovaricin C	19990	95	92	102	62	93	66	29
Streptovaricin C triacetate	210760	103	101	94	95	86	88	82
Streptovaricin D	156216	119	99	100	99	81	65	45
Streptovaricin E	156217	78	91	97	91	90	68	60
Streptovaricin Fc	182858	99	97	95	59	89	58	84
Streptovaricin G	156219	77	80	95	97	91	54	32
Streptovaricin J	182857	101	95	89	112	83	55	40
Streptovaricin fraction a	189793	67	85	91	64	83	58	30
Streptovaricin fraction b	189794	98	82	99	67	74	56	26
Streptovaricin fraction c	189795	94	99	100	57	93	53	43
Streptovaricin fraction d	189796	114	102	102	100	88	106	64
Streptovaricin complex	189792	91	95	101	83	87	78	27
Damavaricin	210762	96	100	104	93	88	84	61
Dapmavarone	210763	72	70	99	78	63	57	53
Streptoval C	169627	75	96	86	96	84	87	71

[a] Data are expressed as percent of control.

Data taken from References 37 and 93.

Table 5
ACTIVITIES OF STREPTOVARICINS FOR CYTOTOXICITY (BALB 3T3), INHIBITION OF MSV (MLV) TRANSFORMATION OF 3T3 CELLS, AND INHIBITION OF REVERSE TRANSCRIPTASE[a]

Compound	3T3 cytotoxicity HND[b] (μg/mℓ)	MSV (MLV) focus (% control)[c]	Reverse transcriptase Activity 10 μg/mℓ	Reverse transcriptase % control 100 μg/mℓ
Streptovaricin A	>40	92 (40)[d]	101	91
Streptovaricin B	>40	96 (40)	92	51
Streptovaricin C	≥40	100 (40)	89	88
Streptovaricin D	10	78 (10)	73	116
Streptovaricin E	40	stimulation	86	76
Streptovaricin G	20-40	104 (40)	70	62
Streptovaricin J	50	60 (50)	91	96
Streptovaricin Fc	20	stimulation	66	69
Streptoval C	10	stimulation	83	89
Streptovaricin complex	40	67 (40)	108	99
Streptovaricin fraction a	40	stimulation	80	67
Streptovaricin fraction b	10	94 (10)	86	94
Streptovaricin fraction c	20	94 (20)	79	95
Streptovaricin fraction d	40	34 (40)	99	100

[a] Adapted from Aldrich, C. D. and Sethi, V. S., unpublished data.
[b] HND = highest nontoxic dose, is defined as that concentration at which the viable cell numbers in the test plates are between 90 to 100% of untreated control dishes.
[c] Control plates in triplicates had between 50 to 100 foci; data are calculated from the average value from three petri dishes.
[d] In parenthesis are the doses of the antibiotic in μg/mℓ.

10^3 daltons as molecular weights for different nucleic acid polymerases, 1 μg of highly purified enzyme protein and 1 mM streptovaricin concentration in the reaction mixture, the ratio of drug: enzyme molecules are in the range of 10^4 to 10^6. These calculations thus clearly indicate that streptovaricins may have specific inhibitory properties for *E. coli* DNA-dependent RNA polymerase, but not against reverse transcriptase or other mammalian nucleic acid polymerases. Data in Table 4 lend further support to this conclusion.[37,93]

The in vitro and in vivo inhibition of infectivity of murine RNA tumor viruses by streptovaricins have been intensively studied by Carter's[8,88-90] and Li's[91] laboratories. Streptovaricin complex causes inhibition of MSV-induced transformation foci to about 30% at 1 to 2 μg/mℓ concentrations; streptovaricin D causes 32 to 48% reduction of number of foci at 10 to 20 μg/mℓ concentration.[89] In these experiments, the drug was maintained in the media for 2 to 3 days and transformation foci were counted on day 7. It has also been noted that streptovaricin complex is capable of reducing the increase in spleen size in Balb/C male mice which was caused by Rauscher leukemia virus (RLV).[90] Li et al.[91] noted that the highest nontoxic doses of streptovaricins for BALB/3T3 cells exposed for 2 days at 37° were: streptovaricin A = Streptovaricin C = < 20 μg; streptovaricin D = 2.8 μg; streptoval C = 5.2 μg, streptoval Fc = 2.8 μg; and streptovarone 5.2 μg, per mℓ of culture medium. Highest nontoxic dose levels of various streptovaricins for 3T3 cells as measured in our laboratory by counting the number of viable cells on day 5 are somewhat higher than those of Li et al.[91] (Table 5) which could be due to metabolism of the drugs in the longest assay time, or due to different concentration of fetal calf serum used in these laboratories.[92] By preincubating streptovaricins at 20, 50, and 100 μg/mℓ with RLV for 60 min at 37°C followed

Table 6
ANTILEUKEMIC ACTIVITY OF STREPTOVARICINS IN P388 AND L-1210 SYSTEMS[a]

	P388 lymphocytic leukemia		L-1210 ascites cells	
Compound	% T/C	Dose (mg/kg)	% T/C	Dose (mg/kg)
Streptovaricin A	136	150	100	100
Streptovaricin C	150	126	103	110
Streptovaricin D	109	25	50	20
Streptovaricin E	101	100	96	25
Streptovaricin Fc	123	200		
Streptovaricin G	119	50	101	25
Streptovaricin J	106	200		
Streptovaricin complex	131	100	101	100
Streptovaricin fraction a	113	400	104	100
Streptovaricin fraction b	123	200	104	100
Streptovaricin fraction c	108	200	104	100
Streptovaricin fraction d	113	100	96	100
Streptoval C	142	50	102	200

[a] Data provided by Douros, J. and Acierto, A., National Cancer Institute, Bethesda, Md.

by virus dilution and measuring its reduction in infectivity by in vitro transformation assay and by decrease in spleen weight of the mice, Li et al.[91] showed a potent virucidal effect of streptoval C, streptoval Fc, and streptovarone. These agents also inhibited reverse transcriptase activity at 200 μg/mℓ.[91] However, inhibition of viral components other than reverse transcriptase cannot be excluded. Similarly, results from our laboratory show that streptovaricin B and streptovaricin G are moderate inhibitors of reverse transcriptase activity, but these agents do not show significant transformation inhibitory properties when used at the nontoxic concentrations in 5-day tests. Clearly, more work is needed to prove the usefulness of streptovaricins as inhibitors of reverse transcriptase and viral transformation.

Antileukemic activity of streptovaricins has been tested in the P388 and L1210 systems of NCl at 50 to 400 mg/kg dose levels (Table 6). In the P388 system, streptovaricin A, streptovaricin C, streptovaricin Fc, streptovaricin complex, and streptoval C show significant antileukemic activity. However, in the L1210 system, these compounds do not show any antileukemic activity. Since streptovaricins are not toxic in animals, these drugs should be tested more rigorously for their antitumor effects in solid tumor and other systems.

V. TOLYPOMYCINS

Tolypomycins have been described as a complex of antibiotics isolated from the fermentation broths of *Streptomyces tolypophorus.*[94,95] The structure of one member of this group, tolypomycin Y, has been determined by X-ray crystallography[96] and closely resembles that of rifamycin S (Figures 1 and 7). Tolypomycin Y differs from rifamycin S in the *ansa* chain with a keto group at C-4, a cyclopropane ring at C-5 and C-6, and in the naphthalene with a tolyposamine (2,3,4,6-tetradeoxy-4-amino sugar) attached as a quinone imide. Tolypomycin Y is very potent against Gram-positive bacteria and its antibacterial activity is attributed to its inhibition of bacterial DNA-dependent RNA polymerase.[31,80]

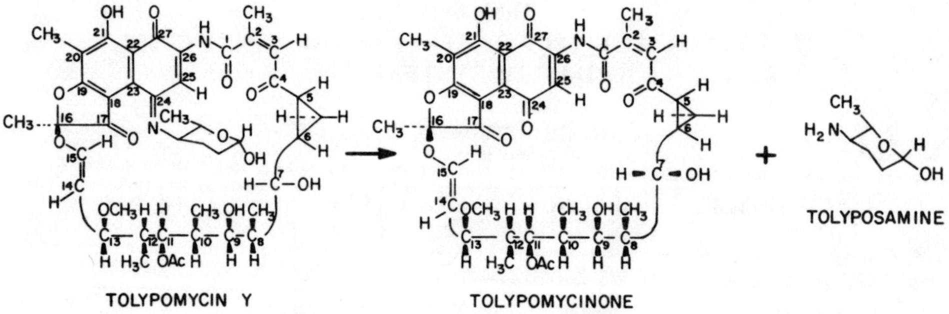

FIGURE 12. Hydrolysis of tolypomycin Y.

On mild acid hydrolysis, tolypomycin Y is converted to tolypomycinone (Figure 12) which has considerably lower antibacterial activity. It has been hypothesized that the lower activity of tolypomycinone could be due to its difference in the spacial conformation of hydroxyl groups at position C-7 and C-9, which in concert with the two oxygen atoms of the hydroxyl groups at C-21 and C-27 of the aromatic chromophore, interact specifically with bacterial RNA polymerase.[98] Analogous to rifamycin S and SV, several derivatives of N-substituted 25-amino tolypomycinone have been synthesized and tested for their activity.[99] Several interesting derivatives with comparable biological activity to rifampicin have been found. There are no published data on the antitumor activity of tolypomycins.

VI. NAPHTHOMYCIN

Naphthomycin was isolated from a *Streptomyces* species.[100] Its structural elucidation was based on proton magnetic resonance studies (Figure 1).[101] It is structurally similar to streptovaricins but it differs as follows: (1) it has a naphthoquinone rather than a hydronaphthoquinone chromophore; (2) it has chloro substitution instead of methyl in the aromatic chromophore; (3) it lacks methylenedioxy and enol acetate groups of the streptovaricins, and (4) the *ansa* chain is longer by six carbon atoms than those of the other naphthalein ansamycin.

Naphthomycin shows antibacterial activity against gram-positive strains, but it does not inhibit bacterial RNA polymerase activity.[10,100] Vitamin K antagonises the activity of the antibiotic[100] and its exact mode of action is not known yet. There is no information on the antitumor activity of this antibiotic.

VII. GELDANAMYCIN

In contrast to the above-described antibiotics, geldanamycin is a benzenic ansamycin, isolated from *Streptomyces hygroscopicus*.[102,103] Inspired by synthesis of clinically active, semisynthetic rifamycins, Rinehart et al.[104,105] have prepared several hydrazone, oxime, phenazine and phenoxazinone derivatives of geldanamycin, and tested their biological activities against several systems. The methods of preparation of these derivatives are similar to those developed for rifamycins, and are briefly depicted in Figures 13 and 14. Geldanamycin is subjected to Mannich condensation to give the t-butylimine derivative of geldanaldehyde, which then undergoes an exchange reaction with a suitable hydrazine or hydroxylamine to yield respective hydrazone or oxime derivatives (Figure 13). Substituents similar to the ones found in active rifamycins have been used to synthesize geldanamycin derivatives.[104] For examples, by condensation of geldanamycin or its 17-O-demethylgeldanamycin with substituted O-phenylenediamine or O-

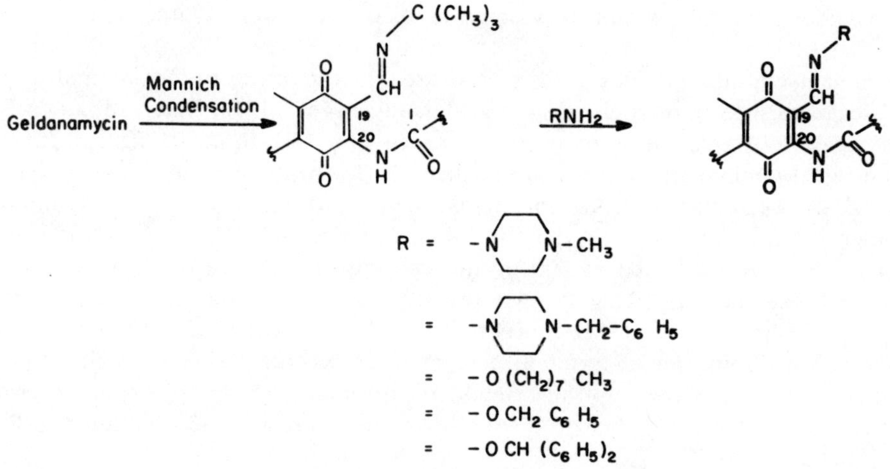

FIGURE 13. Synthesis of hydrazones and oximes of 19-geldanaldehyde.

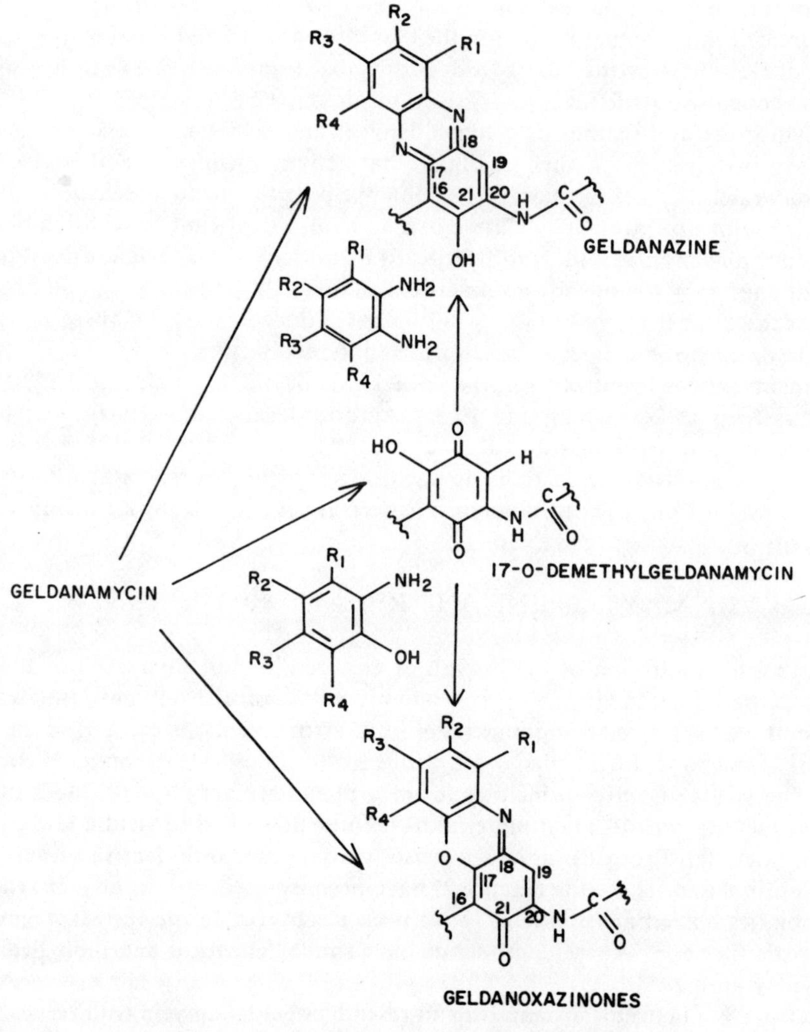

FIGURE 14. Synthesis of geldanazines and geldanoxazinones.

aminophenols, several geldanazines or geldanoxazinones were synthesized[105] (Figure 14).

Geldanamycin inhibits the growth of protozoa *Tetrahymena pyriformis* and *Crithidia fasciculata.* The cytotoxicity (I_{50}) of geldanamycin for L1210, Balb 3T3, and MSV-transformed 3T3 cells in vitro is in the range of 0.001 to 0.004 μg/mℓ concentration.[37,106] Mammalian nucleic acid polymerases (RNA polymerase I, RNA polymerase II, DNA polymerase α, DNA polymerase β, poly (A) polymerase) are not inhibited at 100 μg/mℓ concentration.[93] Avian myeloblastosis virus (AMV) reverse transcriptase activity also is not affected at 100 μg/mℓ concentration.[93] However, inhibition of Rauscher leukemia viral (RLV) reverse transcriptase activity has been reported[106] at 200 to 300 μg/mℓ (2.5×10^{-4}M) concentrations. At 10^{-4} to 10^{-3}*M* (60 to 600 μg/mℓ) concentration, inhibition of terminal deoxynucleotidyl-transferase from human leukemic leukocytes and reverse transcriptase from Simian sarcoma virus has been reported.[87] Because of the higher drug concentrations, these enzyme's inhibitory effects appear to be nonspecific.

Geldanamycin has no significant effect on the production of MSV by freshly-infected 3T3 cells over the 96-hr duration of the experiment at 1 to 4 μg/mℓ antibiotic concentrations as found by Aldrich and Sethi.[37] By treating the virus with 10^{-4} to 10^{-7} *M* concentrations of various geldanamycin derivatives at 37°C for 1 hr, and then testing the treated virus preparations for their in vitro and in vivo infectivity, different levels of inhibition over the control values have been obtained.[106] From these experiments, no conclusive structure-activity relationships have been established.

The hydrazines and oximes of geldanaldehyde, and geldanazines and geldanoxazinones have been tested for their antimicrobial activity against several bacterial and protozoal strains.[104,105] Coversion of geldanamycin to the geldanaldehyde derivatives eliminates antiprotozoal activity. Some oximes show significant growth inhibition of *Diplococcus pneumoniae* and *Staphylococcus hemolyticus.* The geldanazines and geldanoxazinones do not show any antimicrobial activity. Inhibition of *E. coli* RNA polymerase activity at 0.5 μmol (300 to 350 μg/mℓ) concentrations of these derivatives appears to be due to a nonspecific drug-protein interaction.

Geldanamycin has been tested in the P388 lymphocytic leukemia and L1210 leukemia test systems at 2.5 mg/kg and 10 mg/kg dose levels, respectively. The percent treated (T) versus control (C) survival time values were around 107 to 125%, suggesting a slight or no antileukemic activity in these experimental tumor systems. It is not known yet whether any of the geldanamycin derivatives possess any significant antitumor activity.

VIII. MAYTANSINE AND RELATED COMPOUNDS

By testing for inhibition of the growth of cultured human carcinoma of the nasopharynx (KB-cell line) in vitro, Kupchan and his collaborators[107-111] have isolated maytansine and several related compounds (Figure 15) from ethanolic extracts of the fruits, stem barks, and roots of the plants *Maytenus ovatus* Loes, *M. serrata,* and *M. buchananii.* The yields of active principals in these plants are very low (0.2 to 1 mg/kg). Therefore, several tons of plant material have to be processed to yield a few grams of the active material. From *Colubrina texensis,* an oxygenated derivative of maytanbutine, colubrinol and its acetate (Figure 2) have been isolated.[112] Recently, maytansine-like antibiotics, called ansamitocins, have been discovered in the fermentation broth or *Nocardia* species.[113] These compounds have similar chemical and biological properties as maytansine.[113]

Maytansine is a benzenic ansamycin and resembles geldanamycin with respect to the number of C-atoms in the *ansa* ring (Figure 2). However, it differs from geldanamycin

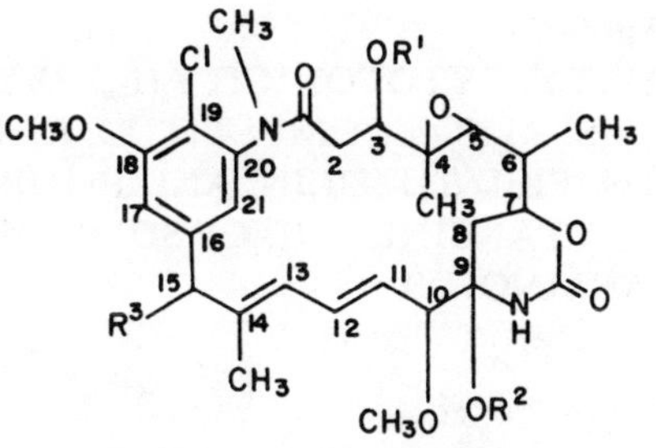

Maytansine	R' = COCH (CH_3) N (CH_3) $COCH_3$; $R^2 = R^3 = H$
Maytanprine	R' = COCH (CH_3) N (CH_3) $COCH_2CH_3$; $R^2 = R^3 = H$
Maytanbutine	R' = COCH (CH_3) N (CH_3) COCH $(CH_3)_2$; $R^2 = R^3 = H$
Maytanvaline	R' = COCH (CH_3) N (CH_3) $COCH_2CH(CH_3)_2$; $R^2 = R^3 = H$
Maytanbutacine	R' = COCH $(CH_3)_2$; $R^2 = H$; $R^3 = OCOCH_3$
Maytanacine	R' = $COCH_3$; $R^2 = R^3 = H$
Maytansinol	R' = $R^2 = R^3 = H$

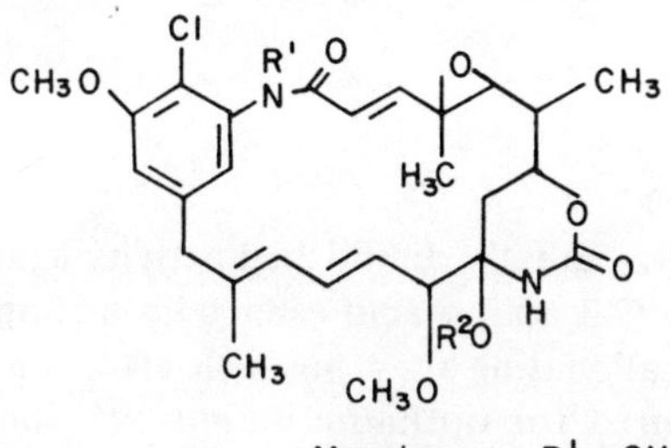

Maysine	R' = CH_3; $R^2 = H$
Normaysine	R' = $R^2 = H$
Maysine methyl ether	R' = $R^2 = CH_3$

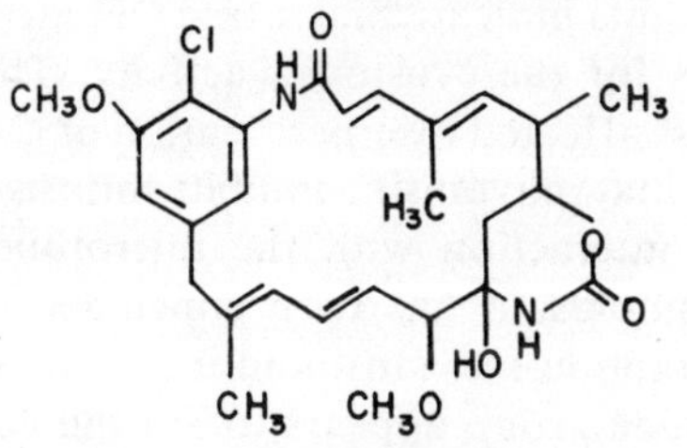

Maysenine

FIGURE 15. Structure of maytansine and related compounds.

in that C-7 and C-9 are linked forming a carbinolamide, and there is an epoxide at C-4 and C-5, a chlorine in the benzene ring, and a large substituent at position C-3. X-ray crystallographic analysis of (3-bromopropyl)-maytansine showed that the 19-membered *ansa*-ring has two parallel sides C-1 to C-6 and C-10 to C-15, which are linked by a benzenic chromophore and a six-membered carbinol-amide ring.[107] The ester group at C-3 and hydroxyl group at C-9 are hydrophilic regions of the molecule in contrast to the rest of the molecule which is hydrophobic.

Among ansamycins, only maytansine and its derivatives have shown considerable promise as antitumor agents. In order to establish structure-activity relationships of maytansinoids, several new C-3 esters and C-9 ethers have been synthesized and tested in several biological systems.[114] Maytansinoids (maytansine, maytanbutine, maytanprine, maytanvaline, maytanacine and maytanbutacine) show excellent antileukemic activity against P388 lymphocytic leukemia in the mouse in μg/kg dose ranges. May-

Table 7
COMPARATIVE ANTITUMOR ACTIVITY, CYTOTOXICITY (I_{50}), (KB CELLS), INHIBITION OF SEA URCHIN MITOSIS, KI FOR COMPETITIVE INHIBITION OF ^{3}H-VINCRISTINE BINDING TO PURIFIED TUBULIN, AND INHIBITION OF TUBULIN POLYMERIZATION OF MAYTANSINE AND OF SOME SELECTED MAYTANSINOIDS

Compound	Antitumor Activity % T/C (μg/kg)	I_{50} (*M*)	Mitosis inhibition (*M*)	Ki (μ*M*)	Tublin polymerization
Maytansine	220 (25)	8.8×10^{-12}	10^{-7}	0.35	1.0
Maytanbutine	190 (0.8)	5.0×10^{-12}	10^{-8}—$^{-9}$	N.T.[a]	1.3
Maytanprine	154 (1.6)	2.0×10^{-13}	10^{-8}	N.T.	1.0
Maytanvaline	187 (12.5)	3.1×10^{-13}	10^{-8}	0.95	0.9
Maysenine	120 (100)	3.9×10^{-9}	10^{-8}	4.83[b]	0.9
Maysine	80 (50)	4.5×10^{-8}	10^{-6}	—[c]	0.3
Normaysine	115 (3.1)	3.5×10^{-8}	10^{-8}	31.76	0.3

[a] N.T. = not tested
[b] Noncompetitive inhibition
[c] No inhibition up to 20 μm

sine, normaysine and maysenine show greatly diminished activity against the P388 system (Table 7). Replacement of the C-3 amino acid ester with a simple alkyl ester, or with an alkyl ester having potential alkylating sites, has little effect on the P388 system, but it does increase the dose required for optimum activity.[114] Maytansinoids show significant cytotoxicity against KB cells, but compounds which lack the C-3 ester are less active than those with esters (Table 7).[114] On conversion of C-9 alcohol of the carbinolamide to an ether, a marked decrease in antileukemic activity and cytotoxicity against KB cells is noted.[114] These results indicate that a free OH group at C-9 is essential for the antileukemic, as well as for the cytotoxic, activity. The antimitotic and antitubulin activities are relatively less affected by etherification of C-9 alcohol.[114]

It has been established previously that maytansine inhibits mitosis in sea urchin eggs and L1210 leukemic cells[115,116] by interaction with the microtubule system of the cells.[115] By testing various maytansinoids, it has been concluded that the C-3 ester linkage appears to be important in tubulin-maytansinoid interaction (Table 7).[114] The presence of the free C-9 carbinolamide group appears to aid the ability of maytansinoids to inhibit tubulin polymerization, suggesting that the compounds may interact with tubulin at two sites.[114] Compounds which do not possess an ester group on C-3 are between 20 to 2000 times less potent as inhibitors of sea urchin cell division[114] than compounds with C-3 ester groups. Esterification of compounds may also facilitate drug transport into the cells.

It has been demonstrated that maytansine and its derivatives bind to tubulin at the vincristine-binding site.[117] Maytansinoids compete for ^{3}H-vincristine-binding sites of tubulin, and the Ki values[118] corroborate the notion that C-3 ester linkage on the drugs is important for binding to tubulin protein (Table 7). In summary, ester groups at C-3 appear to be necessary for potent antitubulin, antimitotic, and antileukemic activity. Blockage of the carbinolamide group by etherification of the C-9 OH group decreases but does not abolish antitubulin activity, has no discernable effect on antimitotic activity, and decreases the in vivo antileukemic effect.

Maytansinoids at 100 μg/mℓ concentration do not inhibit *E. coli* RNA polymerase, mammalian RNA polymerase I and II, DNA polymerase α and β, poly(A) polymerase, and oncornavirus reverse transcriptase activities.[93] At 0.1 to 0.4 ng/mℓ maytansine concentration, the yield of murine sarcoma virus from productively MSV-transformed

3T3 cells is inhibited slightly after 24 hr drug exposure. This inhibition is followed by a slight (and probably not significant) increase in virus production of the maytansine-treated cultures at later time periods. Preliminary results suggest that maytansine decreases virus production of newly-infected cells by only two- to three-fold at days 2, 3, and 4.[37]

Due to pronounced antitumor activity of maytansine in experimental animal model systems, phase I clinical studies of the drug have been conducted at several cancer centers.[119-121] Maytansine doses in the range of 0.01 to 0.8 $mg/m^2/day$, with a total dose of 2.5 mg with different dose schedules, have been applied. Of the evaluable patients, therapeutic benefits have been detected in a limited number with breast cancer, ovarian cancer, non-Hodgkin's lymphoma, melanoma, and head and neck cancer. Drug-related myelosuppression has not been observed. Hepatic toxicity has been found to be subclinical and reversible. Frequently encountered toxic effects are nausea, vomiting, diarrhea, stomatitis, and alopecia. Neurotoxicity of the central and peripheral type has been noted and may be profound. This drug may be useful in the treatment of tumors which are resistant to vincristine. Phase II studies are underway to evaluate fully the therapeutic potential of this interesting drug. However, some recent studies[122,123] indicate the ineffectiveness of this agent in patients with melanoma and breast cancer. Within a few years, we should expect to have more conclusive data on the efficacy of these potential antitumor agents.

IX. SUMMARY

Ansamycins are compounds in which the aromatic chromophore is joined at two nonadjacent positions by an aliphatic *ansa* chain. The naphthalenic (rifamycins, streptovaricins, tolypomycins, and naphthomycin) and the benzenic (geldanamycins and maytansins) ansamycins are produced by fermentation. Maytansins are also produced by plants. The structural similarities among these antibiotics suggest a common pathway of biosynthesis.

Naphthalenic antibiotics are potent antibacterial agents. Some semisynthetic derivatives of rifamycins, rifamycin SV, rifamide, and rifampicine, are used clinically. These drugs exert their antibacterial effect by inhibition of bacterial DNA-dependent RNA-polymerase at the initiation reaction of RNA synthesis. Mammalian nucleic acid polymerase activities are inhibited only at very high concentrations. Some semisynthetic derivatives of rifamycins and streptovaricins, at very high concentrations, inhibit RNA tumor viral reverse transcriptase activity. These effects seem to be nonspecific. Several laboratories have investigated the antiviral properties of ansamycins. Antitransformation and antiviral activities have been found only at high concentrations which are toxic to host cells. These effects also appear to be nonspecific.

Cytotoxic doses of different ansamycins vary drastically against cultured mammalian cells. For example, the cytotoxic dose levels for cultured 3T3 mouse fibroblasts are: maytansins, 10^{-3} to 10^{-4} $\mu g/m\ell$; geldanamycins, 10^{-2} to 10^{-3} $\mu g/m\ell$; rifamycins, 1 to 10 $\mu g/m\ell$, and streptovaricins, 10 to 50 $\mu g/m\ell$. The reason for this wide variation in the cytotoxic levels of these agents is not completely clear, but it could possibly lie in their structure, cellular uptake, metabolism, and different modes of action.

The cellular receptors for all ansamycins are not known. Like the vinca alkaloids, vincristine and vinblastine, maytansine arrests the cells in mitosis. It binds to tubulin at the vincristine binding site and prevents its polymerization to microtubules. It is probably through this mechanism that maytansine exerts its cytotoxic effects. It is not known whether structurally related geldanamycins have a mechanism of action similar to that of maytansine.

Among known ansamycins, maytansine and its derivatives have shown great potential as antitumor agents. Phase I and II trials of this drug on cancer patients show considerable promise. Wide variation in the cytotoxicity of ansamycins offers a challenge to medicinal chemists to modify these structures in order to achieve optimum antitumor activity.

The spacial distribution of hydrophobic and hydrophilic groups on the ansamycin molecule imparts unique physicochemical properties to the molecule. On solubilization, for example, rifamycin forms a gel at higher concentrations and it behaves like a detergent. There are no comparative data on the physicochemical properties (solubility, partition coefficient, viscosity, surface tension, and pK_a) of anasamycins. Therefore, to what extent the biological activities of these agents are influenced by physicochemical properties cannot be evaluated.

Most of the ansamycins are capable of undergoing reversible oxidation-reduction in the quinone-hydroquinone part of the aromatic chromophore moiety. Whether these molecules generate radicals and thereby cause peroxidation of lipids or cleavage of DNA strands analogous to anthracycline antibiotics, is also not known. Moreover, the spacial distribution of keto and hydroxyl functions and the nitrogen atom, seems to be such that these molecules could form chelates with metal ions. The contribution of these properties to the biological activity is also not known. In conclusion, ansamycins are antibiotics with very complex structures and interesting pharmacological properties. It can be predicted that modification of benzenic ansamycins will yield promising antitumor agents.

ACKNOWLEDGMENTS

I am most grateful to Dr. Charles L. Spurr for his continued encouragement and support. I am thankful to Dr. John Douros and Ms. A. M. Acierto for sending me the NCI data on the antitumor activity of ansamycins. I am also thankful to Drs. Don Jackson and Douglas Lyles for a critical reading of the manuscript. I would also like to express my appreciation for David Anderson and Carol Johnson for their library and secretarial assistance. This work was supported by a Center grant CA 12197 to our institution.

REFERENCES

1. **Lüttringhaüs, A. and Gralheer, H.,** *Justus Liebigs Ann. Chem.*, 550, 67, 1942.
2. **Prelog, V. and Oppolzer, W.,** *Helv. Chim. Acta*, 56, 2279, 1973.
3. **Rinehart, K. L., Jr. and Shield, L. S.,** *Fortschr. Chem. Org. Naturst.*, 33, 231, 1976.
4. **Lester, W.,** *Annu. Rev. Microbiol.*, 26, 85, 1972.
5. **Riva, S. and Silvestri, L. G.,** *Annu. Rev. Microbiol.*, 62, 199, 1972.
6. **Rinehart, K. L. Jr.,** *Acc. Chem. Res.*, 5, 57, 1972.
7. **Moss, B.,** in *Selective Inhibitors of Viral Functions*, Carter, W. A., Ed., CRC Press, Boca Raton, Fla., 1973, 313.
8. **Byrd, D. M. and Carter, W. A.,** in *Selective Inhibitors of Viral Functions*, Carter, W. A., Ed., CRC Press, Boca Raton, Fla., 1973, 329.
9. **Sensi, P.,** *Pure Appl. Chem.*, 41, 15, 1975.
10. **Wehrli, W.,** *Topics in Current Chemistry*, Boschke, F., Ed., Springer-Verlag, New York, 1977, 22.
11. **Brufani, M., Kluepfel, D., Lancini, G., Leitish, J., Mesentser, A. S., Prelog, V., Schook, F. P., and Sensi, P.,** *Helv. Chim. Acta*, 56, 2315, 1973.
12. **White, R. J., Martinelli, E., Gallo, G. G., and Lancini, G.,** *Nature*, 243, 273, 1973.
13. **Karlsson, A., Sartori, G., and White, R. J.,** *Eur. J. Biochem.*, 47, 251, 1974.

14. White, R. J., Martinelli, E., and Lancini G., *Proc. Natl. Acad. Sci. U.S.A.*, 71, 3260, 1974.
15. Johnson, R. D., Haber, A., and Rinehart, K. L., *J. Am. Chem. Soc.*, 96, 3316, 1974.
16. Haber, A., Johnson, R. D., and Rinehart, K. L., *J. Am. Chem. Soc.*, 99, 3541, 1977.
17. Kishi, T., Yamana, H., Muroi, M., Harada, S., Asai, M., Hasegawe, T., and Mizuno, K., *J. Antibiot.*, 25, 11, 1972.
18. Deshmukh, P. V., Kakinuma, K., Ameel, J. J., Rinehart, K. L., Wiley, P. F., and Li, L. H., *J. Am. Chem. Soc.*, 98, 870, 1976.
19. Rinehart, K. L., Antosz, F. J., Deshmukh, P. V., Kakinuma, K., Martin, P. K., Milavetz, B. J., Sasaki, K., Witty, T. R., Li, L. H., and Reusser, F., *J. Antibiot.*, 29, 201, 1976.
20. Ghisalba, O., Traxler, P., and Nüesch, J., *J. Antibiot.*, 31, 1124, 1978.
21. Sensi, P., Greco, A. M., and Ballotta, G., *Antibiot. Annu.*, 262, 1959.
22. Margalith, P. and Pagani, H., *Appl. Microbiol.*, 9, 325, 1961.
23. Sensi, P., Ballotta, G., and Gallo G. G., *Farmaco Ed. Sci.*, 16, 165, 1961.
24. Sensi, P., Timbal, M. T., and Maffii, G., *Experientia*, 16, 412, 1960.
25. Oppolzer, W., Prelog, V., and Sensi, P., *Experientia*, 20, 336, 1964.
26. Oppolzer, W. and Prelog, V., *Helv. Chim. Acta*, 56, 2287, 1973.
27. Brufani, M., Cerrini, S., Fedeli, W., and Vaciago, A., *J. Mol. Biol.*, 87, 409, 1974.
28. Sensi, P., Maggi, N., Ballotta, R., Füresz, S., Pallanza, R., and Arioli, V., *J. Med. Chem.*, 7, 596, 1964.
29. Quinn, F. R. and Driscoll, J. S., *J. Med. Chem.*, 18, 4, 1975.
30. Sensi, P., Maggi, N., Füresz, S., and Maffii, G., *Antimicrob. Agents Chemother.*, 699, 1967.
31. Hartmann, G., Honikel, K. O., Knüsel, F., and Nüesch J., *Biochim. Biophys. Acta*, 145, 843, 1967.
32. Wehrli, W., Knüsel, F., Schmid, K., and Staehlin, M., *Proc. Natl. Acad. Sci. U.S.A.*, 61, 667, 1968.
33. Wehrli, W. and Staehelin, M., *Proc. 1st Lepetit Colloq. RNA Polymerase*, Silvestri, L., Ed., North-Holland, Amsterdam, 1970, 65.
34. Wehrli, W., Handschin, J., and Wunderli, W., in *RNA Polymerase*, Losick, R., and Chamberlin, M., Eds., Cold Spring Harbor Laboratory, New York, 1976, 397.
35. Thompson, F. M., Libertini, L. J., Joss, U. R., and Calvin, M., *Science*, 178, 505, 1972.
36. Sethi, V. S. and Okano, P., *Biochim. Biophys. Acta*, 454, 230, 1976.
37. Sethi, V. S., Final Report, National Cancer Institute, Contract NO1-CM-33741, National Institute of Health, Bethesda, Md., 1976, 1.
38. Sethi, V. S., *Prog. Biophys. Mol. Biol.*, 23, 67, 1971.
39. Chamberlin, M. J., in *RNA Polymerase*, Losick, R., and Chamberlin, M., Eds., Cold Spring Harbor Laboratory, New York, 1976, 17.
40. Burgess, R. R., *J. Biol. Chem.* 244, 6168, 1969.
41. Zillig, W., Fuchs, E., Palm, P., Rabussay, D., and Zechel, K., in *1st Lepetit Colloq. RNA Polymerase*, Silvestri, L., Ed., North-Holland, Amsterdam, 1970, 151.
42. Burgess, R. R., in *RNA Polymerase*, Losick, R., and Chamberlin, M., Eds., Cold Spring Harbor Laboratory, New York, 1976, 69.
43. Heil, A. and Zillig, W., *FEBS Lett.*, 11, 165, 1970.
44. Zillig, W., Palm, P., and Heil, A., in *RNA Polymerase*, Losick, R., and Chamberlin, M., Eds., Cold Spring Harbor Laboratory, New York, 1976, 101.
45. Sippel, A. and Hartmann, G., *Biochim. Biophys. Acta*, 157, 218, 1968.
46. Sippel, A. and Hartmann, G., *Eur. J. Biochem.*, 16, 152, 1970.
47. Hinkle, D. C., Mangel, W. F., and Chamberlin, M. J., *J. Mol. Biol.*, 70, 209, 1972.
48. Bähr, W., Stender, W., Scheit, K. H., and Jovin, T. M., in *RNA Polymerase*, Losick, R., and Chamberlin, M., Eds., Cold Spring Harbor Laboratory, New York, 1976, 369.
49. McClure, W. R. and Cech, C. L., *J. Biol. Chem.*, 253, 8949, 1978.
50. Temin, H. M. and Mizutani, S., *Nature*, 226, 1211, 1970.
51. Baltimore, D., *Nature*, 226, 1209, 1970.
52. Green, M. and Gerard, G. F., *Prog. Nucl. Acid Res. Mol. Biol.*, 14, 188, 1974.
53. Wu, A. M. and Gallo, R. C., *CRC Crit. Rev. Biochem.*, 289, 1975.
54. Gurgo, C., Ray, R., and Green, M., *J. Natl. Cancer Inst.*, 49, 61, 1972.
55. Gallo, R. C., Abrell, J. W., Robert, M. S., Yang, S. S., and Smith, R. G., *J. Natl. Cancer Inst.*, 48, 1185, 1972.
56. Tischler, A. N., Thompson F. M., Libertini, L. J., and Calvin, M., *J. Med. Chem.*, 17, 948, 1974.
57. Riva, S., Fietta, A., and Silvestri, L. G., *Biochem. Biophys. Res. Commun.*, 49, 1263, 1972.
58. O'Connor, T. E., Stansly, P., Sethi, V. S., Hadidi, A., and Okano, P., *Intervirology*, 3, 63, 1974.
59. O'Connor, T. E., Aldrich, C. D., and Sethi, V. S., *Ann. N.Y. Acad. Sci.*, 284, 544, 1977.
60. Meilhac, M., Tysper, Z., and Chambon, P., *Eur. J. Biochem.*, 28, 291, 1972.
61. Milavetz, B. I., Horoszewicz, J. S., Rinehart, K. L., and Carter, W. A., *Antimicrob. Agents Chemother.*, 435, 1978.

62. Buss, W. C., Morgan, R., Guttman, J., Barela, T., and Stalter, K., *Science*, 200, 432, 1978.
63. Winicov, I., *Biochemistry*, 18, 1575, 1979.
64. Calvin, M., Joss, V. R., Hackett, A. J., and Owens, R. B., *Proc. Natl. Acad. Sci.*, 68, 1441, 1971.
65. Ting, R. C., Yang, S. S., and Gallo, R. C., *Nat. New Biol.*, 236, 163, 1972.
66. Shannon, W. M., Westbrook, L., and Schabel, F. M., *Intervirology*, 3, 84, 1975.
67. Fralova, L. Y., Meldrays, Y. A., Kochkina, L. L., Giller, S. A., Ermeyev, A. V., Grayevskaya, N. A., and Kisselev, L. L., *Nucl. Acid Res.*, 4, 523, 1977.
68. Szabo, C., Bissell, M. H., and Calvin, M., *J. Virol.*, 18, 445, 1976.
69. Heller, E., Argaman, M., Levy, H., and Goldblum, N., *Nature*, 222, 273, 1969.
70. Subak-Sharpe, J. H., Timbury, M. C., and Williams, J. F., *Nature*, 222, 341, 1969.
71. Moss, B., Rosenblum, E. N., and Grimley, P. M., *Virology*, 45, 135, 1971.
72. Moss, B., Rosenblum, E. N., Katz, E., and Grimley, P. M., *Nature*, 224, 1280, 1969.
73. Pennington, T. H. and Follett, E. A. C., *J. Virol.*, 7, 821, 1971.
74. Szilagyi, J. F. and Pennington, T. H., *J. Virol.*, 8, 133, 1971.
75. Furesz, S., *Antibiot. Chemother.*, 16, 316, 1970.
76. Siminoff, P., Smith, R. M., Sokoloski, W. T., and Savage, G. M., *Am. Rev. Tuberc. Pulm. Dis.*, 75, 576, 1957.
77. Rinehart, K. L., Maheshwari, M. L., Antosz, F. J., Mathur, H. H., Sasaki, K., and Schacht, R. J., *J. Am. Chem. Soc.*, 93, 6273, 1971.
78. Wang, A. H. J., Paul, I. C., Rinehart, K. L., and Antosz, F. J., *J. Am. Chem. Soc.*, 93, 6275, 1971.
79. Yamazaki, H., *J. Antibiot. (Tokyo)*, 21, 209, 1968.
80. Mizuno, S., Yamazaki, H., Nitta, K., and Umezawa, H., *Biochim. Biophys. Acta*, 157, 322, 1968.
81. Yura, T. and Igaraski, K., *Proc. Natl. Acad. Sci. U.S.A.*, 61, 1313, 1968.
82. Yura, T. and Igarashi, K., *Prog. Antimicrob. Anticancer Chemother.*, 2, 405, 1971.
83. Mizuno, S., Yamazaki, H., Nitta, K., and Umezawa, H., *Biochem. Biophys. Res. Commun.*, 30, 379, 1968.
84. Rinehart, K. L., Antosz, F. J., Sasaki, K., Martin P. K., Maheshwari, M. L., Reusser, F., Li, L. H., Moran, D., and Wiley, P. F., *Biochemistry*, 13, 861, 1974.
85. Brockman, W. W., Carter, W. A., Li, L. H., Reusser, F., and Nichol, F. R., *Nature*, 230, 249, 1971.
86. Li, L. H., Cowie, C. H., Gray, L. G., Moran, D. M., Clark, T. D., and Rinehart, K. L., *J. Natl. Cancer Inst.*, 58, 239, 1977.
87. Srivastava, S. B. I., DiCioccio, R. A., Rinehart, K. L., and Li, L. H., *Mol. Pharmacol.*, 14, 442, 1978.
88. Carter, W. A., Brockman, W. W., and Borden, E. C., *Nat. New Biol.*, 232, 212, 1971.
89. Borden, E. C., Brockman, W. W., and Carter, W. A., *Nat. New Biol.*, 232, 214, 1971.
90. Borden, E. C., Carter, W. A., and Sensenbrenner, L. L., *Int. J. Cancer*, 14, 817, 1974.
91. Li, L. H., Clark, T. D., Cowie, C. H., Swenberg, J. A., Renis, H. E., and Rinehart, K. L., *J. Natl. Cancer Inst.*, 58, 245, 1977.
92. Horoszewicz, J. S., Leong, S. S., and Carter, W. A., *Antimicrob. Agents and Chemother.*, 4, 1977.
93. Sethi, V. S., 18th Interscience Conf. Antimicrobial Agents Chemother., Atlanta, Georgia, (Abstr. 254), 1978.
94. Hasegawa, T., Higashide, E., and Shibata, M., *Antibiotics*, 24, 817, 1971.
95. Shibata, M., Hasegawa, T., and Higashide, E., *J. Antibiot.*, 24, 810, 1971.
96. Kamiga, K., Sugino, T., Wada, Y., Nishihawa, M., and Kishi, T., *Experientia*, 25, 901, 1969.
97. Kondo, M., Oishi, T., and Tsuchiya, K., *J. Antibiot.*, 25, 16, 1972.
98. Brufani, M., Cellai, L., Cerrini, S., Fedeli, W., and Vaciago, A., *Mol. Pharmacol.*, 14, 493, 1978.
99. Bellomo, P., Brufani, M., Marchi, E., Mascellani, G., Melloni, W., Montecchi, L., and Stanzani, L., *J. Med. Chem.*, 20, 1287, 1977.
100. Balerna, M., Keller-Schierlein, W., Martius C., Wolf, H., and Zähner, H., *Arch. Mikrobil.*, 65, 303, 1969.
101. Williams, T. H., *J. Antibiot.*, 28, 85, 1975.
102. Sasaki, K., Rinehart, K. L., Slomp, G., Grostic, M. F., and Olson, E. C., *J. Am. Chem. Soc.*, 92, 7591, 1970.
103. DeBoer, C., Meulman, P. A., Wnuk, R. J., and Peterson, D. H., *J. Antibiot.*, 23, 442, 1970.
104. Rinehart, K. L., Sobiczewski, W., Honegger, J. F., Enanoza, R. M., Witty, T. R., Li, V. J., Shield, L. S., Li, L. H., and Reusser, F., *Bioorg. Chem.*, 6, 341, 1977.
105. Rinehart, K. L., McMillan, M. W., Witty, T. R., Tipton, C. D., Shield, L. S., Li, L. H., and Reusser, F., *Bioorg. Chem.*, 6, 353, 1977.
106. Li, L. H., Clark, T. D., Cowie, C. H., and Rinehart, K. L., *Cancer Treat. Rep.*, 61, 815, 1977.

106. Li, L. H., Clark, T. D., Cowie, C. H., and Rinehart, K. L., *Cancer Treat. Rep.*, 61, 815, 1977.
107. Kupchan, S. M., Komoda, Y., Court, W. A., Thomas, G. J., Smith, R. M., Karim, A., Gilmore, D. J., Haltiwanger, R. C., and Bryan, R. F., *J. Am. Chem. Soc.*, 94, 1354, 1972.
108. Kupchan, S. M., Komoda, Y., Thomas, G. J., and Hintz, H. P. J., *J. Chem. Soc. Chem. Commun.*, 1065, 1972.
109. Kupchan, S. M., Komoda, Y., Branfman, A. R., Dailey, R. G., Jr., and Zimmerly, V. A., *J. Am. Chem. Soc.*, 96, 3706, 1974.
110. Kupchan, S. M., Branfman, A. R., Sneden, A. T., Verma, A. K., Dailey, R. G., Jr., Komoda, Y., and Nagao, Y., *J. Am. Chem. Soc.*, 97, 5294, 1975.
111. Kupchan, S. M., Komoda, Y., Branfman, A. R., Sneden, A. T., Court, W. A., Thomas, G. J., Hintz, H. P. J., Smith, R. M., Karim, A., Howie, G. A., Verma, A. K., Nagao, Y., Dailey, R. G., Jr., Zimmerly, V. A., and Sumner, W. C., Jr., *J. Org. Chem.*, 42, 2349, 1977.
112. Wani, M. C., Taylor, H. L., and Wall, M. E., *J. Chem. Soc. Chem. Commun.*, 390, 1973.
113. Higashide, E., Asai, M., Ootsu, K., Tanida, Y., Hasegawa, T., Kishi, T., Sugino, Y., and Yoneda, M., *Nature*, 270, 721, 1977.
114. Kupchan, S. M., Sneden, A. T., Branfman, A. R., Howie, G. A., Rebhun, L. I., McIvor, W. E., Wang, R. W., and Schnaitman, T. C., *J. Med. Chem.*, 21, 31, 1978.
115. Remillard, S., Rebhun, L. I., Howie, G. A., and Kupchan, S. M., *Science*, 189, 1002, 1975.
116. Wolpert-DeFilippes, M. K., Bono, V. H., Dion, R. L., and Johns, D. G., *Biochem. Pharmacol.*, 24, 1735, 1975.
117. Shavit, F., Wolpert-DeFilippes, M. K., and Johns, D. G., *Biochem. Biophys. Res. Comm.*, 72, 47, 1976.
118. York, J., Wolpert-DeFilippes, M. K., Sethi, V. S., and Johns, D. G., *Am. Assoc. Cancer Res. Proc.*, 19, (Abstr. #438), 1978.
119. Cabanillas, F., Rodriguez, V., Hall, S. W., Burgess, M. A., Bodey, G. P., and Freireich, E. J., *Cancer Treat. Rep.*, 62, 425, 1978.
120. Chabner, B. A., Levine, A. S., Johnson, B. L., and Young, R. C., *Cancer Treat. Rep.*, 62, 429, 1978.
121. Blum, R. H. and Kahlert, T., *Cancer Treat. Rep.*, 62, 435, 1978.
122. Eagan, R. T., Creagan, E. T., Ingle, J. N., Frytak, S., and Rubin, J., *Cancer Treat. Rep.*, 62, 1577, 1978.
123. Cabanillas, F., Bodey, G. P., Burgess, M. A., and Freireich, E. J., *Cancer Treat. Rep.*, 63, 507, 1979.

Chapter 4

THE ROLE OF RETINOIDS (VITAMIN A AND ITS ANALOGS) IN THE PREVENTION OF EPITHELIAL CANCER

Charles A. Frolik and Peter P. Roller

TABLE OF CONTENTS

I. INTRODUCTION

Over half of the total cancer deaths in the U.S. are due to malignancies in epithelial tissues including lung, pancreas, colon, bladder, breast, and ovary.[1] Approaches to treatment of this disease in the past have depended on detecting the tumors after the development of invasive cancer and then treating them with cytotoxic chemotherapy, surgery, or radiation. In spite of some advances that have occurred in this area, this method of treatment has met with only limited success. It is now known, however, that the development of invasive cancer is a prolonged process that may take up to 20 yr or more in man before reaching its terminal invasive stage.[2] An alternative approach, therefore, would be to treat the disease during its period of progression, i.e., after the tissue has been exposed to a carcinogenic insult but before the lesion produced has become invasive. There are many innate protective mechanisms, including mutational selection[3] and endocrine and immunological responses,[4,5] that appear to suppress potentially carcinogenic lesions. However, these mechanisms in themselves do not appear to be adequate for elimination of the epithelial cancer problem in humans. Techniques are therefore needed that are able to enhance these protective mechanisms. One possible approach to this problem is the use of vitamin A and its synthetic analogs (retinoids), compounds that are able to modulate epithelial differentiation.

Vitamin A has long been known to be involved in the normal development of epithelial tissue.[6] It has also been demonstrated that in the absence of retinoids in the diet, the normal, differentiated epithelial cells of the bronchi (or the transitional cell of the bladder) are replaced by a keratinized, squamous metaplastic epithelial cell morphologically similar to that observed after carcinogen administration.[7,8] Because of this morphological similarity between the vitamin A-deficient and the carcinogen-treated tissue, and because vitamin A is capable of reversing the abnormal differentiation in the deficient tissue (causing reappearance of normal mature cells), retinoids have been implicated as possibly being useful in influencing the abnormal cell differentiation that characterizes the process of carcinogenesis.

It is the purpose of this review to briefly summarize the chemistry and metabolism of retinoids and then to discuss the in vivo and in vitro experimental systems that have been used to demonstrate the ability of these compounds to prevent epithelial cancer. Many of the ideas discussed in this chapter have originated from several recent reviews in this area.[9-17]

II. CHEMISTRY OF NATURALLY OCCURRING RETINOIDS

A. Introduction and Brief History

Vitamin A and its closely related, naturally occurring analogs are 20 carbon diterpenes that have been isolated chiefly from fish liver oils, visceral parts of fish, eggs, and animal kidney, lung, eyes, and intestinal mucosa. They are structurally related to carotenes, 40 carbon containing polyisoprenoid plant pigments, from which the vitamin A compounds are derived by enzymatic degradation in the intestinal mucosa. Curiously, whereas the biological functions of retinoids are well-recognizable (such as their involvement in the maintenance of normal growth, spermatogenesis, and pregnancy, in the control of epithelial tissue differentiation and in the visual process), there is much yet to be defined about the mechanism of action of vitamin A, other than in the visual process. Furthermore, in spite of considerable progress in the field, the form of vitamin A that is responsible for these various biological activities — unaltered, metabolized and/or protein bound — is still open to question. There are a number of extensive reviews in the literature on the molecular aspects of the problem testifying to the advances made.[15,18-25] This portion of the chapter, though not claiming to be exhaustive, will briefly enumerate the chemistry of several of the naturally occurring retinoids.

A point on the nomenclature deserves clarification. In a strict sense of the word, vitamin A refers to all-*trans*-retinol (Figure 1a). In the past, however, the term vitamin A has encompassed the naturally occurring compounds that exhibit some or all of the characteristic retinoid activities. These compounds (see Figure 1) include retinol (a) and its palmitate and acetate esters, dehydroretinol (b), retinal (c), retinoic acid (d), as well as their cis-trans structural isomers. More recently, the word "retinoids" has been used to refer to any natural or synthetic compound closely related to retinol. These compounds may include a metabolite, oxidation or reduction product, or a derivative thereof, either synthetically or naturally derived, which may or may not exhibit the various physiological activities attributed to retinol.

The existence of an essential nutritional factor, later to be termed vitamin A, was postulated and its physiological effects described long before the vitamin itself was isolated or its chemical structure determined. For example, even the Egyptians recognized the remedial effect for night blindness of ingested liver, the storage organ for vitamin A.[26] In early attempts (1909 to 1919), Hopkins[27] and Stepp[28] found that fat-soluble substances in bread and milk were essential for sustenance of life in animals. A fat-soluble factor, which was found to be essential for growth in animals and also for prevention of xerophthalmia, was also extracted from butter and egg yolk by McCullum and Davis[29,30] and by Osborne and Mendel.[31] Finally, in 1920, Drummond named this essential factor vitamin A[32] and, independently, his group,[33] (as well as Takahashi's group in Japan[34]) succeeded in partial purification from fish oil.

Apparent controversy arose, however, when V. Euler et al. observed that the plant product, carotene, clearly a different substance than vitamin A with different stability and chemical properties, was effective in curing the symptoms of vitamin A deficienty in rats when administered in daily doses of about 5 μg.[35] In experiments that followed, Moore[36,37] and Capper[38] demonstrated the appearance of vitamin A in the liver of vitamin A-deficient rats that were given β-carotene (Figure 2a). Indeed, the enzymatic conversion of β-carotene to retinal and then to retinol in the intestinal mucosa and in the liver of animals has now been demonstrated by several investigators (for a review, see Goodman and Olson[39]). In 1930, the chemical structure of β-carotene was established by Karrer and co-workers in Zurich.[40] Following this, using highly enriched vitamin A obtained from halibut liver oil, this group also established the carbon skel-

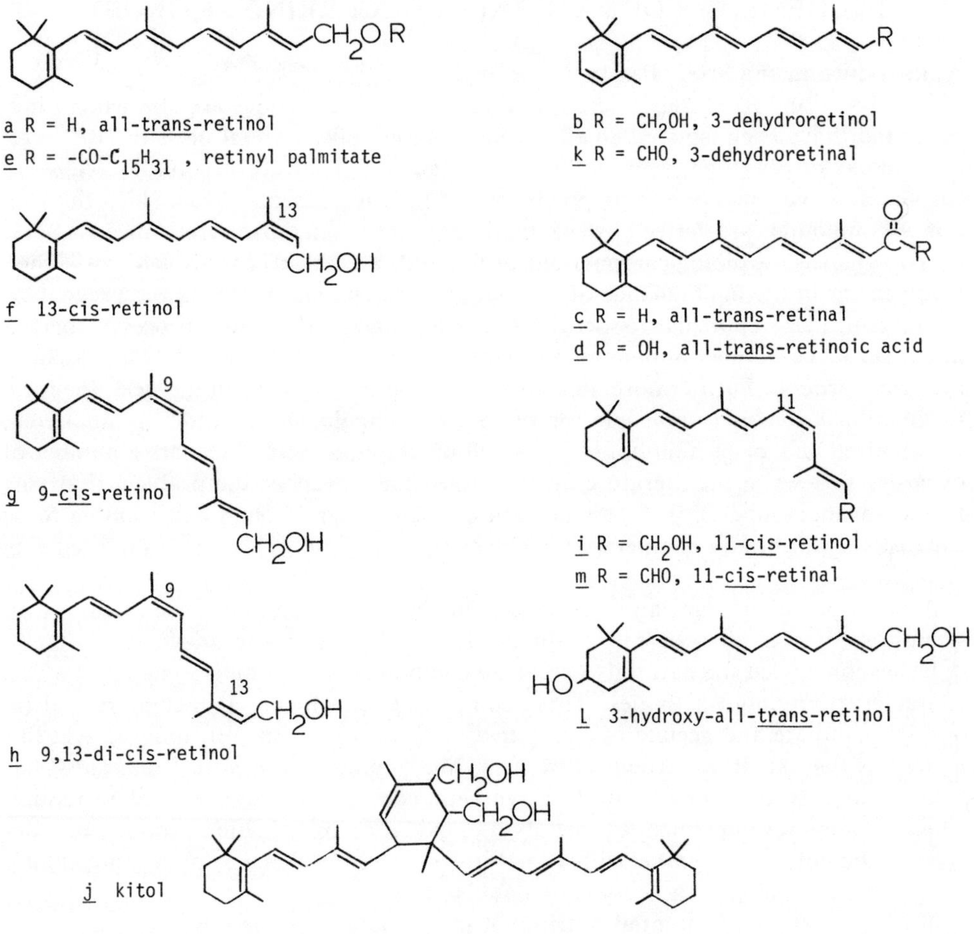

a R = H, all-trans-retinol
e R = -CO-$C_{15}H_{31}$, retinyl palmitate

b R = CH_2OH, 3-dehydroretinol
k R = CHO, 3-dehydroretinal

f 13-cis-retinol

c R = H, all-trans-retinal
d R = OH, all-trans-retinoic acid

g 9-cis-retinol

i R = CH_2OH, 11-cis-retinol
m R = CHO, 11-cis-retinal

h 9,13-di-cis-retinol

L 3-hydroxy-all-trans-retinol

j kitol

FIGURE 1. Naturally occurring retinoids.

eton of retinol by conversion to perhydroretinol.[41] Using this and other chemical information they finally arrived at the now accepted structure of vitamin A (retinol, Figure 1a).[41] Corbet was the first to succeed in obtaining crystalline retinol from natural sources.[42] Chemical synthesis of pure retinol was achieved between 1946 and 1947 by Isler and co-workers[43,44] and independently by Arens and Van Dorp.[45,46]

B. Occurrence and Isolation

1. Retinol and Derivatives

In terms of nutritional requirements, the recommended daily vitamin A intake for adults is 5000 IU,[47] where 1 IU is defined as 0.344 μg retinyl acetate. In as much as certain carotenoids are metabolized to retinoids in the animal intestine, they are also able to contribute to the total vitamin A intake. Animal products rich in vitamin A include fish intestine, fish, pork and beef liver, eggs, milk, and butter. Provitamins, or carotenoids, are especially abundant in carrots, but are also found in other vegetables, milk, and butter. According to a recent estimate,[48] the average diet of a person living in the U.S. is estimated to provide about twice the recommended daily intake of vitamin A, and fruits and vegetables provide almost half of that in the form of provitamins.

Although there are theoretically 16 possible cis-trans isomers of retinol, the most abundant naturally occurring compound is the all-*trans* isomer found in the esterified

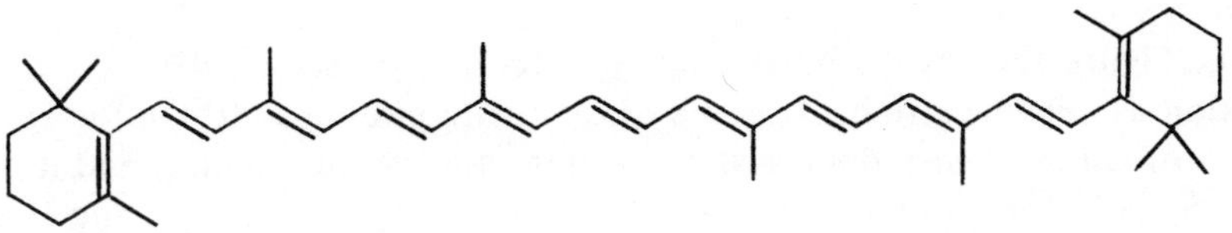

FIGURE 2. Relevant carotenoids.

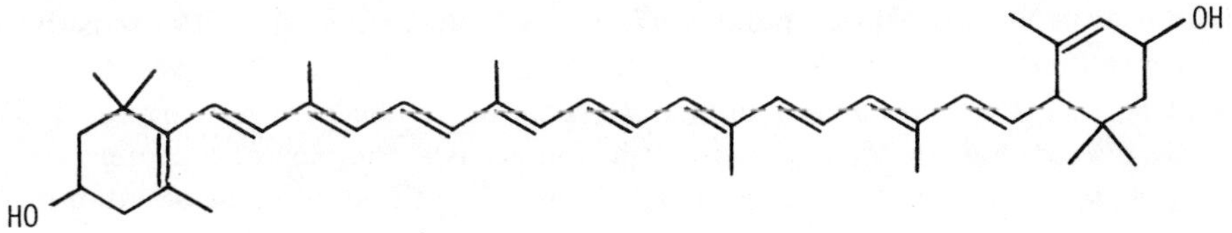

form (for example in salt water fish liver or intestine). Other isomers, however, do occur. The first naturally occurring cis compound, 13-*cis*-retinol (Figure 1f) was isolated by Robeson and Baxter in 1945[49,50] and was found to account for 35% of the total vitamin A in shark-liver oil. The 9-*cis* (Figure 1g) and the 9,13-di-*cis* (Figure 1h) isomers of retinol have also been isolated from fish liver oils.[51] In addition, the sterically hindered 11-*cis*-retinol (Figure 1i) is found in the eye. Most of the retinol in the eye is believed to be stored in the pigment epithelium in the form of retinyl esters,[52,53] and according to Krinsky,[54] the 11-*cis*-retinyl ester may account for up to 65% of this storage depot.

One intriguing, naturally occurring retinoid is the compound kitol (Figure 1j). This compound possesses no vitamin A-like biological activity. It is a stable dimer of retinol and is abundantly encountered in the esterified form in whale liver oil.[55] Its structure has been recently determined with extensive use of NMR spectroscopy.[56] The compound can be considered as a Diels-Alder addition product of the 13,14-double bond of 1 molecule of retinol to the 11,12 - 13,14 diene moiety of a second equivalent of retinol, and the monomer can be regained by molecular vacuum distillation at higher temperatures.[57] Because the natural product is optically active, it is speculated that its formation is enzymatically catalyzed in animals. However, no enzymes have yet been isolated that are able to catalyze either its formation or cleavage. Nevertheless, it has been postulated[58] that the kitol may serve as the storage form of retinol in the whale.

Vitamin A_2, or all-*trans*-3,4-dehydroretinol (Figure 1b), is closely related to retinol but contains an additional double bond in the cyclohexyl moiety of the molecule. The possible existence of this retinoid was first demonstrated in the 1930s, when the liver of fresh water fish appeared to contain an additional vitamin A-like compound that exhibited a high wavelength absorption maximum at 693 nm on treatment with the Carr-Price reagent (antimony trichloride in chloroform),[59,60] in contrast to retinol that has a maximum at 620 nm. Vitamin A_2 was found to be even more sensitive to air oxidation than retinol. It was finally purified in 1948[61] and synthesized in 1952[62] from all-*trans*-retinoic acid (Figure 1d). Vitamin A_2 aldehyde, or 3,4-dehydroretinal (Figure 1k) is known to occur in the retina of fresh water fish, and was isolated from light bleached fractions of porphyropsin by Wald.[63] Vitamin A_2 appears to be biosynthesized from the carotenoid, lutein (Figure 2b).[64] Recently, a related compound 3-hydroxy-all-*trans*-retinol (Figure 1l) has also been isolated from fresh water fish livers.[64]

2. Retinal

Retinal, (Figure 1c), the aldehyde analog of retinol, possesses, the same degree of unsaturation as dehydroretinol. It is the biosynthetic intermediate[65] in the enzymatic conversion of carotene to retinol in the intestinal mucosa of animals, and it has also been detected in trace amounts in the liver of vertebrates.[66] Winterstein and Hegedüs[66] claim identification of trace amounts of retinal in plants, such as clover, spinach, kale, cress, rose hips, and citrus fruits, using the sensitive rhodamin reaction. This reagent forms highly colored compounds with polyenic aldehydes, and the derivative with retinal exhibits a specific absorption band at 472 nm with an estimated limit of sensitivity of 30 ng for retinal.

Retinal performs a well-defined biological function in the eye. As early as 1933, Wald[67] discovered that the "visual purple" in the eye is a complex of a protein *opsin* and the polyene *retinene*. Ten years later, Morton identified *retinene* as vitamin A aldehyde or retinal.[68,69] The details of the mechanism of vision have been impressively worked out by Wald,[70] and others.[71,72] According to the most widely accepted theory, rhodopsin, or visual purple, is composed of opsin, covalently bound to 11-*cis*-retinal (Figure 1m) through a protonated Schiff base, formed between the ∈-amino group of a lysine unit in the protein and the aldehyde functionality of retinal.[73] Upon exposure to light, the 11-*cis*-retinal moiety of the pigment is isomerized to the all-trans form, whereupon the Schiff base is hydrolyzed to the two separate molecules of all-*trans*-retinal and the apoprotein opsin. In two consecutive dark reactions, the all-*trans*-retinal is enzymatically isomerized to the 11-*cis*-form, followed by the nonenzymatic Schiff base formation between the 11-*cis*-retinal and opsin, resulting in regeneration of rhodopsin. Both isomers of retinal are in an equilibrium with their respective retinol isomers though the action of a reversible alcohol dehydrogenase-DPN, enzyme system. Recently, studies by Nakanishi's and Honig's group have shed light on one of the remaining major controversies surrounding the seemingly anomalous absorption properties of rhodopsin.[74-76] Chemically prepared Schiff base, or protonated Schiff base model systems of 11-*cis*-retinal with simpler amines, prepared by previous workers were found to absorb light at at least 50 nm shorter wavelength than rhodopsin itself. In explanation of this phenomenon, Nakanishi and Honig put forward their "external point charge model",[75] supported by experiments and theoretical calculations. Accordingly, the chromophoric unit on rhodopsin in a protonated Schiff base form is in proximity to a spatially well-defined and exactly positioned, negatively charged moiety, approximately 3Å distant through space from carbon 12 of the N-retinylidene group. The negative charge is presumably contributed by a polar functional group attached to the rest of the protein. The existence of a second negative counter ion near the protonated Schiff base nitrogen is postulated as before. These precisely located point charges modulate, or shift to higher wavelength, the absorption spectrum of the retinylidene chromophore through electronic interactions. The case just illustrated presents an impressively well-defined and specific retinoid-protein interaction. At present, it is not clear as to whether a similar interaction will lead to the explanation of other physiological activities of retinoids.

C. Chemical Properties and Synthesis

1. Retinol

Retinol, $C_{20}H_{30}O$, a monocyclic pentaenic primary alcohol, is chemically an unstable molecule. It readily undergoes decomposition by the action of oxygen and heat; it is acid sensitive and, in the presence of light, undergoes cis-trans isomerization. Treatment with an oxidizing agent, such as active manganese dioxide, produces retinal without cis-trans isomerization.[77] Further treatment with a mixture of manganese dioxide and sodium cyanide in 1% acetic acid in methanol produces the methyl ester of retinoic

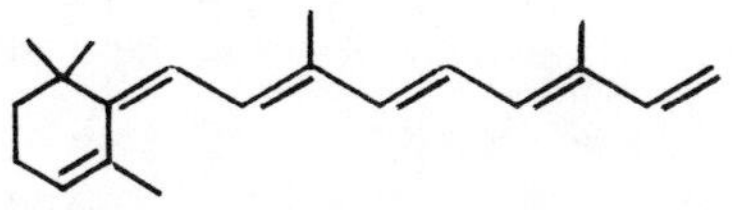

a anhydroretinol

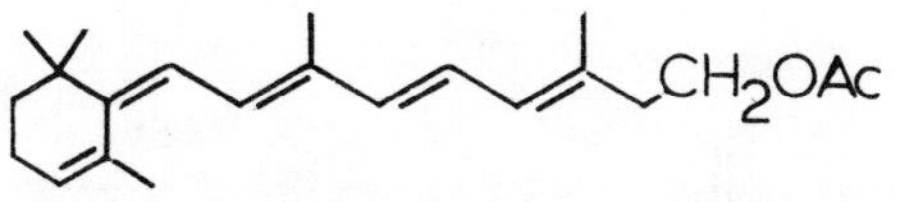

b retrovitamin A acetate

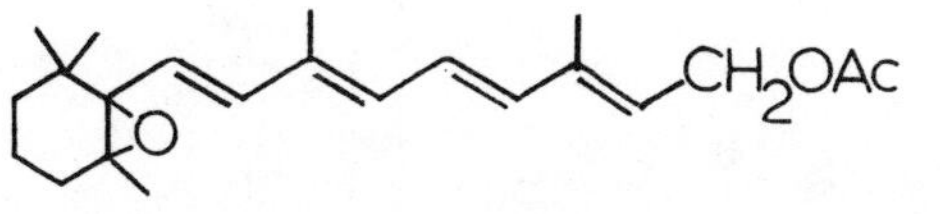

c 5,6-epoxyretinyl acetate

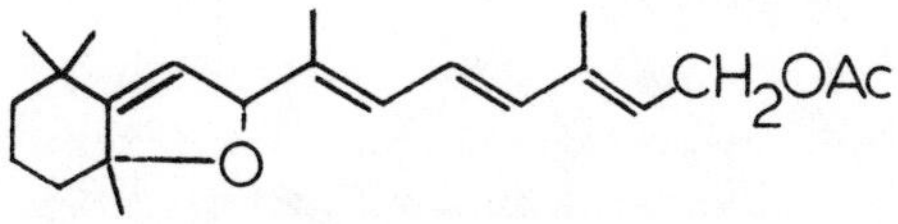

d 5,8-epoxyretinyl acetate

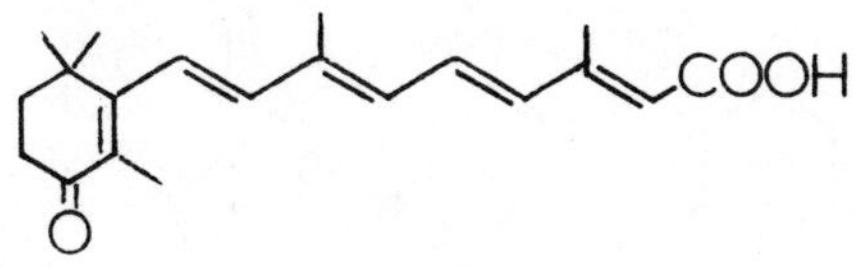

e 4-oxoretinoic acid

f

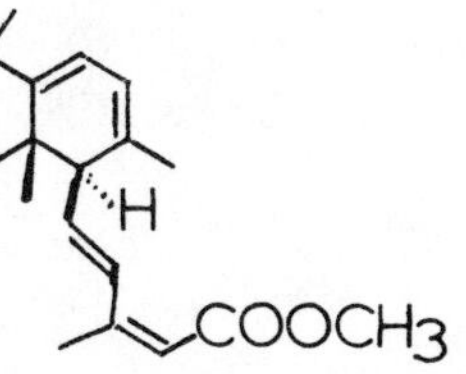

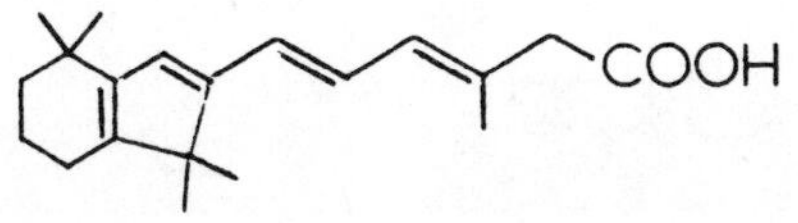

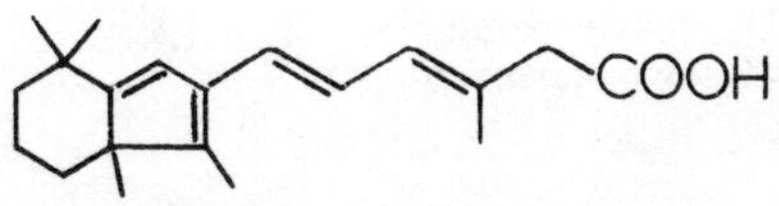

g

FIGURE 3. Various chemical conversion products of retinol, retinyl acetate, and retinoic acid.

acid.[78] Retinol in ethanolic hydrochloric acid undergoes facile dehydration to anhydroretinol, (Figure 3a).[79] Retinyl acetate also undergoes double bond rearrangement in the presence of an acid such as hydrobromic acid in methylene chloride, to yield the product, retrovitamin A acetate, (Figure 3b).[80] Retinol and its esters can undergo Diels-Alder condensation with maleic anhydride with the participation of the 11,12 and 13,14 double bond; in this reaction the all-*trans* isomer is the most reactive.[81] Oxidation of retinyl acetate with monoperphthalic acid effects epoxidation of the endocyclic 5,6 double bond, yielding 5,6-monoepoxyretinyl acetate, (Figure 3c).[82] Acid treatment of the latter product causes isomerization to result in 5,8-monoepoxyretinyl acetate, (Figure 3d).[82] Six of the synthetically prepared cis-trans isomers of retinol, all-*trans*, 9-*cis*, 11-*cis*, 13-*cis*, 9,13-di-*cis*, and 11,13-di-*cis*, are well characterized and physiochemical data on these compounds can be found in the literature.[83,84]

2. *Retinal and Retinoic Acid*

These two retinoids, like retinol, also exhibit susceptibility to air oxidation, and cis-trans isomerization of the side chain double bond by the action of light, iodine, acid, or heat.[85-87] Both retinal and the methyl ester of retinoic acid can be reduced to retinol with lithium aluminum hydride without isomerization of the double bond. In addition to the six synthetically available isomers of retinal and retinoic acid corresponding to those of retinol mentioned above, Liu and co-workers recently have synthesized the hindered isomers 7-*cis*, 7,9-di-*cis*, 7,13-di-*cis*, and the 7,9,13-tri-*cis* isomers of retinal.[88,89] Similar to retinyl acetate, retinoic acid methyl ester can be epoxidized in the 5,6-position of the molecule.[90] Recent studies in Nelson's laboratory have explored the effect of iodine and light on retinoic acid.[91] All-trans-retinoic acid in heptane is rapidly isomerized in the presence of iodine in the dark, giving a mixture of the starting isomer and 13-*cis*-retinoic acid. Under the influence of light, and presumably oxygen, the above mixture also produced good yields of all-*trans*-4-oxoretinoic acid (Figure

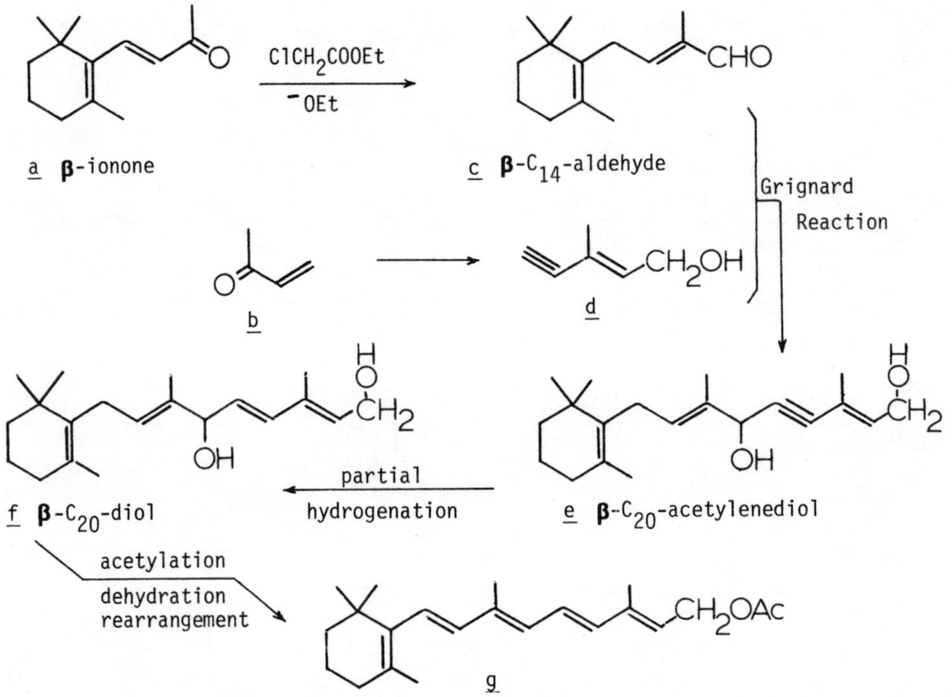

FIGURE 4. Synthesis of retinyl acetate according to Isler et al.[43,44]

3e), and 13-*cis*-4-oxoretinoic acid. This same group has utilized high-pressure liquid chromatography (HPLC), and nuclear magnetic resonance (NMR) spectroscopy for the analysis of 11 isomers of retinoic acid obtained by the action of fluorescent light on a solution of all-*trans*-retinoic acid methyl ester in dimethyl sulfoxide.[92] These compounds were identified as the all-*trans*, 7-*cis*, 9-*cis*, 11-*cis*, 13-*cis*, 7,13-di-*cis*, 9,13-di-*cis*, 11,13-di-*cis*, and the 9,11,13-tri-*cis* isomers of methyl retinoate in addition to the 13-*cis*-(5 →10)-photocyclized isomer (Figure 3f), and the 13-*trans*-(5→10)-photocyclized isomer. After prolonged irradiation, the two photocyclized isomers were shown to accumulate. In less polar solvents such as heptane or acetonitrile, the isomerization rate diminished. Two novel cyclized carbon skeleton rearrangement products, Figures 3g and 3h, were reported recently by Tsukida and co-workers[93] as a result of treatment of all-*trans*-retinoic acid with strong acid. A mechanism for the skeletal rearrangements was also proposed.

3. Synthetic Approaches

A considerable amount of work has gone into development of synthetic methods for preparation of the natural retinoids as well as of their structural analogs, and the literature offers several excellent reviews on this topic.[15,83,94-96] Although increased efforts have recently gone into the preparation of various synthetic aromatic analogs of retinoic acid that might possess better therapeutic activities in dermatology and possibly in cancer prevention,[15] there will be no attempt to review the synthesis of these compounds here. Instead, several of the key synthetic approaches to the natural retinoids will be illustrated. Most syntheses utilize *β*-ionone (Figure 4a) as the starting material. Although *β*-ionone is a naturally occurring compound, it is now mainly synthetically produced.

The original synthetic sequence developed by Isler's group[43,44] (Figure 4) has withstood the test of time and has been developed into a commercial procedure.[97] This

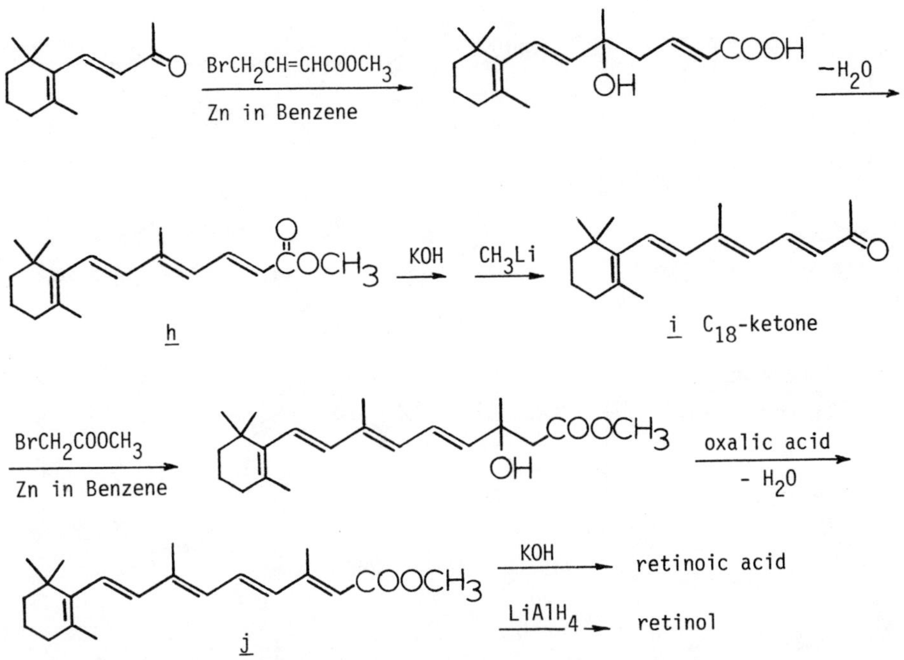

$BrCH_2CH{=}CHCOOCH_3$ / Zn in Benzene
COOH
OH
$-H_2O$
COCH₃ → $COCH_3$
KOH CH_3Li
h
i C_{18}-ketone
$BrCH_2COOCH_3$ / Zn in Benzene
$COOCH_3$
OH
oxalic acid / $- H_2O$
$COOCH_3$
KOH → retinoic acid
$LiAlH_4$ → retinol
j

FIGURE 5. Synthesis of retinoic acid and retinol by Arens and Van Dorp.[45,46]

sequence utilizes as the starting material β-ionone, (a) and methylvinylketone, (b). Thus β-ionone is first converted into β-C_{14}-aldehyde (c) via a glycidic ester synthesis using ethylchloroacetate, treatment with base, and followed by decarboxylation. Methyl vinyl ketone is chain elongated to *cis*-3-methylpent-2-en-4-yn-1-ol, (d). Grignard reaction of the latter compound with the β-C_{14}-aldehyde gives β-C_{20}-acetylenediol (e). Partial hydrogenation gives the β-C_{20}-10,15-diol (f). Acetylation of the primary hydroxyl group, followed by dehydration and isomerization in presence of iodine, yields crystalline all-*trans*-retinyl acetate (g). The procedure is also adaptable to the synthesis of 11-*cis*-retinol and 11,13-di-*cis*-retinol.[44]

The synthesis of retinoic acid[45] and of retinol[46] by Arens and Van Dorp relies on the repeated application of the Reformatskii reaction for building up the side chain (Figure 5). Thus, condensation of β-ionone with γ-bromocrotonic ester in the presence of zinc powder, followed by dehydration with oxalic acid, yields the C_{17}-ionylidene-crotonic ester (h). Saponification to the free acid, followed by methyllithium treatment, gives the key C_{18}-ketone (i). Condensation with bromoacetic ester under Reformatskii conditions, followed by dehydration with oxalic acid, yields retinoic acid methyl ester (j). Reduction of the methyl ester with lithium aluminum hydride yields retinol.[46] The cis/trans confirguration of the products was not delineated at the time of the original synthesis.

Matsui's group[98] in 1958 described the synthesis of four specific isomers of ethyl retinoate and of retinol (Figure 6). The key intermediates are the trans and the cis isomers of the β-C_{15}-aldehyde, (k) and (l), both obtainable from β-ionone. Condensation of the *trans*-β-C_{15}-aldehyde, (k) with ethyl senecioate, (m), with potassium amide in liquid ammonia yields all-*trans*-retinoic acid ethyl ester. The same starting material with sodium amide or with lithium amide gives the 13-*cis*-retinoic acid ethyl ester. In parallel synthetic schemes, the 9-*cis*-β-C_{15}-aldehyde, (l) with potassium amide gives the 9-*cis*-retinoic acid ester, whereas with sodium or lithium amide, the product is the 9,13-di-*cis* ester. The 4 isomeric ethyl retinoates could be reduced to the corresponding retinols with lithium aluminum hydride without rearrangements.

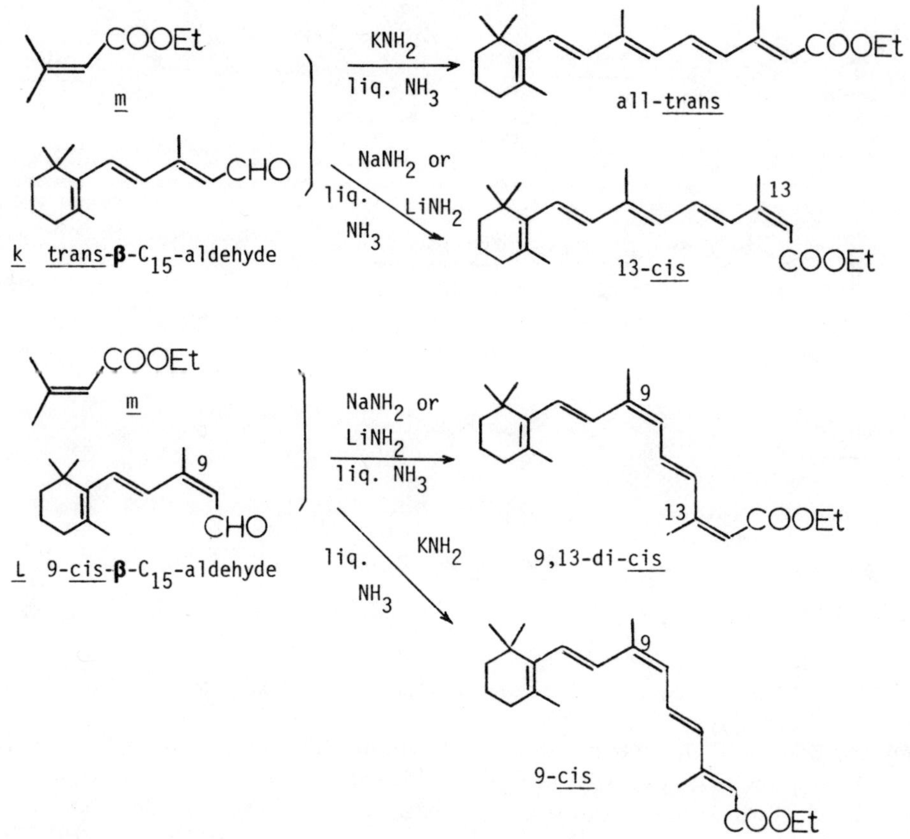

FIGURE 6. Synthesis of all-*trans*, 9-*cis*, 13-*cis*, and 9,13-di-*cis* isomers of retinoic acid ester, according to Matsui et al.[98]

A number of successful synthetic sequences have applied the Wittig reaction for the condensation step. For example, in a recent synthesis of ^{14}C-10 labeled all-*trans*-retinol and retinoic acid, the Horner modification of the Wittig reaction was utilized (Figure 7).[99] Trimethylphosphonoacetate (n) was condensed with β-ionone in the presence of sodium hydride in hexamethyl phosphoric triamide. The resulting methyl *trans*-[10-^{14}C]-β-ionylidene acetate (o) was converted to the *trans*-C_{18}-ketone (p) in 3 steps, which also included a base-catalyzed condensation with acetone. Further condensation of the latter product with another mole of trimethylphosphonoacetate yielded mainly the all-*trans* isomer of [^{14}C-10] methyl retinoate, which could then be reduced to [^{14}C-10]-all-*trans*-retinol. Over 20% yield of the all-*trans*-retinol was achieved based on bromoacetic acid, the precursor to the starting phosphonate ester. Liu's group utilized a similar approach recently to prepare the four hindered 7-*cis* isomers of retinol by condensation of the various 7-*cis*-C_{18}-ketones, analogous to the *trans*-C_{18}-ketone (p) above, with diethyl cyanomethylphosphonate followed by reduction of the cyano group with di-isobutylaluminum hydride, to result in the various retinal isomers.[88,89]

A recently reported synthesis of all-trans-retinoate makes use of the natural product, citral (g), as starting material (Figure 8).[100] The all-*trans*-C_{10} side chain is prepared by first converting citral to the ethyl carboxylate (r) in a 4-step sequence. Selenium dioxide oxidation of the latter produces the aldehyde (s) which, on bromination followed by dehydrobromination, yields the key C_{10}-aldehyde ester (t). Finally, condensation with cyclogeranylphosphonium ylide (u) gives the all-*trans*-retinoic acid ethyl ester (v).

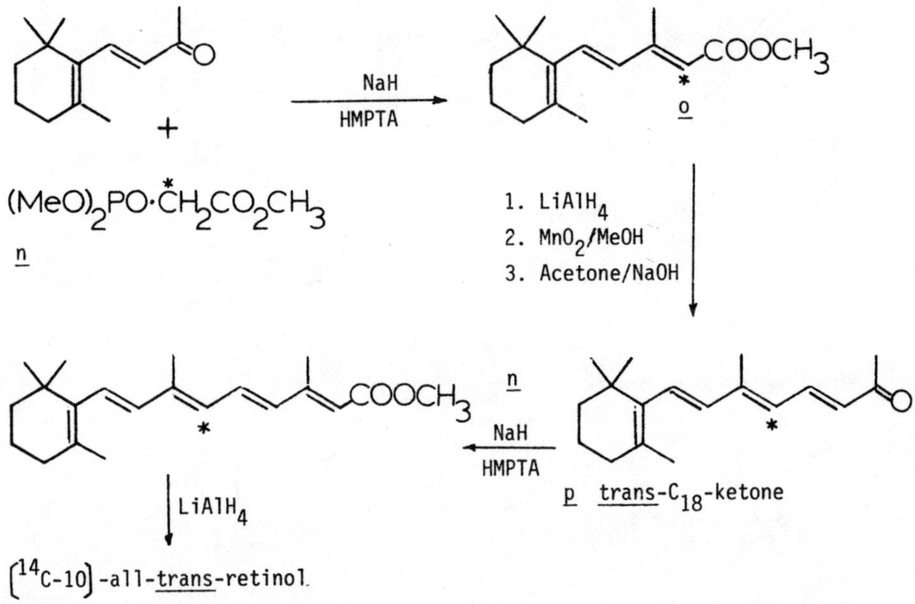

$(MeO)_2PO\cdot\overset{*}{C}H_2CO_2CH_3$ n

NaH / HMPTA → $COOCH_3$ o

1. $LiAlH_4$
2. MnO_2/MeOH
3. Acetone/NaOH

p trans-C_{18}-ketone

n, NaH / HMPTA → $COOCH_3$

$LiAlH_4$ → [^{14}C-10]-all-trans-retinol.

FIGURE 7. Synthesis of methyl [^{14}C-10]-all-*trans*-retinoate and of [^{14}C-10]-all-*trans*-retinol according to BuLock et al.[99]

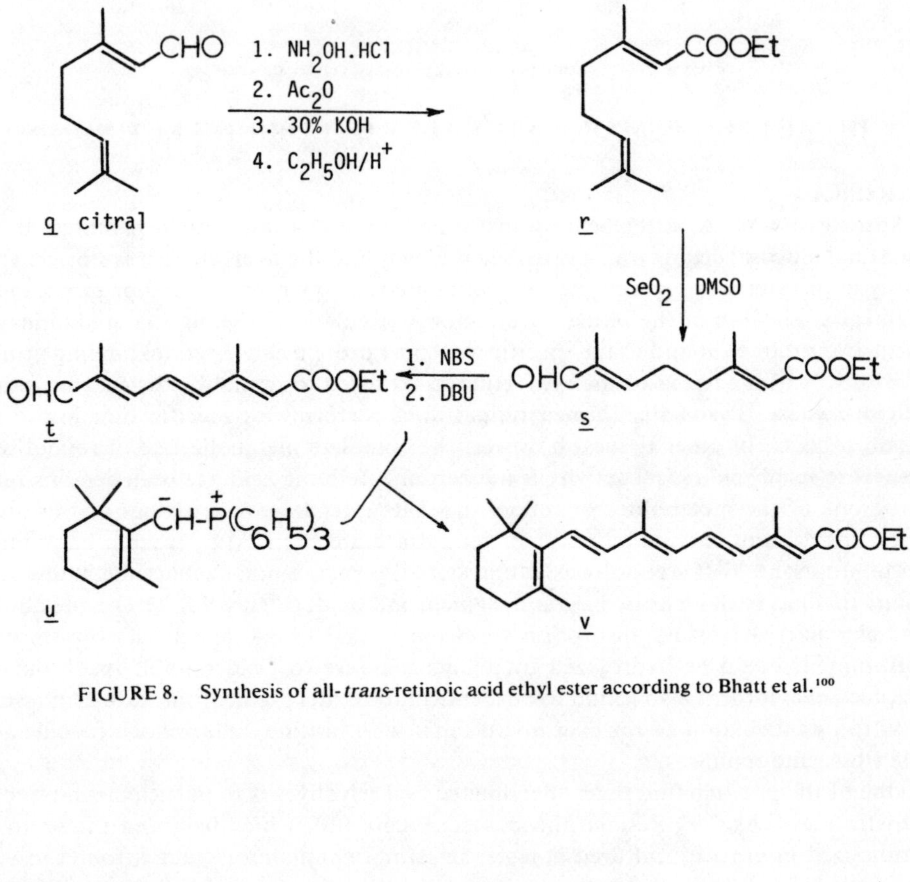

q citral (CHO)

1. $NH_2OH.HCl$
2. Ac_2O
3. 30% KOH
4. C_2H_5OH/H^+

r (COOEt)

SeO_2 DMSO

s OHC … COOEt

1. NBS
2. DBU

t OHC … COOEt

u $\bar{C}H$-$\overset{+}{P}(C_6H_5)_3$

v COOEt

FIGURE 8. Synthesis of all-*trans*-retinoic acid ethyl ester according to Bhatt et al.[100]

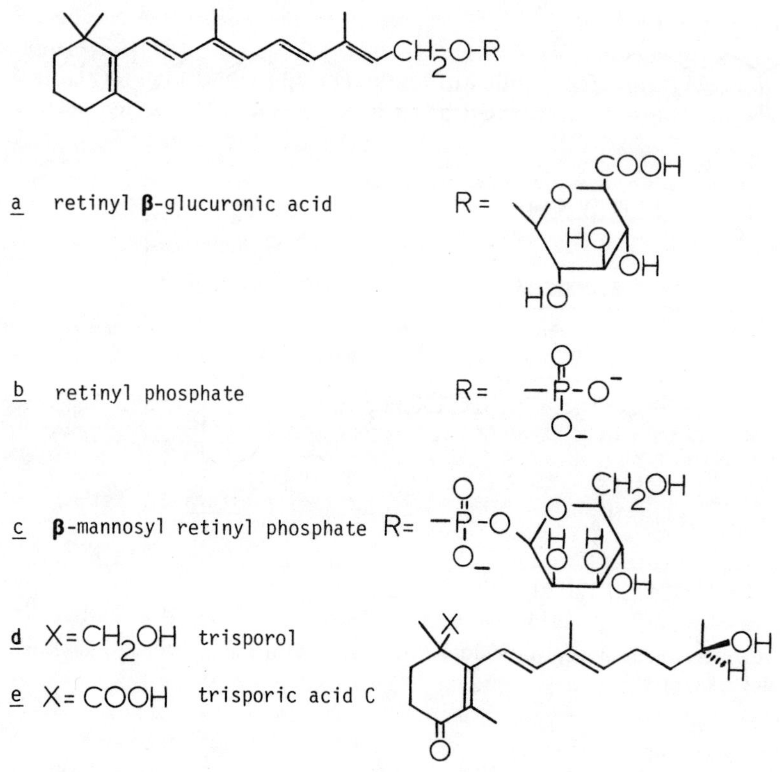

CH_2-O-R

a retinyl β-glucuronic acid R =

b retinyl phosphate R =

c β-mannosyl retinyl phosphate R =

d X = CH_2OH trisporol

e X = COOH trisporic acid C

FIGURE 9. Various biologically formed retinol derivatives.

III. METABOLISM OF NATURALLY OCCURRING RETINOIDS

A. Retinol

Dietary vitamin A, after being transformed primarily into retinyl palmitate in the intestinal mucosal cell, is transported via the lymph to the liver where it is stored extensively in its esterified form.[101] Retinyl palmitate is also stored to a minor extent in the pigment epithelium of the retina.[102] Vitamin A circulates in the plasma predominately as all-*trans*-retinol bound to the specific transport protein called retinol binding protein (RBP).[19,103] In the retina of the eye, retinol is oxidized to retinal by the enzyme retinol dehydrogenase,[104] and the formed retinal then performs its specific function in the vision process. In other tissues, however, the complete metabolic fate of retinol with respect to its physiological activity is not certain. Retinoic acid has been demonstrated to be one of the metabolites of retinol or of its ester in several tissues of rats given physiological amounts of retinol[105] or retinyl acetate.[106,107] Also, when a large intraportal dose of 6,7-^{14}C-retinol was administered to rats, a polar biliary metabolite was found that has been identified as retinyl glucuronic acid, (Figure 9a).[108] This metabolite was characterized by its absorption spectrum (λ_{max}330 nm), and by its fluorescence spectrum. It could be hydrolyzed by β-glucuronidase to yield retinol. Its chromatographic behavior was also found to be identical to retinyl β-glucuronic acid synthesized in vitro by incubation of rat liver microsomes with uridine diphosphoglucuronic acid (UDPGA) and retinol.

One of the possible functions of retinol or its derivatives is to participate in glycosyl transfer reactions.[20,109] Retinyl phosphate, (Figure 9b), which has been shown to be synthesized in vitro by cultured hamster intestinal epithelium[110] and is found in vivo in the rat liver,[111] can act as a carrier in the transfer of sugar moieties from nucleotide sugars to the growing oligosaccharide chains of glycoproteins. The biological synthesis of mannosyl- and galactosyl-retinylphosphate has been reported by several groups.[111-113] The mannosyl transferring ability of mannosyl-retinylphosphate, (Figure 9c) to protein has recently been demonstrated in an in vitro system[109] Retinyl acetate was also found to be metabolized efficiently by the mated cultures of the fungus *Blakeslea trispora*.[114] The metabolites isolated included trisporol (Figure 9d) and trisporic acid C (Figure 9e).

B. Retinoic Acid

Retinoic acid was chemically synthesized long before it was demonstrated to occur in nature. The observation that this synthetic compound was active in promoting the growth of retinoid-deficient rats[45,115] stimulated further interest in its biological activity and metabolism. Although fulfilling the growth-promoting function of retinol, retinoic acid was shown not to be able to restore vision when given to vitamin A-deficient rats,[116] and could not substitute for retinol in maintaining spermatogenesis or pregnancy.[117] It soon was observed that retinoic acid could not be reduced either to retinal or retinol in the body,[118] although both retinal[119,120] and retinol or its acetate ester,[105-107,121] have been found to be metabolized to retinoic acid. That retinoic acid may be a normal intermediate in the metabolism of retinol, has been indicated by the similar tissue metabolic profiles after i.v. administration of either retinoic acid, or retinol or its ester, to retinoid-deficient rats.[107,122]

Extensive metabolic studies have been carried out in the past 15 yr in search of a metabolite (or metabolites) of retinoic acid which would be more active than the parent compound. A number of early studies, using radioactively labeled substrates, have established metabolic profiles but have not elucidated the chemical structures of the products (as reviewed by Olson[25] and Rietz et al.[21]). These studies, however, have demonstrated, that retinoic acid is rapidly cleared from the body[116] and that in some cases, decarboxylation of the terminal carboxyl moiety occurs.[25] The structures of several urinary metabolites have been determined recently. These include four metabolites from rats and humans reported by Rietz et al,[21] where limited spectral evidence was presented to support the probable structures (Figure 10a-d). In addition, urinary metabolites (Figure 10e-g), isolated from rats given an injection of retinoic acid i.p., were reported by Hanni et al.[123] Studies on similarly treated rats also lead to the identification of three fecal metabolites (Figure 10c, h, i), in addition to unchanged retinoic acid.[124] However, the above studies were done on animals that had received large doses of retinoic acid and their physiological significance has not yet been demonstrated. Administration of retinoic acid to vitamin A-deficient rats recently has also allowed isolation of 5,6-epoxyretinoic acid (Figure 11a) from the intestinal mucosa of these animals.[125] The physiological importance of this compound, however, is still open to question.[126] Among biliary metabolites, unchanged retinoic acid, and retinoyl β-glucuronic acid (Figure 11b) could be identified after intraportal injection of retinoic acid.[127] Earlier investigators have also reported isolation of a significant amount of 13-*cis*-retinoic acid from the liver of rats fed large doses of the all-*trans* isomer,[128] although some, if not all, of this isomer was generated as an artifact of the isolation procedure.

Recently, the in vitro metabolism of retinoic acid has been studied using either a hamster tracheal organ-culture system,[129] or a hamster liver cell-free system.[130] From these studies, two metabolites of retinoic acid have been isolated, the 4-hydroxy (Figure 11c) and the 4-oxo (Figure 10c) analogs of the parent compound.[131] These compounds, however, appear to be less active than all-*trans*-retinoic acid[132] and it is thought that metabolism at the C-4 position of the ring may be the initial step in one of the elimination pathways of retinoic acid from the body.

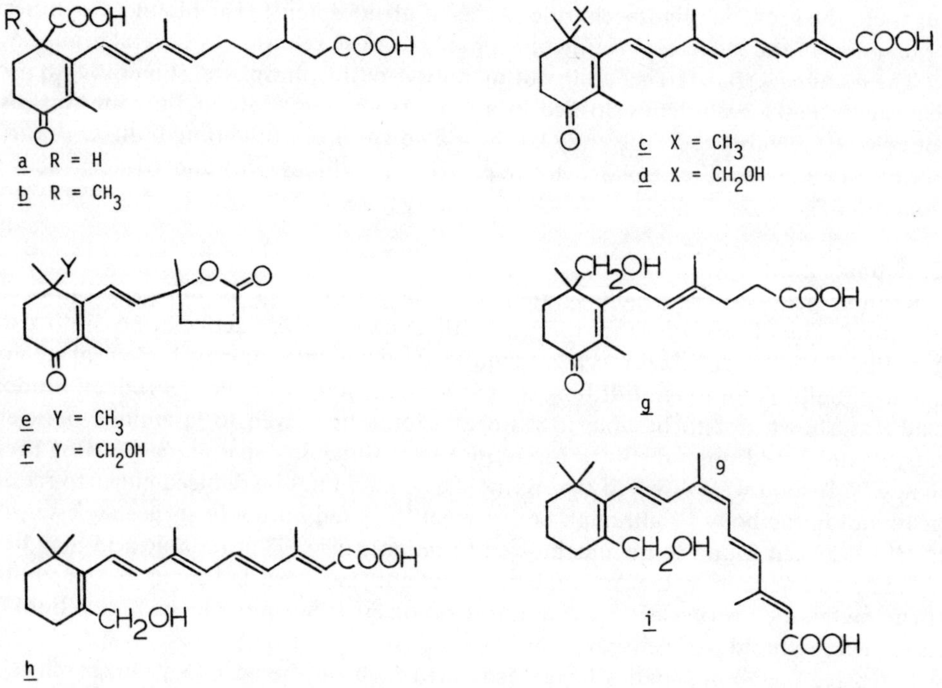

a R = H
b R = CH_3

c X = CH_3
d X = CH_2OH

e Y = CH_3
f Y = CH_2OH

g

h

i

FIGURE 10. Urinary (a to g) and fecal (c, h, i) metabolites of all-trans-retinoic acid.

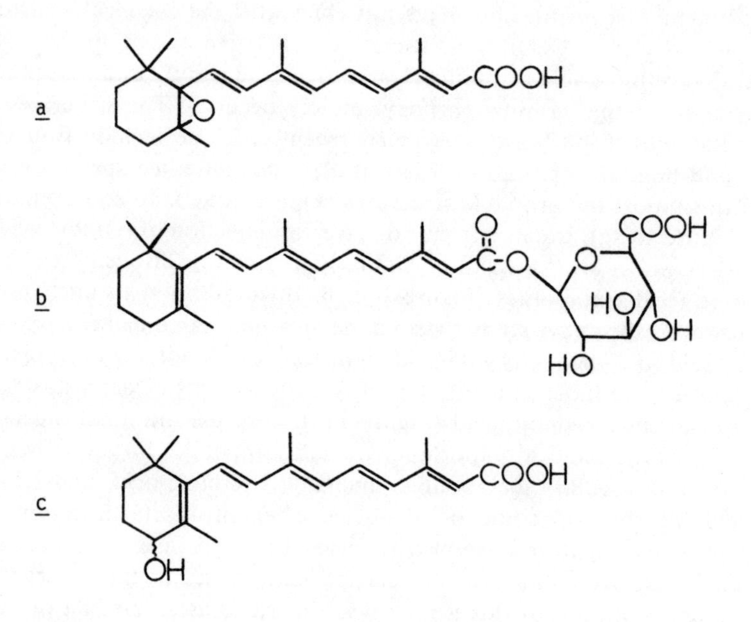

FIGURE 11. Metabolites of all-*trans*-retinoic acid: 5,6-epoxyretinoic acid (1), retinoyl β-glucuronic acid (b), and 4-hydroxy-retinoic acid (c).

IV. HISTORY OF RETINOIDS AS ANTIPROMOTING AGENTS

As early as 1926, a relationship between a retinoid-deficient diet and a greater susceptibility to the formation of certain carcinomas was suggested.[133,134] However, this idea rapidly came under fire when Sugiura[135] reported in 1930 an inability to detect a difference in the development of tumors in rats on a marginally retinoid-deficient diet when compared to rats on a normal diet. Although numerous studies have since demonstrated a relationship between retinoid deficiency and an increased susceptibility to the formation of tumors initiated either through natural causes[136] or through experimental exposure to the action of various carcinogens,[137-140] this controversy still prevails today.

If one accepts the postulate that retinoid deficiency, which causes abnormal differentiation of epithelial tissue, can lead to an increased risk of tumor formation, a disease of abnormal cell differentiation, the idea naturally follows that administration of retinoids — compounds that encourage normal cell differentiation — may help to suppress this tumor formation. In 1932, Kuh,[141] one of the first investigators to test this hypothesis, administered large doses of carotene to tumor-bearing mice and did, indeed, report an inhibition of tumor growth. Since then, many investigations have focused on the possible role of natural and synthetic retinoids in the prevention of epithelial cancers. As expected, a variety of conclusions have been reached. The remainder of this review will deal with the results of these studies and will address the question as to whether retinoids will be able to play an important role in the search for compounds that are capable of preventing carcinogenesis.

V. ACTIVITY OF RETINOIDS IN CANCER PREVENTION

A. Whole Animal Studies

As mentioned earlier, retinoids are involved in the maintenance of a normal epithelium in such tissues as the skin, mammary gland, lung, colon, and urinary bladder. Because malignancies in these tissues account for a high percentage of the total cancer deaths in man,[1] it was only natural to investigate the effects of retinoids on these tissues after exposure to various carcinogens. Although the ultimate goal of this research is the prevention of epithelial cancer in the human population, the usefulness and safety of retinoids must first be proven in several animal systems before investigations in man can take place. During the past 15 years, the effect of retinoids in preventing or reversing tumors of the skin, mammary gland, lung, colon, and urinary bladder, as well as the effect of retinoids on transplantable tumors (especially chondrosarcomas and L1210 leukemia, and oncogenic viruses), have been investigated in the whole animal. In this section the in vivo effects of the natural retinoids, as opposed to the synthetic retinoids, on these various tumor systems will be briefly discussed.

1. Skin Cancer

A possible effect of retinoids on skin papillomas was first suggested by McMichael in 1965.[142] He observed that when rabbits were placed on a diet supplemented with large amounts of retinyl palmitate, there was reduction in the size of papillomas that were induced by inoculation with Shope rabbit virus. However, if the retinoid treatment was discontinued, the papillomas regrew. This observation has been verified by Davies[143] and Shamberger[144] using 7,12-dimethylbenz [α] anthracene-induced papillomas on mouse skin. The technique of inducing skin tumors in mice using topical application of 7,12-dimethylbenz [α] anthracene and croton oil has been extensively utilized by Bollag and co-workers for determining the prophylactic and therapeutic value of retinoids.[15,145-148] In this system, biologically active retinoids given systemically after the administration of the carcinogen will delay the appearance and reduce the number and volume of the papillomas. By comparing the weekly dose needed to cause a 50% inhibition of papillomas, the relative biological activity of various retinoids can be determined (Table 1).

The action of retinoids on a particular type of skin tumor, the keratoacanthoma, has also been investigated.[149-151] The keratoacanthoma, induced in rabbit ears by topical administration of 7,12-dimethylbenz [α] anthracene, is a dry, keratotic tumor that will normally regress by itself. Topical application of retinoic acid during the maturation or regression stages of the keratoacanthoma induced a mucous secretion along with an accelerated rate of regression. Cessation of retinoic acid treatment brought about a reversion of the tumor to its usual keratotic appearance.

2. Mammary Gland Tumors

The action of retinoids on mammary gland carcinogenesis has been under investigation for almost as long as the studies concerning their action on skin tumors. Early work in this area concentrated on the idea that retinoids might enhance the therapeutic value of certain known chemotherapeutic compounds, such as the antineoplastic drug Cytoxan® (cyclophosphamide).[152,153] These studies met with only limited success but did suggest that retinoids may have some effect on mammary gland carcinoma. It was not until 1976 that Moon and co-workers first published a report demonstrating a direct effect of natural retinoids on the prevention of breast carcinoma.[154,155] They demonstrated both a decrease in the total number of mammary adenocarcinomas observed, as well as an increase in the latency period, i.e., the length of time between carcinogen treatment and the appearance of the first palpable tumor after administration of retinyl acetate in the diet of rats given either 7,12-dimethylbenz [α] anthracene[154] or N-methyl-N-nitrosourea.[155] Further studies on the action of retinoids on mammary carcinogenesis have utilized various synthetic retinoids and will be discussed later.

3. Lung Cancer

Lung cancer is another epithelial neoplasia currently being investigated in regard to a possible prophylactic effect of retinoids. Most experiments in this area have utilized either benzo[a]pyrene or 3-methylcholanthrene as carcinogen. Saffiotti et al.,[156] induced squamous metaplasia and squamous tumors in the columnar mucous epithelium of the respiratory tract of the Syrian gold hamster by intratracheal instillation of benzo[a]pyrene. Oral administration of retinyl palmitate after the last dose of carcinogen, was found to inhibit the appearance of such squamous cell tumor development. However, Smith and co-workers,[157,158] were unable to demonstrate any significant suppression by oral retinyl acetate of respiratory tract tumors also induced in hamsters by benzo[a]pyrene. This apparent discrepancy is perhaps partially explained by the fact that, in the latter case, a more effective dose of benzo[a]pyrene was administered (as shown by a nearly twofold higher incidence of respiratory tract tumors.) This more effective dose may have overcome any potential protective effect of the retinoid.

Use of 3-methylcholanthrene for tumor induction has also produced varied results. Employing a tissue transplant method together with 3-methylcholanthrene for induction of carcinomas in mice, Smith, et al.[159] could find no protective effect of intramuscular injections of retinyl palmitate in delaying the appearance of pulmonary epithelial squamous cell carcinoma. In contrast to this, Cone and Nettesheim[160,161] have reported that the development of squamous metaplasia and squamous cell carcinoma in the respiratory tract is significantly reduced in rats that have received oral retinyl acetate prior to 3-methylcholanthrene injections. In a further study, the retinyl acetate administration was begun at 3, 5, and 10 weeks after the last 3-methylcholanthrene dose.

Again, the incidence of metaplastic lung nodules was significantly lower in the retinoid-treated rats than in the control group on a low-retinoid diet, suggesting that the observed protective effect of retinyl acetate was occurring in the postinitiation phase of lung tumor formation in the rat.[162] However, further experimentation, again using as an endpoint marker the development of metaplastic lung nodules in rats induced by 3-methylcholanthrene, revealed that it was the mildly retinoid-deficient state of the control group that had led to a condition of increased susceptibility of the lung to 3-methylcholanthrene. This increased susceptibility was readily abolished by dietary supplementation with moderate amounts of retinyl acetate. Administration of additional retinyl acetate provided no further protection against carcinogenic insult.[139,140] It was concluded, therefore, that at least in this species and tumor system, there was no inhibitory effect of excessive amounts of retinyl acetate on tumor development. Thus, at this point in time it still remains to be positively demonstrated that an excess of the natural retinoids will provide any protection against carcinogenic insult in the lung, provided the dietary retinoid level is sufficient to maintain normal lung differentiation.

4. Colon Cancer

Colon cancer has only recently been investigated with regard to the possible inhibitory effect of retinoids on tumor development. Initial studies seemed to indicate that rats on a deficient or a marginal vitamin A diet, when given aflatoxin B_1, were more susceptible to colon carcinoma than rats on an adequate vitamin A diet.[163] In contrast to this, Narisawa, et al.[164] reported that the incidence of large bowel tumor induction by N-methyl-N^1-nitro-N-nitrosoguanidine in rats on a retinoid-deficient diet was reduced when compared to rats on a retinoid-adequate diet. These latter results would suggest that the retinoid-deficient large intestine has a different susceptibility to carcinogenic insult than the respiratory[139,140] and urinary tracts.[137,165] In addition, administration either of retinyl palmitate to rats in which intestinal tumors had been induced by intragastric administration of 1,2-dimethylhydrazine,[166] or of 13-*cis*-retinoic acid to rats given intrarectal administration of N-methyl-N-nitrosourea[167] did not significantly modify the carcinogenic response in the colon. It therefore appears that the natural retinoids, under the conditions tested to date, do not have any inhibitory action on the appearance of colon carcinogenesis induced by various chemical carcinogens. Further studies concerning the possible beneficial effects of synthetic retinoids on colon cancer will be discussed later.

5. Urinary Bladder Cancer

The effects of retinoids on urinary bladder carcinoma appear to be more promising than those observed for colon carcinoma. It was initially reported in 1960 that the bladder of a retinoid-deficient rat was more susceptible to the action of carcinogens than the bladder of a normal rat.[137] This conclusion was again reached in 1976 when it was demonstrated that urinary bladder tumorigenesis was accelerated in retinoid-deficient rats given a urothelial dose of N-[4-(5-nitro-2-furyl)-2-thiazolyl] formamide.[165] This experiment, however, failed to show any positive effect of excess dietary retinyl palmitate either on the prevention of bladder epithelial hyperplasia or on tumor incidence. Such a beneficial effect, nevertheless, was first reported using the synthetic retinoid, 13-*cis*-retinoic acid,[168-171] and will be discussed later. These early experiments have recently been confirmed by Miyata and co-workers[172] and by Dawson, et al.[173] using natural retinoids. Miyata's group has demonstrated that high doses of dietary retinyl acetate will inhibit keratinization and squamous metaplasia in bladder lesions from rats induced with N-butyl-N-(4-hydroxybutyl)nitosamine, while the second group has shown a similar prophylactic effect of dietary retinyl acetate against transitional cell and squamous cell bladder tumors in mice given N-[4-(5-nitro-2-furyl)-2-thiazolyl] formamide.

6. Transplantable Tumors

Several investigations have been made into the possible use of retinoids to interfere with the proliferation of fully transformed malignant cells. In an initial study by Bollag in 1971, no inhibitory effect of all-*trans*-retinoic acid was observed on the growth of six transplantable tumors that included Ehrlich solid carcinoma, Ehrlich ascites carcinoma, Crocker sarcoma 180, and leukemia L-1210 in mice and Walker 256 carcinosarcoma and uterus epithelioma T-8 (Guerin) in rats.[174,175] However, Felix et al.[176] were able to show an inhibitory effect of retinyl palmitate on the growth and development of S-91 (Cloudman) murine melanoma. This inhibitory effect was attributed to the highly immunogenic nature of the S-91 melanoma and a possible enhancement of immunity against the tumor by the retinoid. A direct effect of natural retinoids on a transplantable adenocarcinoma cell line has also been demonstrated.[177] Mice inoculated with the adenocarcinoma cells showed increased survival and a decreased rate of breast tumor growth when fed a diet supplemented with retinyl palmitate as compared to control mice on a normal diet. The inhibition of growth by various synthetic retinoids of the transplantable rat chondrosarcoma, a nonepithelial tumor, has likewise been described.[178-180] These reports therefore demonstrate that retinoids in certain situations are able to modify the growth of malignant, transformed cells.

Although retinoids by themselves have displayed no protective effect against L-1210 leukemia in vivo,[174,175,181] they do appear to enhance the activity against the tumor of certain chemotherapeutic agents such as 1,3-bis(2-chloro-ethyl)-1-nitrosourea,[181,182] cyclophosphamide[181,183] and X-ray treatment.[184] This effect is believed to be due to an action of retinoids on the tumor cell membrane, thereby making the cell more susceptible to the chemotherapeutic agent.[185] Retinyl palmitate treatment has also been reported to enhance the tumor suppression mediated by *Mycobacterium bovis* (strain BCG) against a transplantable, syngeneic, 3-methylcholanthrene-induced sarcoma in mice. Again, the retinoid by itself had no effect.[186]

7. Oncogenic Viruses

Finally, mention should be made of several reports that have demonstrated an action of retinoids on tumors induced by oncogenic viruses. Mice were placed on a diet supplemented with retinyl palmitate or on a normal control diet, and were inoculated with the Moloney strain of murine sarcoma virus. The experimental animals showed an increased latency period, a reduction in both the number of tumors and the growth rate of the tumors, and a faster resorption rate of the tumor compared to the control mice.[187-189] As will be discussed later, these effects have again been suggested to be operating through an interaction with the immune system. There has been little follow-up work on these initial observations.

B. In Vitro Studies

Several in vitro systems have recently been described for determining both the biological activity of retinoids and their possible use as chemopreventive agents. These systems have an advantage over the in vivo experiments in that they require smaller quantities of the compound for testing and data can be obtained in a relatively short period of time. These techniques may also be useful for the elucidation of the mode of action of retinoids in the suppression of malignant transformation. The systems thus far described include a hamster tracheal organ culture,[190-194] a mouse prostate organ culture,[195-201] an L-1210 leukemia cell culture,[184,185] an ornithine decarboxylase assay from mouse epidermis[202-204] or bovine lymphocytes,[205-207] a restoration of density and anchorage-dependent growth to a transformed fibroblast or melanoma cell,[208-216] a chick embryo metatarsal skin explant[217,218] and, most recently, a sarcoma growth factor assay[219] and a mouse mammary gland cell culture.[220]

1. Action of Retinoids on Organ Cultures

The mouse prostate organ culture was the first in vitro system used to demonstrate the ability of retinoids to reverse the hyperplastic and anaplastic epithelial lesions normally found after exposure of the tissue to chemical carcinogens. In 1955, Lasnitzki demonstrated this antineoplastic effect of retinoids in the cultured mouse prostate gland when she reported that the presence of excess vitamin A in the culture medium prevented keratinization and significantly decreased the incidence and degree of epithelial hyperplasia in glands that had been treated with 3-methylcholanthrene.[195] This initial observation has been verified, both by Lasnitzki using natural and synthetic retinoids,[197,198] and by Chopra and Wilkoff.[199] The latter investigators have exposed the mouse prostate tissue to N-methyl-N'-nitro-N-nitrosoguanidine, as well as to 3-methylcholanthrene, and have observed that retinoic acid will inhibit and reverse the hyperplastic changes induced by either of the carcinogens in vitro.[199] This technique has also been used as an assay system to evaluate the relative antihyperplastic activity of various retinoids[200] (Table 2). Finally, retinoic acid is likewise able to inhibit and reverse the hyperplasia induced by exposure of the mouse prostate gland in vitro to testosterone,[201] a compound that has been linked with human prostatic cancer.[221-223]

A second system that has been utilized in the study of two different problems in the area of the prevention of epithelial cancer by retinoids, is the hamster tracheal organ culture system. The first area of investigation concerns the effect of retinoids on the neoplastic changes caused by exposure of the tissue to a carcinogenic insult. This technique was initially employed by Crocker and Sanders to demonstrate that 13-*cis*-retinol could prevent the formation of a squamous metaplastic epithelium in trachea that had been exposed in vitro to benzo[a]pyrene.[190] Similar results have been obtained by Lane and Miller using retinyl acetate and retinoic acid.[191] Mossman and Craighead have also reported recently that the metaplastic changes caused by exposure of the tracheas to asbestos fibers could be prevented by addition of retinyl methyl ether to the culture medium.[224]

The second area of research that utilizes the hamster trachea in organ culture is concerned with the determination of the biological activity of both the natural and synthetic retinoids. It has been shown that a retinoid-deficient trachea in organ culture will undergo squamous metaplasia with formation of keratohyaline granules and keratin. Addition of a biologically active retinoid to the culture medium will cause a reversal of this keratinization. The concentration of the retinoid being tested that will cause this reversal in 50% of the tracheal organ cultures gives an indication of its relative biological activity.[192-193] Using this assay system, it has been possible to evaluate the biological activity of numerous natural and synthetic retinoids (Table 3)[192,194,225]

Another in vitro technique used for determining the biological activity of retinoids employs chick embryo metatarsal skin explant cultures. This system, similar to the tracheal organ culture assay, tests the ability of different compounds to influence epithelial differentiation as measured by the degree of keratinization and mucous metaplasia.[217,218] The effectiveness of a series of retinoids in modulating epithelial differentiation is shown in Table 4.

2. Action of Retinoids on Ornithine Decarboxylase

Two of the in vitro techniques just described, the hamster tracheal and mouse prostate organ cultures, have demonstrated that retinoids are able to interact with epithelial tissues previously exposed to carcinogens to allow a return of the epithelium to normalcy. A possible mechanism for this interaction of retinoids with the initiated cell has recently been suggested by Verma et al.[202,203] using 12-O-tetradecanoyl-phorbol-13-acetate-induced mouse epidermal ornithine decarboxylase as a marker enzyme for the promoting or antipromoting ability of a test compound. Epidermal ornithine decarboxylase activity, determined in vitro, is dramatically induced after topical application to mouse skin of the phorbol ester 12-O-tetradecanoyl-phorbol-13-acetate, a skin tumor-promoting agent.[202] This enzyme induction has been proposed, although as yet unproven, to be essential for skin tumor promotion.[226,227] Retinoic acid is a specific and potent inhibitor of this induction in a dose- and time-dependent manner.[202] This inhibition of phorbol ester-induced ornithine decarboxylase activity could be correlated with inhibition of mouse skin papilloma formation suggesting a relationship between the two events.[204] Both natural and synthetic retinoids have been tested in this in vivo/in vitro system.[205] To date, all-*trans*-retinoic acid is the most potent retinoid for inhibition of the induced decarboxylase activity. Similar data have been obtained using phytohemagglutinin-treated bovine lymphocytes. In this second system, as before both the 12-0-tetradecanoyl-phorbol-13-acetate-induced ornithine decarboxylase activity[205] and the DNA synthetic activity[206] of the cells are maximally inhibited in the presence of all-*trans*-retinoic acid and less so by other retinoids.[206,207] It has therefore been suggested that retinoids may inhibit the formation of certain epithelial carcinomas through the inhibition of ornithine decarboxylase, an enzyme possibly involved in the early stages of tumor promotion.

3. Action of Retinoids on Transformed Cell Lines

The ability of retinoids to restore a normal morphology to a transformed cell, as occurs in the hamster tracheal and the mouse prostate epithelia in vitro after exposure to a carcinogen, has also been demonstrated in several fibroblast and melanoma cell lines. For example, retinoic acid is able to restore density-dependent growth to a transformed mouse L-929 cell line.[208] Growth inhibition by retinoic acid also occurs in several other transformed and tumor cell lines from mice, rats, and humans.[210,211] However, the ability of retinoic acid to inhibit the growth of human melanoma and breast carcinoma cells has been shown to vary from cell line to cell line.[213] This suggests that cells, even though they are from a similar histopathological type, may respond differently to retinoid treatment. Therefore, a variation in the response would be anticipated in the treatment of such tumors with retinoids. Other systems that display an induction of density-dependent growth by retinoids include fibroblast cultures of hamster Nil and Nilpy cells,[214] mouse 3T3 cells,[214] and 3T12-3 cells.[215]

The effect of retinoids on another marker of cell normalcy, anchorage-dependent growth, has also been investigated. Loss of anchorage-dependent growth is believed to be associated with neoplastic transformation.[228] Retinoic acid is capable of restoring this normal cell function both in a murine melanoma cell line[209] and in transformed mouse fibroblasts grown in monolayer.[215,216]

Finally, two additional in vitro systems have very recently been reported that demonstrate an effect of retinoids on cell transformation. The first technique utilizes rat fibroblast cells that have been treated with sarcoma growth factor. The sarcoma growth factor stimulates cell division and promotes the growth of the cells in soft agar, a function not characteristic of the normal fibroblast cell.[229] Retinoids are able to block this transforming effect of sarcoma growth factor.[219] In the second technique, mouse mammary glands when exposed in vitro to 7,12-dimethylbenz[a]anthracene, display marked epithelial transformation. This transformation is reduced by over 90% in the presence of retinoids administered either throughout the time of carcinogen treatment or immediately following removal of carcinogen,[220] once again demonstrating a protective effect of retinoids on epithelial transformation and carcinogenesis.

4. Conclusion

Studies utilizing in vitro systems have demonstrated that retinoids are required for maintenance of normal epithelial cell differentiation, that they can prevent the transformation of the epithelial cell after exposure to a carcinogen, and that they can cause certain transformed cell lines to revert to normal cell characteristics. However, further investigation into the practical use of retinoids for treatment of invasive malignancies is still needed. In addition to using in vitro systems to study the modification of cellular function by retinoids, this technique has also been employed for determining the biological and therapeutic activity of retinoids and should be of great help in the future for studying the mechanism whereby retinoids are able to control or modify both normal and abnormal cellular differentiation.

C. Human Studies

While the data gathered from the whole animal and the in vitro studies are inherently of interest, the practicality of the use of retinoids in the treatment of cancer resides in its applicability to the human population. Vitamin A and its derivatives have been used with varying degrees of success on numerous skin disorders in humans including acne,[230-232] psoriasis,[233] keratosis follicularis (Drier's disease),[234-238] lamellar ichthyosis,[235,239,240] congenital ichthyosiform erythroderma,[235,241] and erythrokeratoderma variabilis.[242] The scope of this review, however, will deal mainly with the effect of retinoids on various human neoplasias.

1. Vitamin A Deficiency and Cancer

As mentioned earlier, several experiments with whole animals have suggested that an increased risk of tumor formation occurs in those animals on a retinoid-deficient diet. A number of investigators have likewise reported that plasma vitamin A levels are lower in patients with gastrointestinal cancer,[243] keratosis follicularis,[234] alimentary tract tumors,[244] bronchial and lung carcinoma (especially squamous and oat cell carcinoma),[245,246] and cancer of the oral cavity and oropharnyx[247] than the levels observed in control patients. Conversely, humans on a diet high in vitamin A displayed a lower incidence of lung cancer.[248] The importance of this observation, that there may be an increased risk for various carcinomas in humans with low vitamin A levels, is amplified by data indicating that a relatively high percentage of the normal human population may have abnormally low liver levels of vitamin A[249-251] (the liver being the major site of vitamin A storage in the body), and therefore may have an increased susceptibility to certain epithelial neoplasias. It should, however, be kept in mind that the decreased vitamin A levels in patients with epithelial cancer may be a secondary effect of neoplasia and not a primary cause.

2. Retinoids and Cancer Treatment

The investigations into the effects of retinoids on tumors in man center mainly on various skin abnormalities because of both the accessibility of the site and the ease and greater safety of topical treatment. Bollag and Ott have treated premalignant (senile or actinic keratoses) and malignant (basal cell carcinoma) lesions of the skin with topical all-*trans*-retinoic acid and have observed a partial or complete regression of the disorder in a majority (59 out of 67) of the cases.[148,252-254] Although this demonstrates a favorable influence of a noncytotoxic agent on premalignant and malignant epithelial tumors, the regression rate obtained was inferior to that observed for conventional methods of treatment. Therefore, on the basis of the data, retinoic acid was not recommended as routine treatment. Oral all-*trans*-retinoic acid, by itself, has also been found to have a beneficial effect on epidermal skin tumors, although the greatest improvement was obtained if given in combination with cytostatic agents or X-ray treatment.[255] Oral 13-*cis*-retinoic acid has been noted to cause improvement of basal cell carcinoma[239] and also to slow tumor progression and to stop metastatic spread in

Table 1
RELATIVE THERAPEUTIC ACTIVITY OF RETINOIC ACID AND ITS ANALOGS IN REGRESSION OF PAPILLOMAS IN MICE[a]

Trivial name	Structure	ED_{50}[b]
All-trans-Retinoic acid	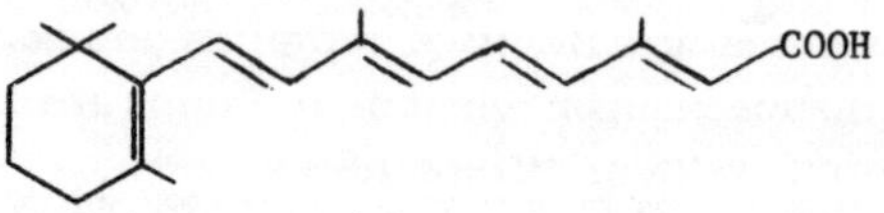 COOH	400
alpha-Retinoic acid	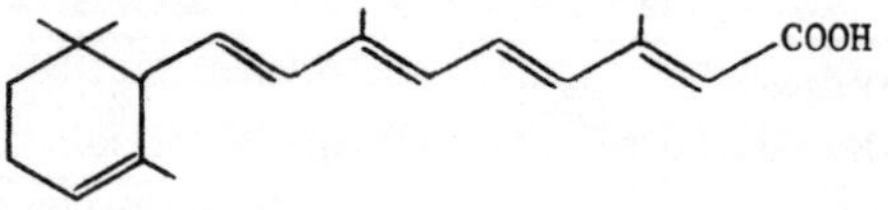 COOH	200
TMMP[c] analog of retinoic acid	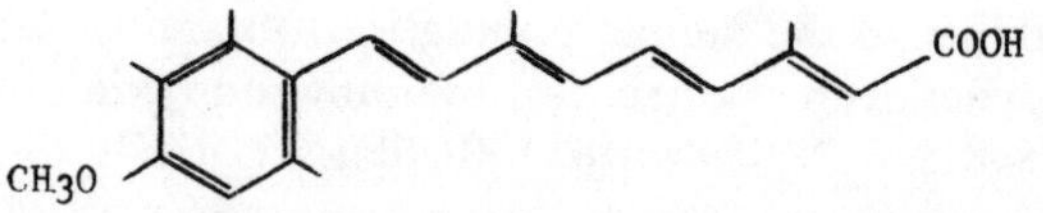 CH_3O; COOH	50
TMMP[c] analog of retinoic acid ethyl ester	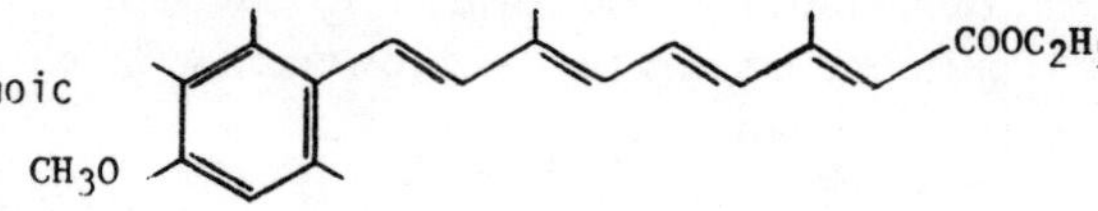 CH_3O; $COOC_2H_5$	25

[a] Adapted from Reference 15.
[b] i.p. dose (μg) given once a week for 2 weeks that causes a 50% regression of papillomas.
[c] Trimethylmethoxyphenyl.

Table 2
ACTIVITY OF RETINOIDS IN THE REVERSAL OF N-MEHTYL-N'-NITRO-N-NITROSOGUANIDINE-INDUCED HYPERPLASIA IN MOUSE PROSTATE ORGAN CULTURES[a]

Trivial name	Structure, $R_1 = R_2 =$ [b]	% Inhibition[c]
Retinyl methyl ether	I, $-H_2$, $-OCH_3$	76
Cyclopentenyl analog of retinoic acid	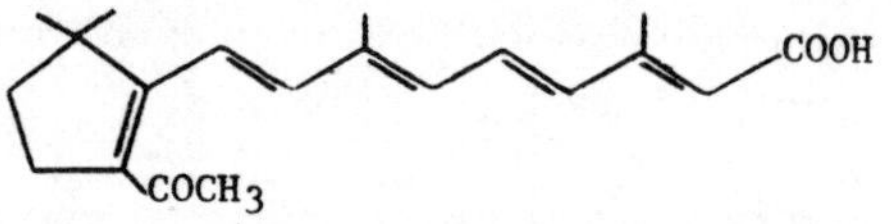 COOH; $COCH_3$	71
all-*trans*-Retinoic acid	I, =O, -OH	37
TMMP[d] analog of retinoic acid	II, =O, -OH	36
N-Acetyl retinyl amine	I, $-H_2$, $-NHCOCH_3$	29

[a] Adapted from Reference 200.
[b] I, CR_1R_2
II, CH_3O, CR_1R_2
[c] Percent mitotic inhibition of carcinogen-treated mouse prostate gland caused by 96 hr of exposure to 6.1 to 6.6×10^{-7} *M* retinoid.
[d] Trimethylmethoxyphenyl.

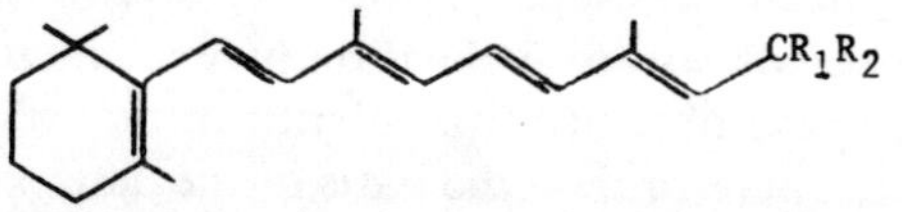

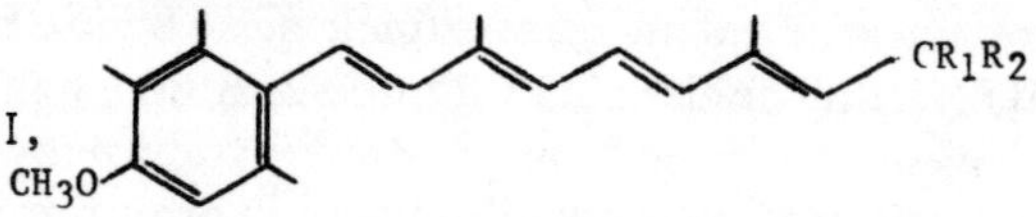

Table 3
ACTIVITY OF RETINOIDS IN INHIBITION OF KERATINIZATION IN HAMSTER TRACHEAL ORGAN CULTURES[a]

Trivial name	Structure, R_1 = , R_2 = [b]	ED_{50}(M)[c]
Natural retinoids		
all-*trans*-Retinoic acid	I, =O, –OH	3×10^{-10}
Retinal	I, –H, =O	3×10^{-10}
Retinol	I, $-H_2$, –OH	2×10^{-9}
Retinyl acetate[d]	I, $-H_2$, $-OCOCH_3$	3×10^{-9}
Synthetic retinoids		
Retinoic acid ethyl ester	I, =O, $-OC_2H_5$	5×10^{-10}
Retinoic acid ethyl amide	I, =O, $-NHC_2H_5$	2×10^{-9}
Retinyl methyl ether	I, $-H_2$, $-OCH_3$	3×10^{-9}
TMMP[e] analog of retinoic acid	II, =O, –OH	6×10^{-9}
N-Acetyl retinyl amine	I, $-H_2$, $-NHCOCH_3$	9×10^{-9}
TMMP[e] analog of retinoic acid ethyl ester	II, =O, $-OC_2H_5$	2×10^{-8}
TMMP[e] analog of retinoic acid ethyl amide	II, =O, $-NHC_2H_5$	3×10^{-8}
Retinyl butyl ether	I, $-H_2$, $-OC_4H_9$	3×10^{-8}
N-retinyl phthalimide	I, $-H_2$, $-N(CO)_2C_6H_4$	5×10^{-7}

[a] Modified from Reference 225.

[b]

I. CR_1R_2

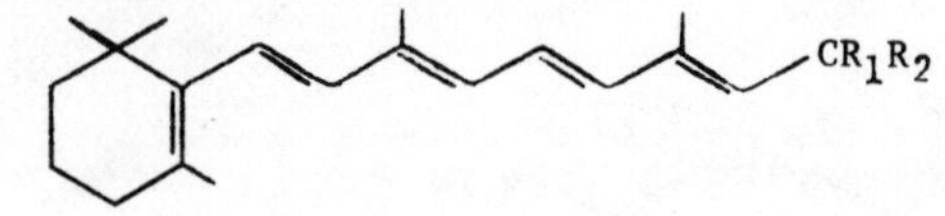

II. CH_3O CR_1R_2

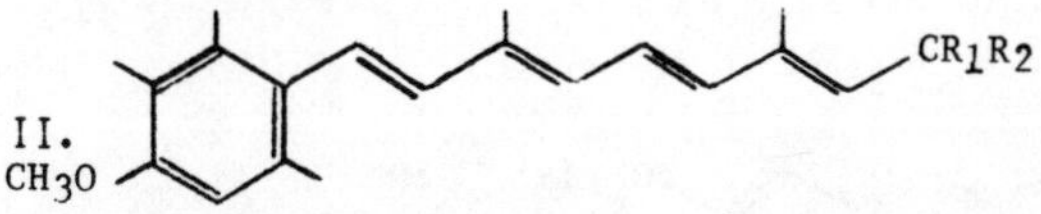

[c] ED_{50}(M) is the dose that causes a reversal of keratinization in the epithelium of 50% of the retinoid-deficient hamster tracheas using standard in vitro assay conditions.[192-193]

[d] Not a true natural retinoid but, because it is readily converted to retinol,[121] it is being classified as such for the purpose of this review

[e] Trimethylmethoxyphenyl.

Table 4
ACTIVITY OF RETINOIDS IN THE INHIBITION OF KERATINIZATION IN CHICK EMBRYO METATARSAL SKIN EXPLANT CULTURES[a]

Trivial Name	Structure, R_1 = , R_2 = [b]	M[c]
Natural retinoids		
all-*trans*-Retinoic acid	I, =O, –OH	4.3×10^{-6}
Retinyl acetate[d]	I, $-H_2$, $-OCOCH_3$	1.6×10^{-5}
Retinol	I, $-H_2$, –OH	1.8×10^{-5}
Synthetic retinoids		
Cyclopentenyl analog of retinoic acid	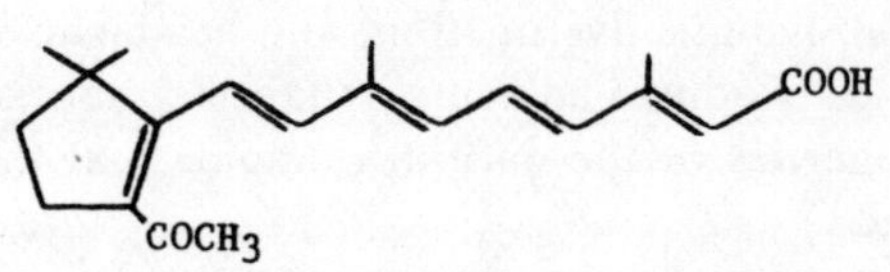 COOH; $COCH_3$	5.5×10^{-7}

Table 4 (continued)
ACTIVITY OF RETINOIDS IN THE INHIBITION OF KERATINIZATION IN CHICK EMBRYO METATARSAL SKIN EXPLANT CULTURES[a]

Trivial Name	Structure, $R_1 =, R_2 =$ [b]	M[c]
Retinoic acid ethyl ester	I, =O, $-OC_2H_5$	1.1×10^{-6}
Retinoic acid ethyl amide	I, =O, $-NHC_2H_5$	2.2×10^{-6}
alpha-Retinoic acid	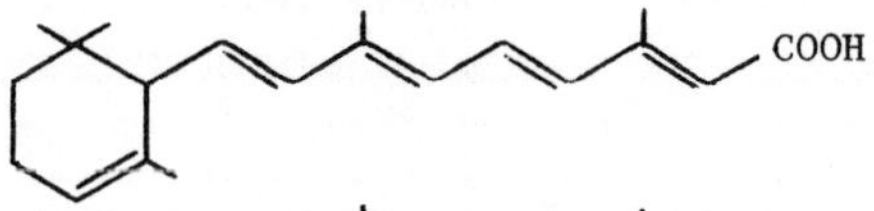 COOH	
13-*cis*-Retinoic acid	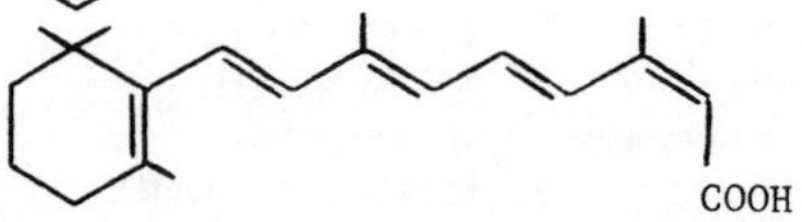 COOH	
		2.2×10^{-6}
		2.2×10^{-6}
TMMP[e] analog of retinoic acid	II, =O, $-OH$	3×10^{-6}
TMMP analog of retinoic acid ethyl ester	II, =O, $-OC_2H$	1.4×10^{-5}
TMMP analog of retinoic acid ethyl amide	II, =O, $-NHC_2H_5$	1.4×10^{-5}

[a] Modified from Reference 217

[b] I, CR_1R_2

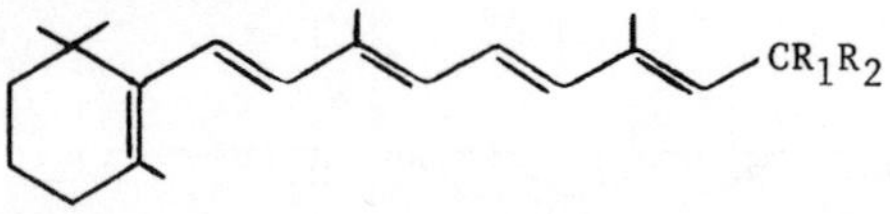

II, CH_3O, CR_1R_2

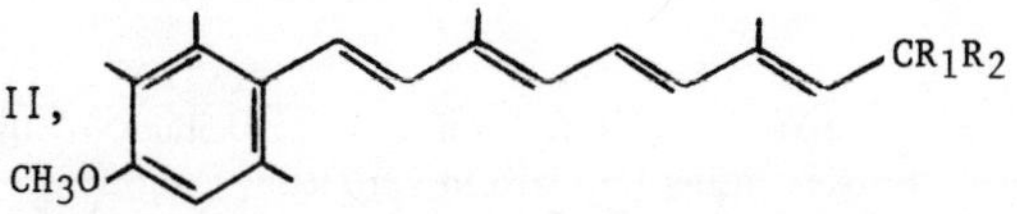

[c] Concentration of retinoid required to inhibit approximately 80% of 13-day-old chick embryo metatarsal skin explants from keratinizing in organ culture.

[d] Not a true natural retinoid but because it is readily converted to retinol[121] it is being classified as such for the purpose of this review.

[e] Trimethylmethoxyphenyl.

nine males who had unresectable lung cancer with metastases.[256] Finally, oral retinyl palmitate together with 5-fluorouracil and cobalt-60 radiation has produced very good 3 to 4 years survival rates in patients with head and neck tumors.[257]

3. Conclusion

Although the reports just described do indicate a beneficial effect of retinoids on various human malignancies, a carefully controlled study is needed. However, the inherent toxicity of the natural retinoids is also a major problem in their use for cancer prophylaxis in humans. The main symptoms of retinoid toxicity include alterations of the skin (e.g., hair loss, erythema, desquamation) and mucous membranes (e.g., conjunctivitis, cheilitis), hepatic dysfunction, and headache.[258,259] In addition, topical application of retinoids produces an acute irritant dermatitis. It is in the attempt to modify these toxic properties of the natural retinoids that the area of synthetic retinoids has been investigated.

D. Natural vs. Synthetic Retinoids

As just discussed, the toxicity of the natural retinoids is a serious problem that must be overcome prior to the practical application to humans of any beneficial tumor-preventing properties these compounds may possess. For this reason, numerous chemically modified analogs of vitamin A have been synthesized and tested in various systems to determine their biological activity in terms of their growth, differentiation, and antipromoting characteristics. One possible method for obtaining a compound with high biological activity and low toxicity rests on the idea that the various biological properties of retinoids (growth, reproduction, visual cycle, and toxicity) can be separated from one another. That this separation of the various biological properties can occur was first suggested in the early 1960s when it was discovered that rats maintaining on dietary retinoic acid grew normally but were incapable of maintaining the visual cycle[115] and either spermatogenesis in male rats[260] or reproduction in female rats.[261] Similarly, the separation of the anticarcinogenic properties of retinoids from their toxic[262-264] and their growth-promoting activities[198] has been suggested. Finally, recent synthesis of various 15-methylated retinoids has produced a group of compounds that will support reproduction and spermatogenesis but will not prevent degeneration of the retina.[265] It thus appears that by proper modification of the retinoid molecule, the various aspects of its biological activity can be emphasized or deemphasized.

A second method for developing a clinically useful drug is based on the hope that by appropriate changes in the basic retinoid structure, compounds can be synthesized that will concentrate in specific target tissues. Administration of the natural compounds — retinal, retinol, and its esters — causes a marked accumulation of retinoids in the liver and does not result in a dose-dependent increase in plasma vitamin A levels. A step away from this accmulation of retinoids in the liver is evident in studies involving the compound axerophthene. After administration of this retinoid to a rat, there was less deposition of retinyl palmitate in the liver and a greater concentration of total retinoid in the mammary gland when compared to a rat receiving an equivalent dose of retinyl acetate.[266] As will be discussed later, similar results have been obtained both for N-(4-hydroxyphenyl)retinamide[270] and for retinyl methyl ether.[225]

1. In Vivo and In Vitro Studies

Of the numerous analogs of vitamin A that have been synthesized to date, only a few have reached the stage of testing for their in vivo ability to inhibit tumor promotion by carcinogens. These compounds include 13-*cis*-retinoic acid, the aromatic series of retinoids and the amide and ether analogs of retinoic acid and retinol, respectively (Figure 12). The synthetic analog that has been most extensively investigated is the 13-*cis* isomer of retinoic acid (Figure 12a). Although this compound has a lower toxicity than all-*trans*-retinoic acid in both mice[267] and rats,[139,268] it provides no additional protective qualities in the rat compared to the all-*trans* isomer for 3-methylcholanthrene-induced lung cancer[139] or for N-methyl-N-nitrosourea-induced colon cancer.[167] However, there is one report in the literature that demonstrates that 13-*cis*-retinoic acid treatment is able to reduce both the incidence of colon carcinoma in rats and the number of tumors per animal after dimethylhydrazine treatment.[269] Since this is the first reported beneficial effect of excess retinoids on colon carcinogenesis, additional evidence will have to be obtained before any definite conclusions can be reached. The possible protective effects of the 13-*cis* isomer on urinary bladder carcinoma are more extensively demonstrated. Dietary 13-*cis*-retinoic acid reduces both the incidence and severity of transitional and squamous cell carcinomas of the urinary bladder after treating either rats with N-methyl-N-nitrosourea[168,169] or rats and mice with N-butyl-

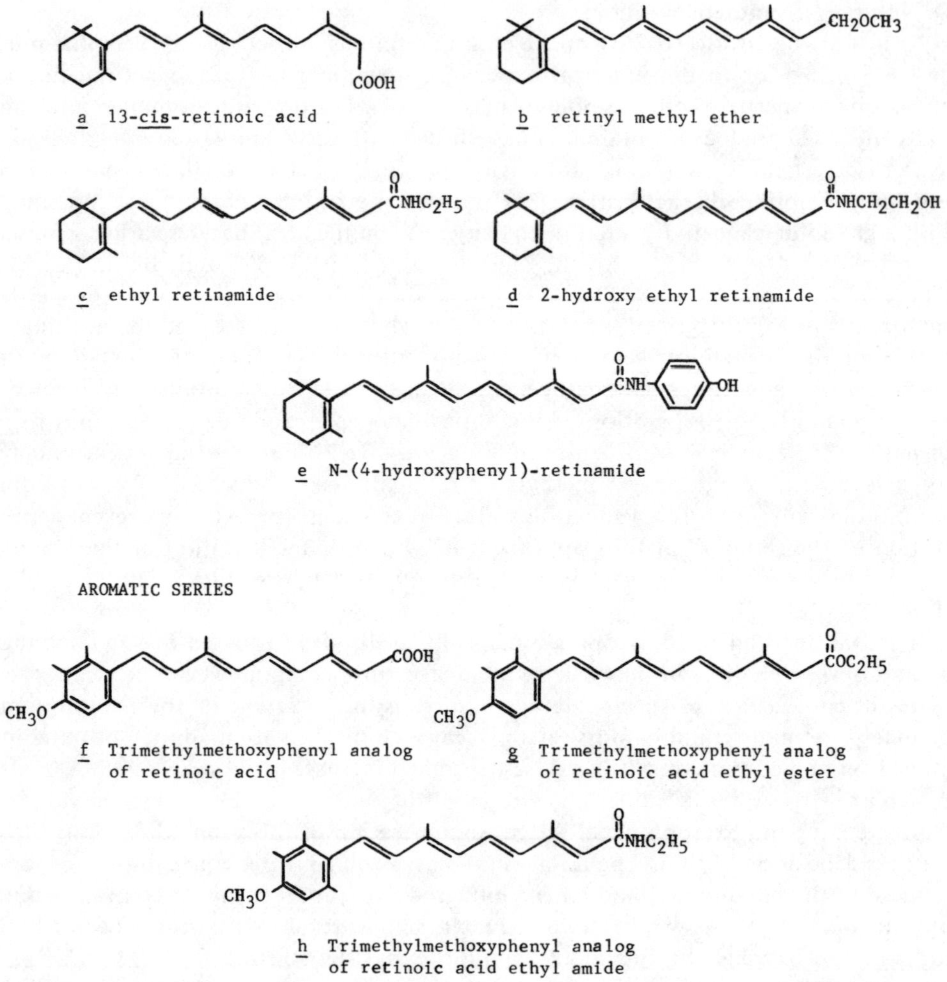

FIGURE 12. Structures of synthetic retinoids being used in studies concerning inhibition of tumor promotion.

N-(4-hydroxybutyl)nitrosamine.[170,171] These experiments therefore demonstrate a positive effect of this analog on urinary bladder cancer in two different animal species using two different carcinogens.

As mentioned previously, several other analogs of retinoic acid have received recent attention in animal studies. N-(4-hydroxyphenyl)-all-*trans*-retinamide (Figure 1e) has been shown to be an effective agent for inhibition of breast cancer induced in rats by N-methyl-N-nitrosourea.[270] Although this compound was less active than retinyl acetate in this test system, its lower toxicity made it a superior chemopreventive agent. Part of the reason for its lower toxicity compared to retinyl acetate appears to be due to its altered tissue distribution. The retinoid concentration in breast tissue was much higher in the N-(4-hydroxyphenyl)retinamide-dosed rats compared to the retinyl acetate-dosed animals, while, in the liver, the reverse was true. Retinyl acetate-treated rats showed a marked deposition of retinyl esters in the liver while the retinamide-treated animals showed little accumulation.[270] Similar results have been obtained for retinyl methyl ether (Fibure 12b). This compound, when administered to rats, concentrates in the mammary gland when compared to retinyl acetate[225] and is superior to retinyl acetate in the inhibition of rat mammary carcinogenesis induced by 7,12-

dimethylbenz[a]anthracene.[271] These results, therefore, give examples of the ability to alter the tissue distribution of retinoids by structural modifications. Two other amide analogs, ethyl and 2-hydroxyethyl retinamide (Figure 12c and d), have recently been reported to be effective in reducing both the incidence and severity of transitional cell carcinoma of the rat urinary bladder after N-butyl-N-(4-hydroxybutyl)nitrosamine treatment.[272]

The final group of synthetic retinoids that have been investigated for their antitumor properties in animal models fall into the general group of the aromatic analogs (Figure 12 f to h). The aromatic ethyl ester analog of retinoic acid is more active and slightly less toxic than all-*trans*-retinoic acid in retarding the growth of papillomas and in reducing the incidence of skin carcinoma induced in mice by 7,12-dimethyl-benz[a]anthracene and croton oil.[262,263,273] The aromatic free acid, the ethyl ester, and the ethyl amide will all inhibit tumor growth of the transplantable rat chondrosarcoma and, at sufficiently high concentrations, will cause regression of already established tumors.[274] 13-*cis*-retinoic acid is also effective in this chondrosarcoma system.[275] From the results of these in vivo studies, it thus appears that it is indeed possible to modify the basic retinoid structure to either alter its biological activity or to change its tissue distribution.

As has already been discussed, there are numerous in vitro systems that are being used for the determination of the biological activity and potential chemopreventive usefulness of synthetic retinoids. The results from these in vitro assays plus the data gathered from the in vivo experiments just discussed support several major conclusions.[225] These conclusions include (1) in all in vitro systems the most active natural retinoid is all-*trans*-retinoic acid, (2) the presence of a terminal carboxyl group is associated with increased toxicity, and (3) modification of the terminal polar group can lessen the toxicity and can affect the pharmacokinetic properties with only a slight decrease in activity. Using these ideas as stepping stones, it may be possible in the future to synthesize even more potentially useful retinoids that may be safely used for prevention of epithelial tumors in man.

2. Human Studies

Although the ideal retinoid (one with high activity and low toxicity) has not yet been found, several analogs are currently being tested in man. In most cases, these analogs are being used in various nonneoplastic skin disorders in which the efficacy of vitamin A treatment has previously been demonstrated. Since this review is concerned with the antineoplastic properties of retinoids, only brief mention will be made of a few of these studies — mainly to demonstrate that synthetic analogs may play a role in decreasing the toxicity of vitamin A in man.

As in animals, 13-*cis*-retinoic acid has received the most extensive testing in humans. For such skin disorders as severe and treatment-resistant acne, lamellar ichthyosis, Darier's disease, and other keratinizing dermatoses, this isomer appears to be an effective alternative to the all-*trans* compound with a minimal, short-term toxicity.[232,239,240] The 13-*cis* analog is also being used for preliminary studies in the treatment of two neoplastic lesions, basal cell carcinoma[239] and urinary bladder cancer. Other compounds that demonstrate a reduced, short-term toxicity in clinical trials are again found in the aromatic series of retinoids. The aromatic ethyl amide (Figure 12h) has proved to be much less effective than all-*trans*-retinoic acid in the reduction of acne but it is also much less toxic.[276] The aromatic ethyl ester (Figure 12g) has been tested either by itself in patients with Drier's disease[238] or psoriasis,[233,277-279] or in combination with a photosensitizer (methoxsalen) and long-wave UV radiation,[280] usually resulting in an improvement of the disorder without many of the serious side effects often observed after all-*trans*-retinoic acid treatment. This limited number of reports emphasizes that much work still needs to be done in order to find a compound with retinoid-like activity that is both effective as an antineoplastic agent and yet not toxic after either short or long term administration.

3. Conclusion

From the various in vivo and in vitro studies just described, as well as from limited human trials, it is clear that the future direction of research in the use of retinoids for the prevention of cancer lies in the area of synthetic, rather than natural, compounds. The goal of this research will be to find the most effective and least toxic synthetic retinoid that displays a high degree of tissue specificity for protection against cancer at a particular organ site.

VI. MECHANISM OF ACTION OF RETINOIDS IN CANCER PREVENTION

A. Effects on Carcinogen Metabolism

There have been numerous mechanisms proposed for the action of retinoids in the prevention of epithelial cancer. The idea that retinoids may block or somehow alter carcinogen metabolism seems, at first, an attractive hypothesis. Indeed, the activity of cytochrome P-450, an enzyme system involved in the activation of certain carcinogens,[281] has been found to vary in vitamin A-deficient rats.[282-284] Also, benzo[a]pyrene and 7,12-dimethylbenz[a]anthracene metabolism has been reported to be altered by retinoid treatment.[285-288] However, evidence has accumulated from various sources to suggest that retinoids do not exert their anticarcinogenic properties through the alteration of carcinogen metabolism. First, Bornstein et al. were unable to observe any changes in benzo[a]pyrene metabolism in lung microsomes from vitamin A-deficient compared to normal hamsters.[289] Second, and perhaps more important, retinyl acetate was able to suppress 7,12-dimethylbenz[a]anthracene induced mammary tumorigenesis even when administration of the retinoid was not begun until 7 days after exposure to the carcinogen,[154] a time when 7,12-dimethylbenz[a]anthracene uptake and binding to DNA was completed.[290] Finally, retinoids are able to alter tumor induction by direct-acting carcinogens that do not require metabolic activation.[164,199] These and other data,[162,170] therefore, imply that retinoids are not preventing carcinogenesis by direct interaction with the carcinogen but rather are acting at some step during the postinitiation phase of tumorigenesis.

B. Immunological Effects

One such step in the progression of tumorigenesis that has been reported to be affected by retinoids is an enhancement of the host-immune response against preneoplastic tissue. Vitamin A has been shown to have an adjuvant effect.[291-294] This effect is believed to be due either to its ability to stimulate cell-mediated cytotoxicity[294-296] or to its ability to destabilize the lysosomal membrane.[291,292] The destabilization of the membranes is thought to stimulate cell division at a time when the antigen is available in the cell, thus resulting in an induction of the immune response.[291,292] Hogan-Ryan and Fennelly have also suggested recently that vitamin A may cause certain cell surface membrane changes that eventually will lead to an increased immunogenicity of the cells.[297] Other reports suggesting a possible stimulation of a cell-mediated immune response by retinoids have demonstrated that either retinoic acid[298] or retinyl palmitate[299,300] can increase the susceptibility of skin transplants to immunological rejection. However, Boss, et al.[301] have postulated that this graft rejection is due to a disturbance in the wound-healing process by hypervitaminosis A and not to an effect on the immune system.

This enhancement of the immune rejection response has been suggested as being involved in the mechanism of action of retinoids against several tumor systems. For example, retinol or retinyl palmitate caused a reduction of the radiation dose needed to control an antigenic fibrosarcoma in a normal mouse, but each was without effect in an immune-suppressed animal.[302] Also, although ineffective by itself, retinyl palmitate is effective in enhancing tumor suppression by *Mycobacterium bovis* (strain BCG), an immunostimulant that suppresses tumor growth by an immunological mechanism.[303] This enhanced suppression is observed both for a transplantable 3-methylcholanthrene induced sarcoma[304] and for a transplantable, murine Lewis lung tumor.[305]

In the experiments just described, the retinoid used alone was without effect on tumor progression. Several reports, however, have demonstrated that the inhibitory effect on tumor development of vitamin A, by itself, may in certain cases be immune mediated. Retinyl palmitate will inhibit the growth of a transplantable, highly immunogenic murine melanoma while it is ineffective against two other less immunogenic 3-methylcholanthrene induced murine tumors.[176] Also, oral administration of either retinyl palmitate or 13-*cis*-retinoic acid has slowed tumor progression and stopped metastatic spread in patients with metastatic, unresectable squamous cell carcinoma of the lung.[256] These same patients also displayed an immune potentiating effect after the vitamin A therapy. These studies, however, were not carefully controlled clinical programs and, therefore, any definite conclusions must await further experimentation.

Although the above effects of retinoid therapy can be explained by an interaction of the retinoid with the immune system, many of these effects could also possibly be mediated by an increased lability of the lysosomal membrane in cells after vitamin A treatment. Another indication that mechanisms other than the immune response may be responsible for the action of retinoids on tumor cells is that retinoids are effective in the prevention of tumors in vitro where no immune mechanism exists.[197,198,200,208-211,220] This suggests that the enhancement of the immune response by retinoids, although perhaps being an important mechanism in certain situations, is not the only means whereby carcinogenesis is inhibited by retinoids and that, as suggested above, there may be a more direct inhibition of tumorigenesis.

C. Interaction with Cell Membranes

There are two different mechanisms proposed for a direct interaction of vitamin A with the preneoplastic epithelial cell. One theory postulates that the presence of vitamin A causes membrane changes, particularly in the cell and lysosomal membranes, and that these changes are then responsible for the destruction of the tumor cell. The second mechanism involves an interaction of vitamin A with the genetic machinery of the cell, thereby regulating the synthesis of specific proteins. It is not the purpose of this chapter to extensively describe the work that has gone into the evolution of these two theories. Rather, the reader will be referred to reviews that very adequately cover these areas. Only the experiments that involve carcinogenesis will be mentioned here.

The idea that vitamin A alters membrane properties has been advanced by Lucy and Dingle (for a review see References 306 and 307) who have demonstrated that vitamin A in excess will cause lysis of the lysosome[308,309] and erythrocyte.[210] They have hypothesized that vitamin A, under physiological conditions, functions by controlling membrane permeability.[310] In related work, De Luca and co-workers (for a review see Reference 20) have demonstrated that retinol is converted to a phosphate derivative.[311,312] They have suggested that this retinyl phosphate is then involved in the transfer of glycosides (especially mannose) to membrane glycoproteins.[313]

That retinoids may interact with tumor cells by changing the cell surface membrane characteristics has been proposed by several investigators. Brandes et al.[185] noted that L-1210 leukemia cells treated in vitro or in vivo with retinol had an altered glycoprotein surface coat as detected by histochemical staining techniques. This change appeared to be similar to that observed after treatment of the cells with neuraminidase. The neuraminidase-like effect of retinol in vitro was verified by Hogan-Ryan and Fennelly[297] who noted a release of sialic acid from ascites tumor cells after exposure to retinol resulting in a change in the net charge of the surface membrane. Other cell surface changes have recently been demonstrated in the hamster fibroblast NIL 2K and NILpy cell lines.[214] In these studies the glycolipid hematoside was greatly enhanced by addition of retinol to NIL 2K cultures and a decrease in the cell surface glycoprotein Gap a (or LETS) was accompanied by the appearance of a new surface membrane glycoprotein. Therefore, cell surface changes have been observed after vitamin A treatment although the primary or secondary nature of these effects is not yet clear.

The neuraminidase-like effect of vitamin A can easily be explained by the action of retinoids in labilizing lysosomes with a subsequent release of acid hydrolases. Lysosomal enzyme release has been postulated to be a general mechanism for tissue-regressive changes.[314] That this process may indeed be occurring after retinoid treatment, at least in the murine L-1210 leukemia cell line, has been demonstrated by several investigators. Brandes and co-workers[183] initially noted that the involution of the spleen found after transplantation of L-1210 leukemia into mice was reduced by cyclophosphamide and retinol treatment. This reduction was accompanied by an increase in the activity of lysosomal acid phosphatase. Similar results were obtained after retinoid treatment of mammary gland carcinoma in mice.[152] It has also been demonstrated that in vitro treatment of L-1210 leukemia cells with retinol produces a dose- and time-dependent release of β-glucuronidase from lysosomes.[315] Finally, there is also a report of retinoids affecting lysosomal stability in the papilloma tumor system.[144] As has been mentioned previously, when mice, initiated with 7,12-dimethylbenz[a]anthracene and promoted with croton oil, are treated with retinoids, the number of papillomas formed is greatly reduced.[143-148] However, filipin, a lysosomal labilizer, will similarly reduce tumor incidence while two lysosomal stabilizers, chloroquine and hydrocortisone, will increase the number of tumors.[144] Also, the activity of two lysosomal enzymes, acid phosphatase and aryl sulfatase, was increased after long-term treatment with filipin or retinol and decreased with chloroquine or hydrocortisone treatment.[144] From these data, it appears likely that retinoids, at least in the skin papilloma system, may be inhibiting tumorigenesis, at least in part, by labilizing the lysosomes of the premalignant cell.

However, these surface-active properties of vitamin A do not seem to account for all of the observed tumor-preventive properties of retinoids. For example, retinoids will inhibit the squamous metaplasia found in the alveolar epithelium of the mouse prostate after in vitro exposure to 3-methylcholanthrene.[195-199] However, the surface-active detergent, sodium dodecyl sulfate, has no inhibitory effect under identical conditions.[197] Also, hydrocortisone, a lysosomal stabilizer,[316] will similarly inhibit the effects of 3-methylcholanthrene on rat prostate epithelium.[317] Therefore, since both a lysosomal labilizer, retinol, and a lysosomal stabilizer, hydrocortisone, display the same antitumor properties, it is unlikely that the lysosomes play a primary role in tumor prevention, at least in the mouse prostatic epithelium. A similar conclusion has been reached using a malignant murine melanoma cell line whose in vitro proliferation is inhibited by retinoids.[211] In this system, both cortisone and hydrocortisone, steroids that prevent the release of lysosomal enzymes by retinoids,[318,319] failed to reduce the inhibitory effect of retinoic acid on cell proliferation. Also, exposure of the melanoma cell to retinoic acid did not cause a significant change in the cytosol level of two lysosomal enzymes, acid phosphatase and aryl sulfatase, when compared to controls.[211] It must therefore be concluded that the interaction of vitamin A with membrane surfaces, especially lysosomal, while perhaps being an important mechanism for tumor prevention by retinoids in certain systems, does not explain the observations reported in other models.

D. Interaction with the Genetic Machinery of the Cell

The second proposed theory for the direct interaction of retinoids with the initiated cell is an extension of the envisioned role of vitamin A in the normal control of growth and differentiation of epithelial tissues. Although the exact mechanism whereby vitamin A performs this task is still unknown, a number of investigators have postulated (for a general review see Reference 19) that vitamin A may control differentiation in a manner similar to steroid hormones. The steroid hormones are currently believed to bind to a specific cytosol-receptor protein after entering the target cell, followed by a translocation of this protein-hormone complex to the nucleus where it intereacts with the chromatin to change genome expression.[320] Indeed, a number of reports have demonstrated altered RNA synthesis in vitamin A-deficient compared to vitamin A-normal animals, either in vivo[321-325] or in vitro.[321,326-328] Again, this review will concentrate principally on how possible manipulation of gene expression by vitamin A may account for some of the observed effects of retinoids on preneoplasia.

Recently, Blalock and Gifford have reported that retinoic acid may be involved in the regulation of gene expression for the induced glycoprotein, interferon. They found that interferon production in mouse L-cells, stimulated with Newcastle disease virus, was suppressed by all-*trans*-retinoic acid along with a change in total RNA synthesis.[329] By following the time course of interferon production along with the time course of actinomycin D and cycloheximide inhibition of RNA and protein synthesis, they concluded that retinoic acid induces the production of a regulatory protein that, in turn, inhibits transcription of the interferon genome.[330,331] It is tempting to speculate that this induction of a regulatory protein by retinoic acid is a return to a "normal" state for a transformed cell.

As has previously been mentioned, any interaction of retinoids with the genetic machinery of the cell is believed to follow a pathway similar to that observed for steroid hormones. This hypothesis would necessitate a series of cytosol and nuclear retinoid-binding proteins. Indeed, such cytosol-binding proteins specific for either retinol (CRBP) or retinoic acid (CRABP) have been detected in many normal tissues (see, for example, References 332-334). That these binding proteins may also occur in the cell nucleus has likewise been demonstrated.[19,335-338] The question then, for the purpose of this review, is whether these binding proteins are involved in the interaction of retinoids with the initiated cell.

A number of investigators have been able to find retinoid-binding proteins in tumor tissue. For example, Ong et al.[339] have examined both human lung and breast carcinomas and were able to detect CRABP. Interestingly, this protein was not observed in normal lung or breast tissue from the same patients. This inability to detect the binding protein in normal tissue could either indicate that it is absent or merely that its concentration is below the sensitivity level of the binding assay. Similar data have been obtained using the mouse where CRABP was not detected in normal mouse lung, but was observed in both Lewis lung tumors and lungs with metastatic Lewis lung foci.[332,340] Because CRABP can be detected in many tissues of rat fetus while being absent in those same tissues in the adult,[339,341] it has been postulated that the presence of CRABP in tumor tissue may be yet another example of the occurrence in malignant tissue of a cellular protein which is normally present in that tissue only at the fetal stage.[342,343] The same workers have also found a high concentration of CRBP and CRABP in the mouse skin papilloma,[344] a tumor that is responsive to retinoid treatment,[143-148] while observing lower and more variable concentrations of the proteins in transplantable tumors, the growth of which, in general, does not appear to be inhibited by retinoids.[174,175] Finally, in L-1210 leukemia cells, a system that does not respond to retinoids by themselves, retinoid binding protein can not be detected.[344]

From the basis of the work just presented, it appears that the presence of CRABP or CRBP may be a required but not necessarily sufficient condition for sensitivity to retinoid treatment. This conclusion is supported by observations concerning the presence of retinoid-binding proteins in colon tumors which, as discussed previously, have been shown to be largely insensitive to retinoid treatment.[166,167] Sani and co-workers have examined four chemically induced (with either N-nitroso-N-methylurethane or 1,2-dimethylhydrazine dihydrochloride) transplantable colon tumors from mice and were able to detect CRABP in two highly metastatic tumors whereas two nonmetastatic tumor lines, as well as the normal colon, had no detectable CRABP.[332,338] Similarly, Ong et al.,[345] examining colon tissue from rats treated with 1,2-dimethylhydrazine, have observed that the level of CRBP in the adenocarcinoma was much higher than the level in normal colon from the same animal. The CRABP, however, showed no detectable change between the neoplastic and the normal tissue. Interestingly, the CRBP from the chemically induced adenocarcinoma was almost fully saturated with endogenous retinol. It is, therefore, interesting to speculate that if, indeed, the protective effects of retinoids are mediated by the binding proteins, then the lack of response to retinoids for this system may be due to the fact that the CRBP present in the initiated cell is already fully saturated with retinol and therefore unable to respond to additional retinoid treatment. This not only could explain the lack of response of the colon tumor to retinoid treatment, but also could provide further evidence that the presence of retinoid-binding proteins in the tumor tissue is not a sufficient condition for tumor sensitivity to retinoids.

The question still remains as to whether retinoid-binding proteins are involved in the mechanism of action for both the growth promoting and the possible chemopreventive activity of retinol, retinoic acid and their analogs. In an attempt to answer this question, Chytil and Ong have compared the ability of various retinoids to bind to CRABP isolated from rat testis, mouse papillomas, and human breast tumor with their ability to reverse metaplasia in tracheal organ culture.[19,346] They observed a very good correlation between the binding affinity of the retinoids to these three proteins, and both their ability to reverse metaplasia in the hamster trachea and their reported ability to inhibit carcinogenesis. Although this is only circumstantial evidence, it does support the idea of a possible involvement of these binding proteins in the mechanism of action of retinoids in tumor prevention.

A discussion of a possible mechanism of action of retinoids in tumor prevention would not be complete without mentioning a recent study of Todaro and co-workers[219] involving sarcoma growth factor (SGF). SGF is a polypeptide (mol wt approximately 12,000) that has been isolated from murine sarcoma virus-transformed 3T3 cells.[229] This protein is able to stimulate cell division in normal rat kidney fibroblast cells and produces a reversible, morphological transformation of the cells growing in monolayer that allows them to form multiple cell layers and to grow in a soft agar medium where normal cell growth is prevented.[229] However, when retinyl acetate or retinoic acid is added to the medium simultaneously with SGF, the phenotypic cell transformation no longer takes place.[219] Although the molecular interaction between retinoids and SGF is still not understood, this in vitro system should offer an excellent tool for the investigation of the mechanism of suppression of malignant transformation by retinoids.[17]

E. Conclusion

From the evidence presented, it appears that the mechanism of action of retinoids in preventing tumor formation or growth is still unclear. L-1210 leukemia cells, which do not contain retinoid-binding proteins[344] and therefore presumably do not have a mechanism for the interaction of retinoids with the cell's genetic machinery and which also are not responsive to retinoids alone,[183] appear to respond to retinoid treatment in combination with other therapeutic agents through a lysosomal membrane labilization mechanism. On the other hand, tumors such as the transplantable, highly immunogenic murine melanoma studied by Felix et al.[176] are most likely affected by the adjuvant properties of retinoids. Finally, many initiated cells may interact with retinoids through an enhancement of the normal controls on differentiation that vitamin A exerts in the epithelial cell — perhaps being mediated through a retinoid-binding protein or a growth factor. The only general conclusion that can be reached at this time appears to be that there is no single mechanism which is able to explain all the effects of retinoids in the various systems examined, and that the mechanism or combination of mechanisms employed depends on the tissue or cell being studied.

VII. CAUTIONS

When considering the use of retinoids in cancer prevention, there are two principal areas of concern that need to be dealt with. The first, as has been mentioned several times in this review, is the toxicity of these compounds. The second, is the reports by several investigators that retinoids may increase the carcinogeneic effect of certain substances in selected systems.

The toxicity of vitamin A was noted very early in its history when Takahashi and co-workers,[347] testing the effects of an excess of a crude concentrate of vitamin A given orally to rats and mice, noted loss of hair, emaciation, paralysis of the hind legs and, within several weeks, death. From these initial observations in 1925 to the present, numerous investigators have examined both the symptoms of vitamin A toxicity and the mechanism whereby vitamin A causes these symptoms. As discussed earlier in this review, excess vitamin A in humans manifests itself in alterations of the skin (hair loss, erythema, desquamation) and mucous membranes (conjunctivitis, cheilitis), hepatic dysfunction, and headache.[258,259] One additional effect of retinoid excess is its teratogenicity. Excessive intake of either retinol and its esters[348,349] or retinoic acid[350-353] has been shown to cause abnormal litters in rats,[348,350] mice,[349,350] and hamsters.[351,353] It is these undesirable toxic symptoms of retinoids that, hopefully, can be eliminated by the synthesis of chemically modified compounds that will be able to prevent epithelial neoplasia without causing the undue side effects mentioned above, even when administered over a long period of time.

Several in vivo and in vitro systems have been used to determine the toxicity of various retinoids, and therefore, to assess their potential usefulness as long-term chemopreventive agents. These include the release of chondroitin sulfate from rabbit ear cartilage in organ cultures,[354,355] bone resorption and occurrence of fractures in mice,[267] loss of weight, skin scaling, loss of hair and bone fractures also in mice,[263] release of proteoglycan from tracheal cartilage in vitro,[225] and a recently described in vitro technique using fluorescence polarization methods to measure the change of membrane fluidity.[356] As mentioned earlier, with the use of these systems it has been possible to demonstrate a separation of the toxic properties of vitamin A from its growth-promoting[197,198,357] and its anticarcinogenic[262-264,355] activities, giving rise to the hope that a biologically active synthetic retinoid can be found with lower toxicity. The results of these assays indicate that the presence of a free terminal carboxylic acid moiety leads to a compound with increased toxicity. Modification of this polar end group to obtain a compound with a less polar terminal function decreases the toxicity.[225,355] By appropriate modification of the retinoid molecule, it has been possible to decrease the toxicity inherent in the compound while retaining the ability to affect epithelial tumors.[169,270,273] Work is currently underway to create analogs with maximal activity and minimal toxicity.

The mechanism of retinoid toxicity is still not firmly established. It has been observed in the past that excess vitamin A causes labilization of membranes with subsequent release of lysosomal enzymes.[306,307] Recently, Meeks and Chen[356] have demonstrated that retinoic acid will interact in vitro with several membrane systems to decrease the microviscosity of the membrane as determined by steady state fluorescence polarization. These experiments confirm the idea that retinoids do indeed have a direct effect on membranes. Whether this interaction of retinoids with the cell membrane is able to cause the multitude of symptoms observed during retinoid toxicity is a problem currently being investigated by Harrison and co-workers. These investigators noted that many of the symptoms of vitamin A toxicity also occur after prostaglandin administration. They have postulated that the symptoms of retinoid intoxication may be mediated by prostaglandin production.[358] In support of this hypothesis, Ziboh and co-workers[359] have detected an increase in prostaglandin levels in guinea pig skin after topical application of retinoic acid. This increase in prostaglandin concentration could take place by release of the lysosomal enzyme phospholipase A which has been previously shown to induce prostaglandin synthesis.[360] Elevation of prostaglandin levels by toxic levels of retinoids has also been demonstrated in dog kidney cell cultures.[361] Finally, Harrison and co-workers were able to prevent several of the toxic effects of retinoic acid in mice by administration of aspirin[358] or ibuprofen,[362] both of which have been shown to be inhibitors of prostaglandin synthesis.[363] It is therefore a possibility that prostaglandin synthesis is tied in with retinoid toxicity and, that by controlling the former, the latter may also be brought under control.

As mentioned in the introductory paragraph of this section, several investigators have reported adverse effects of retinoids in certain biological systems. Large excesses of topical retinyl palmitate (10 to 20 mg per application) have been implicated in increasing the incidence of carcinoma induced by either 7,12-dimethylbenz[a]anthracene in hamster cheek pouch[364-367] or by Rous sarcoma virus in chickens.[368,369] Similarly, dietary supplementation with high amounts of retinyl palmitate has been thought to augment the induction of squamous cell carcinoma in the pulmonary epithelium of mice.[370] Retinyl palmitate is not the only retinoid implicated to have an adverse effect on certain epithelial tumors. Topical retinoic acid treatment of hairless mice has also been reportd to promote UV-induced carcinogenesis in the animals.[371] The same investigators, however, have recently demonstrated that at low concentrations of retinoic acid, this promotion no longer occurs.[372] A number of reports have also found retinoids to have neither a beneficial nor a harmful effect on tumorigenesis.[135,157,158,373-375] All of these undesirable effects just mentioned are most likely explained by the irritating, toxic properties that high concentrations of the natural retinoids have been shown to have on epithelial tissues. These reports only serve to emphasize the need to synthesize less toxic analogs.

Recently, it has also been reported that retinoic acid, as well as the tumor promoter phorbol myristate acetate and Rous sarcoma virus, will induce the synthesis of plasminogen activator in chick fibroblasts[376] and in chick myogenic cultures.[377] This evidence has caused some investigators to call retinoic acid a tumor promotor under certain conditions. Plasminogen activator is a protein that is thought to be involved in tissue remodeling with a possible function in the ovulation process.[378] It has been implicated as possibly having a role in tumor promotion.[379] However, the evidence for this is not strong enough to be able to correlate the ability of a compound to induce the synthesis of plasminogen activator to the activity of that compound as a tumor promotor. Therefore, it is not safe to draw conclusions regarding the tumor-promoting properties of retinoic acid based merely on the induction of plasminogen activator synthesis.

In conclusion, taken by themselves, the reports just mentioned suggest that the feeding of excess retinoids to an animal has a harmful, or at the least a nonbeneficial, role in tumorigenesis. However, there are now sufficient reports in the literature demonstrating a beneficial effect of retinoids that it may safely be concluded that certain natural retinoids can prevent the development of epithelial cancer both in animals and in man, but their usefulness as agents for chemoprevention of cancer is limited because of excessive toxicity. In certain experimental cases, these toxic properties may overcome any beneficial effects. As already mentioned, this problem again emphasizes the need for an analog of vitamin A that displays reduced toxicity symptoms.

VIII. CONCLUSIONS

From the data discussed in this review, a number of conclusions can be reached. (1) Animals (and perhaps humans) on a low dietary intake of vitamin A are more susceptible to carcinogenesis than animals on a normal vitamin A diet. (2) Retinoids are useful for the prevention of cancer in the skin, respiratory tract, mammary gland, and urinary bladder of experimental animals after exposure to carcinogens. (3) The mechanism whereby retinoids influence epithelial preneoplasia appears to vary depending on the system being studied. These mechanisms include enhancement of the host's immune system, labilization of membranes (especially lysosomal membranes) and interaction with the genetic machinery of the epithelial cell through retinoid binding proteins, or, perhaps, through modulating the effect of growth factors. (4) A major problem in the use of natural retinoids for cancer prevention is their inherent toxic properties. These toxic properties preclude any long-term administration of the natural retinoids to the human population at a concentration sufficient to prevent cancer. (5) Analogs of vitamin A show promise of being useful in cancer prevention not only because they are less toxic, but also because they can be targeted to specific organs. Future developments in this field, therefore, will depend on the ability to synthesize an active retinoid that does not display any toxic symptoms at pharmacologically active dose levels.

REFERENCES

1. **Silverberg, E. and Holleb, A.,** *Cancer,* 25, 2, 1975.
2. **Farber, E.,** *Cancer Res.,* 36, 2703, 1976.
3. **Cairns, J.,** *Nature,* 255, 197, 1975.
4. **Lemon, H. M.,** *Cancer,* 23, 781, 1969.
5. **Kripke, M. L. and Borsos, T.,** *J. Natl. Cancer Inst.,* 52, 1393, 1974.
6. **Wolbach, S. B. and Howe, P. R.,** *J. Exp. Med.,* 42, 753, 1925.
7. **Wong, Y. C. and Buck, R. C.,** *Lab. Invest.,* 24, 55, 1971.
8. **Harris, C. C., Sporn, M. B., Kaufman, D. G., Smith, J. M., Jackson, F. E., and Saffiotti, U.,** *J. Natl. Cancer Inst.,* 48, 743, 1972.
9. **Sporn, M. B.,** *Cancer Res.,* 36, 2699, 1976.
10. **Sporn, M. B., Dunlop, N. M., Newton, D. L., and Smith, J. M.,** *Fed. Proc.,* 35, 1332, 1976.
11. **Sporn, M. B.,** in *Origins of Human Cancer,* Hiatt, H. H., Watson, J. D., and Winston, J. A. Eds., Cold Spring Harbor Laboratory, New York, 1977, 801.
12. **Sporn, M. B.,** *Nutr. Rev.,* 35, 65, 1977.

13. **Sporn, M. B.**, in *Nutrition and Cancer*, Winnick, M., Ed., John Wiley and Sons, New York, 1977, 119.
14. **Sporn, M. B.**, in *Carcinogenesis*, Vol. 2, Slaga, T. J., Sivak, A., and Boutwell, R. K., Eds., Raven Press, New York, 1978, 545.
15. **Mayer, H., Bollag, W., Hänni, R., and Ruegg, R.**, *Experientia*, 34, 1105, 1978.
16. **Sporn, M. B., Newton, D. L., Smith, J. M., Acton, N., Jacobson, A. E., and Brossi, A.**, in *Carcinogens: Identification and Mechanisms of Action*, Griffin, A. C. and Shaw, C. R., Eds., Raven Press, New York, 1979, 441.
17. **Sporn, M. B. and Newton, D. L.**, *Fed. Proc.*, 38, 2528, 1979.
18. **De Luca, L. M.**, in *Handbook of Lipid Research*, Vol. 2, The Fat Soluble Vitamins, DeLuca, H. F., Ed., Plenum Press, New York, 1978 Chap. 1.
19. **Chytil, F. and Ong, D. E.**, in *Vitamins and Hormones, Vol. 36*, Munson, P. L., Diczfalusy, E., Glover, J., and Olson, R. E., Eds., Academic Press, New York, 1978, 1.
20. **De Luca, L. M.**, in *Vitamins and Hormones*, Vol. 35, Munson, P. L., Diczfalusy, E., Glover, J., and Olson, R. E., Eds., Academic Press, New York, 1977, 1.
21. **Rietz, P., Wiss, O., and Weber, F.**, in *Vitamins and Hormones*, Vol. 32, Harris, R. S., Munson, P. L., Diczfalusy, E., and Glover, J., Eds., Academic Press, New York, 1974, 237.
22. **Cama, H. R. and Sastry, P. S., Eds.**, *World Review of Nutrition and Dietetics*, Vol. 31, S. Karger, Basel, 1978.
23. **Pitt, G. A. J.**, in *Carotenoids*, Isler, O., Ed., Birkhauser Verlag, Basel, 1971, chap. 10.
24. **Morton, R. A., Ed.**, *Fat Soluble Vitamins*, Pergamon Press, Oxford, 1970.
25. **Olson, J. A.**, in *Vitamins and Hormones*, Vol. 26, Harris, R. S., Wool, I. G., Loraine, J. A., and Munson, P. L., Eds., Academic Press, New York, 1978, 1.
26. **Moore, T.**, *Vitamin A*, Elsevier, Amsterdam, 1957, chap. 1.
27. **Hopkins, F. G.**, *Analyst*, 31, 385, 1906.
28. **Stepp, W.**, *Biochem. Z.*, 22, 458, 1909; *Z. Biol.*, 57, 135, 1911; *Z. Biol.*, 59, 366, 1919.
29. **McCullum, E. V. and Davis, M.**, *J. Biol. Chem.*, 15, 167, 1913.
30. **McCullum, E. V. and Davis, M.**, *J. Biol. Chem.*, 23, 181, 1915.
31. **Osborne, T. B. and Mendel, L. B.**, *J. Biol. Chem.*, 15, 311, 1913.
32. **Drummond, J. C.**, *Biochem. J.*, 14, 660, 1920.
33. **Drummond, J. C., Channon, H. J., and Coward, K. H.**, *Biochem. J.*, 19, 1047, 1925.
34. **Takahashi, K.**, *J. Chem. Soc. Jpn*, 43, 826, 1922.
35. **V. Euler, H., V. Euler, B., and Hellstrom, H.**, *Biochem. Z.*, 203, 370, 1928.
36. **Moore, T.**, *Lancet*, 217, 380, 1929.
37. **Moore, T.**, *Biochem. J.*, 24, 696, 1930.
38. **Capper, N. S.**, *Biochem. J.*, 24, 980, 1930.
39. **Goodman, D. S. and Olson, J. A.**, *Methods Enzymol.*, 15, 463, 1969.
40. **Karrer, P., Helfenstein, A., Wehrli, H., and Wettstein, A.**, *Helv. Chim. Acta*, 13, 1084, 1930; 14, 614, 1931.
41. **Karrer, P. and Morf, R.**, *Helv. Chim. Acta*, 16, 557, 625, 1933.
42. **Holmes, H. N. and Corbet, R. E.**, *J. Am. Chem. Soc.*, 59, 2042, 1937.
43. **Isler, O., Kofler, M., Huber, W., and Ronco, A.**, *Experientia*, 2, 31, 1946.
44. **Isler, O., Huber, W., Ronco, A., and Kofler, M.**, *Helv. Chim. Acta*, 30, 1911, 1947.
45. **Arens, J. F. and Van Dorp, D. A.**, *Nature*, 157, 190, 1946.
46. **Arens, J. F. and Van Dorp, D. A.**, *Nature*, 160, 189, 1947.
47. **Marks, J.**, in *The Vitamins in Health and Disease, A Modern Reappraisal*, J. and A. Churchill Ltd., London, 1968, 29.
48. **Greenwood, C. T. and Richardson, D. P.**, *World Rev. Nutr. Diet*, 33, 1, 1979.
49. **Robeson, C. D. and Baxter, J. G.**, *Nature*, 155, 300, 1945.
50. **Robeson, C. D. and Baxter, J. G.**, *J. Am. Chem. Soc.*, 69, 136, 1947.
51. **Brown, P. S., Blum, W. P., and Stern, M. H.**, *Nature*, 184, 1377, 1959.
52. **Dowling, J. E.**, *Nature*, 188, 114, 1960.
53. **Futterman, S. and Andrews, J. S.**, *J. Biol. Chem.*, 239, 81, 1964.
54. **Krinsky, N. I.**, *J. Biol. Chem.*, 232, 881, 1958.
55. **Embree, N. D. and Shantz, E. M.**, *J. Am. Chem. Soc.*, 65, 910, 1943.
56. **Burger, B. V., Garbers, C. F., Pachler, K., Bonnett, R., and Weedon, B. C. L.**, *Chem. Commun.*, 588, 1965.
57. **Kaneko, R.**, *Rep. Gov. Chem. Ind. Res. Inst. Tokyo*, 57, 23, 1962.
58. **Giannotti, C., Das, C. B., and Lederer, E.**, *Bull. Soc. Chim. Fr.*, 3299, 1966.
59. **Heilbron, I. M., Gillam, A. E., and Morton, R. A.**, *Biochem. J.*, 25, 1352, 1931.
60. **Lederer, E., Rosanova, V., Gillam, A. E., and Heilbron, I. M.**, *Nature*, 140, 233, 1937.
61. **Shantz, E. M.**, *Science*, 108, 417, 1948.

62. Farrar, K. R., Hamlet, J. C., Henbest, H. B., and Jones, E. R. H., *J. Chem. Soc.*, 2657, 1952.
63. Wald, G., *Nature*, 139, 1017, 1937.
64. Barna, A. B., Das, R. C., and Verma, K., *Biochem. J.*, 168, 557, 1977.
65. Goodman, D. S. and Huang, H. S., *Science*, 149, 879, 1965.
66. Winterstein, A. and Hegedüs, B., *Z. Physiol. Chem.*, 321, 97, 1960.
67. Wald, G., *Nature*, 132, 316, 1933.
68. Morton, R. A., *Nature*, 153, 69, 1944.
69. Morton, R. A. and Goodwin, T. W., *Nature*, 153, 405, 1944.
70. Wald, G., *Nature*, 219, 800, 1968.
71. Wald, G. and Hubbard, R., in *Fat Soluble Vitamins*, Morton, R. A., Ed., Pergamon Press, Oxford, 1970, chap. 8.
72. Morton, R. A., in *Photochemistry of Vision*, Dartnall, H. J. A., Ed., Springer-Verlag, Berlin, 1972, chap. 2.
73. Callender, R. and Honig, B., *Annu. Rev. Biophys. Bioeng.*, 6, 33, 1977.
74. Arnaboldi, M., Motto, M. G., Tsujimoto, K., Balogh-Nair, V., and Nakanishi, K., *J. Am. Chem. Soc.*, 101, 7082, 1979.
75. Honig, B., Dinur, U., Nakanishi, K., Balogh-Nair, V., Gawinowicz, M. A., Arnaboldi, M., and Motto, M. G., *J. Am. Chem. Soc.*, 101, 7084, 1979.
76. Sheves, M., Nakanishi, K., and Honig, B., *J. Am. Chem. Soc.*, 101, 7086, 1979.
77. Ball, S., Goodwin, T. W., and Morton, R. A., *Biochem. J.*, 42, 516, 1948.
78. Kaneko, R., Seki, K., and Suzuki, M., *Chem. Ind. London*, 1016, 1971.
79. Barnholdt, B., *Acta Chem. Scand.*, 11, 909, 1957.
80. Beutel, R. H., Hinkley, D. F., and Pollak, P. I., *J. Am. Chem. Soc.*, 77, 5166, 1955.
81. Huisman, H. O., Smit, A., Van Leeuwen, P. H., and Van Rij, J. H., *Recl. Trav. Chim.*, 75, 977, 1965.
82. Jungalwala, F. B. and Cama, H. R., *Biochem. J.*, 95, 17, 1965.
83. Schwieter, U. and Isler, O., in *The Vitamins*, Vol. 1, 2nd ed., Sebrell, W. H., Jr. and Harris, R. S., Eds., Academic Press, New York, 1967, 5.
84. Zechmeister, L., *Cis-Trans Isomeric Carotenoids, Vitamin A and Arylpolyenes*, Academic Press, New York, 1962.
85. Liu, R. S. H., Asato, A. E., and Denny, M., *J. Am. Chem. Soc.*, 99, 8095, 1977.
86. Denny, M. and Liu, R. S. H., *J. Am. Chem. Soc.*, 99, 4865, 1977.
87. Tsukida, K., Ito, M., and Kodama, A., *J. Nutr. Sci. Vitaminol.*, 24, 143, 1978.
88. Ramamurthy, V. and Liu, R. S. H., *Tetrahedron*, 31, 201, 1975.
89. Asato, A. E. and Liu, R. S. H., *J. Am. Chem. Soc.*, 97, 4128, 1975.
90. Morgan, B. and Thompson, J. N., *Biochem. J.*, 101, 835, 1966.
91. McKenzie, R. M., Hellwege, D. M., McGregor, M. L., and Nelson, E. C., *Lipids*, 14, 714, 1979.
92. Halley, B. A. and Nelson, E. C., *J. Chromatogr.*, 175, 113, 1979.
93. Tsukida, K., Ito, M., Tomeoka, F., and Kodama, A., *J. Nutr. Sci. Vitaminol.*, 24, 335, 1978.
94. Dyke, S. F., *The Chemistry of Vitamins*, Interscience, New York, 1965, chap. 12.
95. Isler, O., Rüegg, R., Schwieter, U., and Würsch, J., *Vitamin. Horm. N.Y.*, 18, 295, 1960.
96. Mayer, H. and Isler, O., in *Carotenoids*, Isler, O., Ed., Birkhauser Verlag, Basel, 1971, chap. 6.
97. Isler, O., Kläui, H., and Solms, U., in *Vitamins*, Vol. 1, 2nd ed., Sebrell, W. H. Jr., and Harris, R. S., Eds., Academic Press, New York, 1967, 101.
98. Matsui, M., Okano, S., Yamashita, K., Miyano, M., Kitamura, S., Kobayashi, A., Sato, T. and Mikami, R., *J. Vitaminol. (Kyoto)*, 4, 178, 1958.
99. BuLock, J. D., Quarrie, S. A., and Taylor, D. A., *J. Labelled Compd.*, 9, 311, 1973.
100. Bhatt, M. V. and Venkatesh Prasad, H. N., *World Rev. Nutr. Diet.*, 31, 141, 1978.
101. Blomstrand, R. and Werner, B., *Scand. J. Clin. Lab. Invest.*, 19, 339, 1967.
102. Andrews, J. S. and Futterman, S., *J. Biol. Chem.*, 239, 4073, 1964.
103. Goodman, D. S., in *The Fat-Soluble Vitamins*, DeLuca, H. F., and Suttie, J. W., Eds., The University of Wisconsin Press, Madison, 1969, chap. 13.
104. Bliss, A. F., *Arch. Biochem.*, 31, 197, 1951.
105. Kleiner-Bössaler, A., and DeLuca, H. F., *Arch. Biochem.*, 142, 371, 1971.
106. Emerick, R. J., Zile, M., and DeLuca, H. F., *Biochem. J.*, 102, 606, 1967.
107. Ito, Y., Zile, M., DeLuca, H. F., and Ahrens, H. M., *Biochim. Biophys. Acta*, 369, 338, 1974.
108. Lippel, K. and Olson, J. A., *J. Lipid Res.*, 9, 168, 1968.
109. Masushige, S., Rosso, G. C., and Wolf, G., *World Rev. Nutr. Diet.*, 31, 16, 1978.
110. Frot-Coutaz, J. P., Silverman-Jones, C. S., and De Luca, L. M., *J. Lipid Res.*, 17, 220, 1976.
111. Barr, R. M., and De Luca, L. M., *Biochem. Biophys. Res. Commun.*, 60, 355, 1974.
112. Peterson, P. A., Rask, L., Helting, T., Östberg, L., and Fernstedt, Y., *J. Biol. Chem.*, 251, 4986, 1976.

113. Rosso, G. C., De Luca, L. M., Warren, C. D., and Wolf, G., *J. Lipid Res.*, 16, 235, 1975.
114. Austin, D. J., BuLock, J. D., and Drake, D., *Experientia*, 26, 348, 1970.
115. Van Dorp, D. A. and Arens, J. F. *Nature*, 158, 60, 1946.
116. Dowling, J. E. and Wald, G., *Proc. Natl. Acad. Sci., U.S.A.*, 46, 587, 1960.
117. Thompson, J. N., Howell, J. McC., and Pitt, G. A., *Proc. R. Soc. London, Ser. B*, 159, 510, 1964.
118. Arens, J. F. and Van Dorp, D. A., *Nature*, 158, 622, 1946.
119. Dunagin, P. E., Jr., Zachman, R. D., and Olson, J. A., *Biochim. Biophys. Acta*, 90, 432, 1964.
120. Deshmukh, D. S., Malathi, P., and Ganguly, J., *Biochim. Biophys. Acta*, 107, 120, 1965.
121. Ito, Y. L., Zile, M., Ahrens, H., and DeLuca, H. F., *J. Lipid Res.*, 15, 517, 1974.
122. Roberts, A. B. and DeLuca, H. F., *Biochem. J.*, 102, 609, 1967.
123. Hänni, R., Bigler, F., Meister, W., and Englert, G., *Helv. Chim. Acta*, 59, 2221, 1976.
124. Hänni, R. and Bigler, F., *Helv. Chim. Acta*, 60, 2309, 1977.
125. McCormick, A. M., Napoli, J. L., Schnoes, H. K., and DeLuca, H. F., *Biochemistry*, 17, 4085, 1978.
126. Mallia, A. K., John, J., John, K. V., Lakshmanan, M. R., Jungalwala, F. B., and Cama, H. R., *Indian J. Biochem.*, 7, 102, 1970.
127. Dunagin, P. E., Zachman, R. D., and Olson, J. A., *Science*, 148, 86, 1965.
128. Zile, M. H., Emerick, R. J., and DeLuca, H. F., *Biochim. Biophys. Acta*, 141, 639, 1967.
129. Frolik, C. A., Tavela, T. E., Newton, D. L., and Sporn, M. B., *J. Biol. Chem.*, 253, 7319, 1978.
130. Roberts, A. B., Nichols, M. D., Newton, D. L., and Sporn, M. B., *J. Biol. Chem.*, 254, 6296, 1979.
131. Frolik, C. A., Roberts, A. B., Tavela, T. E., Roller, P. P., Newton, D. L., and Sporn, M. B., *Biochemistry*, 18, 2093, 1979.
132. Roberts, A. B. and Frolik, C. A., *Fed. Proc.*, 38, 2524, 1979.
133. Fujimaki, Y., *J. Cancer Res.*, 10, 469, 1926.
134. Burrows, M. F., *J. Cancer Res.*, 10, 239, 1926.
135. Sugiura, K. and Benedict, S. R., *J. Cancer Res.*, 14, 306, 1930.
136. Orten, A. U., Burn, C. G., and Smith, A. H., *Proc. Soc. Exp. Biol. Med.*, 36, 82, 1937.
137. Capurro, P., Angrist, A., Black, J., and Moumgis, B., *Cancer Res.*, 20, 563, 1960.
138. Kaufman, D. G., Genta, V. M., and Harris, C. C., in *Experimental Lung Cancer-Carcinogenesis and Bioassays*, Karbe, E. and Park, J. F., Eds. Springer-Verlag, Berlin, 564, 1974.
139. Nettesheim, P. and Williams, M. L., *Int. J. Cancer*, 17, 351, 1976.
140. Nettesheim, P., Snyder, C., and Kim, J. C. S., *Environ. Health Perspect.*, 29, 89, 1979.
141. Kuh, C., *Yale J. of Biol. Med.*, 5, 123, 1932.
142. McMichael, H., *Cancer Res.*, 25, 947, 1965.
143. Davies, R. E., *Cancer Res.*, 27, 237, 1967.
144. Shamberger, R. J., *J. Natl. Cancer Inst.*, 47, 667, 1971.
145. Bollag, W., *Experientia*, 27, 90, 1971.
146. Bollag, W., *Eur. J. Cancer*, 8, 689, 1972.
147. Bollag, W., *Experientia*, 28, 1219, 1972.
148. Bollag, W. and Ott, F., *Acta Derm. Vener.*, 55 (Suppl. 74), 163, 1975.
149. Prutkin, L., *J. Invest. Derm.*, 49, 165, 1967.
150. Prutkin, L., *Cancer Res.*, 28, 1021, 1968.
151. Prutkin, L., *Cancer Res.*, 33, 128, 1973.
152. Brandes, D., and Anton, E., *Lab. Invest.*, 15, 987, 1966.
153. Nathanson, L., Maddock, C. L., and Hall, T. C., *J. Clin. Pharm.* 9, 359, 1969.
154. Moon, R. C., Grubbs, C. J., and Sporn, M. B., *Cancer Res.*, 36, 2626, 1976.
155. Moon, R. C., Grubbs, C. J., Sporn, M. B., and Goodman, D. G., *Nature*, 267, 620, 1977.
156. Saffiotti, U., Montesano, R., Sellakumar, A. R., and Borg, S. A., *Cancer*, 20, 857, 1967.
157. Smith, D. M., Rogers, A. E., Herndon, B. J., and Newberne, P. M., *Cancer Res.*, 35, 11, 1975.
158. Smith, D. M., Rogers, A. E., and Newberne, P. M., *Cancer Res.*, 35, 1485, 1975.
159. Smith, W. E., Yazdi, E., and Miller, L., *Environ. Res.*, 5, 152, 1972.
160. Cone, M. V. and Nettesheim, P., *J. Natl. Cancer Inst.*, 50, 1599, 1973.
161. Nettesheim, P., Cone, M. V., and Williams, M. L., *Proc. Am. Assoc. Cancer Res.*, 14, 59, 1973.
162. Nettesheim, P., Cone, M. V., and Snyder, C., *Cancer Res.*, 36, 996, 1976.
163. Newberne, P. M. and Rogers, A. E., *J. Natl. Cancer Inst.*, 50, 439, 1973.
164. Narisawa, T., Reddy, B. S., Wong, C. Q., and Weisburger, J. H., *Cancer Res.*, 36, 1379, 1976.
165. Cohen, S. M., Wittenberg, J. F., and Bryan, G. T., *Cancer Res.*, 36, 2334, 1976.
166. Rogers, A. E., Herndon, B. J., and Newberne, P. M., *Cancer Res.*, 33, 1003, 1973.
167. Ward, J. M., Sporn, M. B., Wenk, M. L., Smith, J. M., Feeser, D., and Dean, R. J., *J. Natl. Cancer Inst.*, 60, 1489, 1978.
168. Squire, R. A., Sporn, M. B., Brown, C. C., Smith, J. M., Wenk, M. L., and Springer, S., *Cancer Res.*, 37, 2930, 1977.

169. Sporn, M. B., Squire, R. A., Brown, C. C., Smith, J. M., Wenk, M. L., and Springer, S., *Science*, 195, 487, 1977.
170. Grubbs, C. J., Moon, R. C., Squire, R. A., Farrow, G. M., Stinson, S. F., Goodman, D. G., Brown, C. C., and Sporn, M. B., *Science*, 198, 743, 1977.
171. Becci, P. J., Thompson, H. J., Grubbs, C. J., Squire, R. A., Brown, C. C., Sporn, M. B., and Moon, R. C., *Cancer Res.*, 38, 4463, 1978.
172. Miyata, Y., Tsuda, H., Matayoshi-Miyasato, K., Fukushima, S., Murasaki, G., Ogiso, T., and Ito, N., *Gann*, 69, 845, 1978.
173. Dawson, W. D., Miller, W. W., and Liles, W. B., *Invest. Urol.*, 16, 376, 1979.
174. Bollag, W., *Cancer Chemother. Rep.*, 55, 53, 1971.
175. Bollag, W., *Schweiz. Med. Wochenschr.*, 101, 11, 1971.
176. Felix, E. L., Loyd, B., and Cohen, M. H., *Science*, 189, 886, 1975.
177. Rettura, G., Schittek, A., Hardy, M., Levenson, S. M., Demetriou, A., and Seifter, E., *J. Natl. Cancer Inst.*, 54, 1489, 1975.
178. Heilman, C. and Swarm, R. L., *Fed. Proc.*, 34, 822, 1975.
179. Shapiro, S. S., Bishop, M., Poon, J. P., and Trown, P. W., *Cancer Res.*, 36, 3702, 1976.
180. Trown, P. W., Buck, M. J., and Hansen, R., *Cancer Treat. Rep.*, 60, 1647, 1976.
181. Cohen, M. H. and Carbone, P. P., *J. Natl. Cancer Inst.*, 48, 921, 1972.
182. Cohen, M. H., *J. Natl. Cancer Inst.*, 48, 927, 1972.
183. Brandes, D., Anton, E., and Wai Lan, K., *J. Natl. Cancer Inst.*, 39, 385, 1967.
184. Brandes, D., Rundell, J. O., and Ueda, H., *J. Natl. Cancer Inst.*, 52, 945, 1974.
185. Brandes, D., Sato, T., Ueda, H., and Rundell, J. O., *Cancer Res.*, 34, 2151, 1974.
186. Meltzer, M. S. and Cohen, B. E., *J. Natl. Cancer Inst.*, 53, 585, 1974.
187. Seifter, E., Fisklatt, M., Levine, N., and Rettura, G., *Life Sciences*, 13, 945, 1973.
188. Levine, N. S., Salisburg, R. E., Seifter, E., Walker, H. L., Mason, A. D., Jr., and Pruitt, B. A., Jr., *Experientia*, 31, 1309, 1975.
189. Seifter, E., Rettura, G., Padawer, J., Demetriou, A. A., and Levenson, S., *J. Natl. Cancer Inst.*, 57, 355, 1976.
190. Crocker, T. T. and Sanders, L. L., *Cancer Res.*, 30, 1312, 1970.
191. Lane, B. P. and Miller, S. L., *Proc. Am. Assoc. Cancer Res.*, 17, 211, 1976.
192. Clamon, G. H., Sporn, M., Smith, J., and Saffiotti, U., *Nature*, 250, 64, 1974.
193. Sporn, M. B., Clamon, G. H., Smith, J. M., Dunlop, N. M., Newton, D. L., and Saffiotti, U., in *Experimental Lung Cancer, Carcinogenesis and Bioassays*, Karbe, E., and Park, J. F., Eds., Springer-Verlag, Berlin, 1974, 575.
194. Sporn, M., Clamon, G. H., Dunlop, N., Newton, D., Smith, J., and Saffiotti, U., *Nature*, 253, 47, 1975.
195. Lasnitzki, I., *Br. J. Cancer*, 9, 434, 1955.
196. Lasnitzki, I., *Exp. Cell Res.*, 28, 40, 1962.
197. Lasnitzki, I. and Goodman, D. S., *Cancer Res.*, 34, 1564, 1974.
198. Lasnitzki, I., *Br. J. Cancer*, 34, 239, 1976.
199. Chopra, D. P. and Wilkoff, L. J., *J. Natl. Cancer Inst.*, 56, 583, 1976.
200. Chopra, D. P. and Wilkoff, L. J., *J. Natl. Cancer Inst.*, 58, 923, 1977.
201. Chopra, D. P. and Wilkoff, L. J., *Nature*, 265, 339, 1977.
202. Verma, A. K. and Boutwell, R. K., *Cancer Res.*, 37, 2196, 1977.
203. Verma, A. K., Rice, H. M., Shapas, B. G., and Boutwell, R. K., *Cancer Res.*, 38, 793, 1978.
204. Verma, A. K., Shapas, B. G., Rice, H. M., and Boutwell, R. K., *Cancer Res.*, 39, 419, 1979.
205. Kensler, T. W., Verma, A. K., Boutwell, R. K., and Mueller, G. C., *Cancer Res.*, 38, 2896, 1978.
206. Kensler, T. W. and Mueller, G. C., *Cancer Res.*, 38, 771, 1978.
207. Wertz, P. W., Kensler, T. W., Mueller, G. C., Verma, A. K., and Boutwell, R. K., *Nature*, 277, 227, 1979.
208. Dion, L. D., Blalock, J. E., and Gifford, G. E., *J. Natl. Cancer Inst.*, 58, 795, 1977.
209. Dion, L. D., Blalock, J. E., and Gifford, G. E., *Exp. Cell Res.*, 117, 15, 1978.
210. Lotan, R. and Nicolson, G. L., *J. Natl. Cancer Inst.*, 59, 1717, 1977.
211. Lotan, R., Giotta, G., Nork, E., and Nicolson, G. L., *J. Natl. Cancer Inst.*, 60, 1035, 1978.
212. Lotan, R., *J. Cell Biol.*, 79, 29a, 1978.
213. Lotan, R., *Cancer Res.*, 39, 1014, 1979.
214. Patt, L. M., Itaya, K., and Hakomori, S., *Nature*, 273, 379, 1979.
215. Adamo, S., Akalovsky, I., and De Luca, L. M., *Proc. Am. Assoc. Cancer Res.*, 19, 107, 1978.
216. Adamo, S., De Luca, L. M., Akalovsky, I., and Bhat, P. V., *J. Natl. Cancer Inst.*, 62, 1473, 1979.
217. Wilkoff, L., Peckham, J. C., Dulmadge, E. A., Mowry, R. W., and Chopra, D. P., *Cancer Res.*, 36, 964, 1976.
218. Wilkoff, L. J., Chopra, D. P., and Peckham, J. C., *J. Invest. Dermatol.*, 72, 11, 1979.

219. **Todaro, G. J., DeLarco, J. E., and Sporn, M. B.,** *Nature,* 276, 272, 1978.
220. **Dickens, M. S. and Sorof, S.,** *Proc. Am. Assoc. Cancer Res.,* 20, 71, 1979.
221. **Lasnitzki, I.,** *J. Natl. Cancer Inst. Monogr.,* 12, 1381, 1963.
222. **Siiteri, P. K. and Wilson, J. D.,** *J. Clin Invest.,* 49, 1737, 1970.
223. **Sandburg, A. A., Kirdani, R. Y., Yamanaka, H., Varkarakis, M. J., and Murphy, G. P.,** *Cancer Chemother. Rep.,* 59, 175, 1975.
224. **Mossman, B. T. and Craighead, J. E.,** *Proc. Amer. Assoc. Cancer Res.,* 19, 370, 1978.
225. **Sporn, M. B., Dunlop, N. M., Newton, D. L., and Henderson, W. R.,** *Nature,* 263, 110, 1976.
226. **Boutwell, R. K.,** in *Origins of Human Cancer, Book B,* Hiat, H. H., Watson, J. D., and Winsten, J. A. Eds., Cold Spring Harbor, New York, 1977, 165.
227. **O'Brien, T. G.,** *Cancer Res.,* 36, 2644, 1976.
228. **Shin, S., Freedman, V., Risser, R., and Pollack, R.,** *Proc. Natl. Acad. Sci. U.S.A.,* 72, 4435, 1975.
229. **DeLarco, J. E. and Todaro, G. J.,** *Proc. Natl. Acad. Sci. U.S.A.,* 75, 4001, 1978.
230. **Straumfjord, J. V.,** *Northwest Med.,* 42, 219, 1943.
231. **Heel, R. C., Brogden, R. N., Speight, T. M., and Avery, G. S.,** *Drugs,* 14, 401, 1977.
232. **Peck, G. L., Olsen, T. G., Yoder, F. W., Strauss, J. S., Downing, D. T., Pandya, M., Butkus, D., and Battandier, J. A.,** *New Eng. J. Med.,* 300, 329, 1979.
233. **Orfanos, C. E. and Runne, U.,** *Br. J. Dermatol.,* 95, 101, 1976.
234. **Peck, S. M., Chargin, L., and Sobotka, H.,** *Arch. Dermatol. and Syphilol.,* 43, 1, 1941.
235. **Burgoon, C. F., Graham, J. H., Urbach, F., and Musgnug, R.,** *Arch. Dermatol.,* 87, 63, 1963.
236. **Fulton, J. E., Gross, P. R., Cornelius, C. E., and Kigman, A. M.,** *Arch. Dermatol.,* 98, 396, 1978.
237. **Gunther, S.,** *Acta Derm. Venereol.,* 55 (Suppl. 74), 146, 1975.
238. **Orfanos, C. E., Kurka, M., and Strunk, V.,** *Arch. Dermatol.,* 114, 1211, 1978.
239. **Peck, G. L., Yoder, F. W., Olsen, T. G., Pandya, M. D., and Butkus, D.,** *Dermatologica,* 157 (Suppl. 1), 11, 1978.
240. **Peck, G. L. and Yoder, F. W.,** *Lancet,* 3, 1172, 1976.
241. **Eriksen, L. and Cormane, R. H.,** *Br. J. Dermatol.,* 92, 343, 1975.
242. **Van der Wateren, A. R. and Cormane, R. H.,** *Br. J. Dermatol.,* 97, 83, 1977.
243. **Abels, J. C., Borham, A. T., Pack, G. T., and Rhoads, C. P.,** *J. Clin. Nutr.,* 20, 749, 1941.
244. **Basu, T. K., Raven, R. W., Dickerson, J. W. T., and Williams, D. C.,** *Int. J. Vitamin. Nutr. Res.,* 44, 14, 1974.
245. **Basu, T. K., Donaldson, D., Jenner, M., Williams, D. C., and Sakula, A.,** *Br. J. Cancer,* 33, 119, 1976.
246. **Mettlin, C., Graham, S., and Swanson, M.,** *J. Natl. Cancer Inst.,* 62, 1435, 1979.
247. **Ibrahim, K., Jafarey, N. A., and Zuberi, S. J.,** *Clin. Oncol.,* 3, 203, 1977.
248. **Bjelke, E.,** *Int. J. Cancer,* 15, 561, 1975.
249. **Hoppner, K., Phillips, W. E. J., Murray, T. K., and Campbell, J. S.,** *Can. Med. Assoc. J.,* 99, 983, 1968.
250. **Hoppner, K., Phillips, W. E. J., Erdody, P., Murray, T. K., and Perrin, D. E.,** *Can. Med. Assoc. J.,* 101, 736, 1969.
251. **Underwood, B. A., Siegel, H., Weisell, R. C., and Dolinski, M.,** *Am. J. Clin. Nutr.,* 23, 1037, 1970.
252. **Bollag, W. and Ott, F.,** *Agents and Actions,* 1, 172, 1970.
253. **Bollag, W. and Ott, F.,** *Schweiz. Med. Wochenschr.,* 101, 17, 1971.
254. **Bollag, W. and Ott, F.,** *Cancer Chemother. Rep.,* 55, 59, 1971.
255. **Stuttgen, G., Ippen, H., and Mahrle, G.,** *Int. J. Dermatol.,* 16, 500, 1977.
256. **Micksche, M., Cerni, C., Kokron, O., Titscher, R., and Wrba, H.,** *Oncology,* 34, 234, 1977.
257. **Komiyama, S., Hiroto, J., Ryu, S., Nakashima, T., Kuwano, M., and Endo, H.,** *Oncology,* 35, 253, 1978.
258. **Moore, T.,** in *The Vitamins,* Vol. 1, 2nd Ed., Sebrell, W. H., Jr., and Harris, R. S., Eds., Academic Press, New York, 1967, 280.
259. **Stimson, W. H.,** *New Eng. J. Med.,* 265, 369, 1961.
260. **Thompson, J. N., Howell, J. McC., and Pitt, G. A. J.,** *Biochem. J.,* 80, 25P, 1961.
261. **Thompson, J. N., Howell, J. McC., and Pitt, G. A. J.,** *Biochem. J.,* 80, 16P, 1961.
262. **Bollag, W.,** *Experientia,* 30, 1198, 1974.
263. **Bollag, W.,** *Eur. J. Cancer,* 10, 731, 1974.
264. **Bollag, W.,** *Chemotherapy,* 21, 236, 1975.
265. **Tosukhowong, P. and Olson, J. A.,** *Biochim. Biophys. Acta,* 529, 438, 1978.
266. **Newton, D. L., Frolik, C. A., Roberts, A. B., Smith, J. M., Sporn, M. B., Nurrenbach, A. and Paust, J.,** *Cancer Res.,* 38, 1734, 1978.
267. **Hixson, J. and Denine, E. P.,** *Toxicol. Appl. Pharm.,* 44, 29, 1978.
268. **Hixson, E. J., Burdeshaw, J. A., Denine, E. P., and Harrison, S. D.,** *Toxicol. Appl. Pharm.,* 47, 359, 1979.

269. Newberne, P. M. and Suphakarn, V., *Cancer,* 40, 2553, 1977.
270. Moon, R. C., Thompson, H. J., Becci, P. J., Grubbs, C. J., Gander, R. J., Newton, D. L., Smith, J. M., Phillips, S. L., Henderson, W. R., Mullen, L. T., Brown, C. C. and Sporn, M. B., *Cancer Res.,* 39, 1339, 1979.
271. Grubbs, C. J., Moon, R. C., Sporn, M. B., and Newton, D. L., *Cancer Res.,* 37, 599, 1977.
272. Thompson, H. J. and Becci, P. J., *Proc. Am. Assoc. Cancer Res.,* 20, 96, 1979.
273. Bollag, W., *Eur. J. Cancer,* 11, 721, 1975.
274. Trown, P. W., Buck, M. J., and Hansen, R., *Cancer Treat. Rep.,* 60, 1647, 1976.
275. Heilman, C. and Swarm, R. L., *Fed. Proc.,* 34, 822, 1975.
276. Christiansen, J., Holm, P., and Reymann, F., *Dermatologica,* 154, 219, 1977.
277. Ott, F. and Bollag, W., *Schweiz. Med. Woschenschr.,* 105, 439, 1975.
278. Schimpf, A., *Z. Hautkr.,* 51, 265, 1976.
279. Dahl, B., Mollenbach, K., and Reymann, F., *Dermatologica,* 154, 261, 1977.
280. Fritsch, P. O., Honigsmann, H., Jaschke, E., and Wolff, K., *J. Invest. Dermatol.,* 70, 178, 1978.
281. Yang, S. K., Deutsch, J. and Gelboin, H. V., in *Polycyclic Hydrocarbons and Cancer,* Vol. 1, Gelboin, H. V. and Ts'o, P. O. P., Eds., Academic Press, New York, 1978, 205.
282. Becking, G. C., *Can. J. Physiol. Pharmacol.,* 51, 6, 1973.
283. Colby, H. D., Kramer, R. E., Greiner, J. W., Robinson, D. A., Krause, R. F., and Canady, W. J., *Biochem. Pharmacol.,* 24, 1644, 1975.
284. Hauswirth, J. W. and Brizuela, B. S., *Cancer Res.,* 36, 1941, 1976.
285. Rasmussen, R. E., Wang, I. Y. and Crocker, T. T., *J. Natl. Cancer Inst.,* 59, 693, 1972.
286. Hill, D. L. and Shih, T. W., *Cancer Res.,* 34, 564, 1974.
287. Kkhandudzha, H. L., Esakova, T. D., Semin, B. K., Petrusevich, Y. M., and Tarusov, B. N., *Bull. Exp. Biol. Med.,* 83, 687, 1977.
288. Kohl, F. V., and Rudiger, H. W., *J. Cancer Res. Clin. Oncol.,* 93, 149, 1979.
289. Bornstein, W. A., Lamden, M. P., Chuang, H. L., Gross, R. L., Newberne, P. M., and Bresnick, E., *Cancer Res.,* 38, 1497, 1978.
290. Janss, D. H., Moon, R. C., and Irving, C. C., *Cancer Res.,* 32, 254, 1972.
291. Dresser, D. W., *Nature,* 217, 527, 1968.
292. Spitznagel, J. K. and Allison, A. C., *J. Immunol.,* 104, 119, 1970.
293. Cohen, B. E. and Cohen, I. K., *J. Immunol.,* 111, 1376, 1973.
294. Dennert, G., Crowley, C., Kouba, J., and Lotan, R., *J. Natl. Cancer Inst.,* 62, 89, 1979.
295. Lotan, R. and Dennert, G., *Cancer Res.,* 39, 55, 1979.
296. Dennert, G. and Lotan, R., *Eur. J. Immunology,* 8, 23, 1978.
297. Hogan-Ryan, A. and Fennelly, J. J., *Eur. J. Cancer,* 14, 113, 1978.
298. Floersheim, G. L. and Bollag, W., *Transplantation,* 14, 564, 1972.
299. Jurin, M. and Tannock, I. F., *Immunology,* 23, 283, 1972.
300. Rettura, G., Levenson, S. M., Schittek, A., and Seifter, E., *Surg. Forum,* 26, 301, 1975.
301. Boss, J. H., Bitterman, W., and Gabriel, M., *Transplantation,* 4, 293, 1966.
302. Tannock, I. F., Suit, H. D. and Marshall, N., *J. Natl. Cancer Inst.,* 48, 731, 1972.
303. Bartlett, G. L., Zbar, B., and Rapp, H. J., *J. Natl. Cancer Inst.,* 48, 245, 1972.
304. Meltzer, M. S., and Cohen, B. E., *J. Natl. Cancer Inst.,* 53, 585, 1974.
305. Kurata, T. and Micksche, M., *Oncology,* 34, 212, 1977.
306. Lucy, J. A. and Dingle, J. T., *Nature,* 204, 156, 1964.
307. Dingle, J. T. and Lucy, J. A., *Biol. Rev.,* 40, 422, 1965.
308. Dingle, J. T., Sharman, I. M., and Moore, T., *Biochem. J.,* 98, 476, 1966.
309. Fell, H. B., *Proc. Nutr. Soc. Eng. Scot.,* 24, 166, 1965.
310. Dingle, J. T. and Lucy, J. A., *Biochem. J.,* 84, 611, 1962.
311. De Luca, L., Maestri, N., Rosso, G., and Wolf, G., *J. Biol. Chem.,* 248, 641, 1973.
312. De Luca, L., Rosso, G., and Wolf, G., *Biochem. Biophys. Res. Commun.,* 41, 615, 1970.
313. Rosso, G. C., Masushige, S., Quill, H., and Wolf, G., *Proc. Natl. Acad. Sci. U.S.A.,* 74, 3762, 1977.
314. De Duve, C., in *Int. Symp. Injury, Inflammation and Immunity,* Thomas, L., Uhr, J., and Grant, L., Eds., Williams & Wilkins, Baltimore, 1964, 283.
315. Rundell, J. O., Sato, T., Wetzelberger, E., Ueda, H., and Brandes, D., *J. Natl. Cancer Inst.,* 52, 1237, 1974.
316. Lasnitzki, I., Dingle, J. T., and Adams, S., *Exp. Cell Res.,* 43, 120, 1966.
317. Lasnitzki, I., *J. Natl. Cancer Inst.,* 35, 1001, 1965.
318. Fell, H. B. and Thomas, L., *J. Exp. Med.,* 114, 343, 1961.
319. Weissman, G. and Thomas, L., *J. Clin. Invest.,* 42, 661, 1963.
320. O'Malley, B. W. and Means, A. R., *Science,* 183, 610, 1974.
321. Zachman, R. D., *Life Sci.,* 6, 2207, 1967.

322. Johnson, B. C., Kennedy, M., and Chiba, N., *Am. J. Clin. Nutr.*, 22, 1048, 1969.
323. Zile, M. and DeLuca, H. F., *Arch. Biochem. Biophys.*, 140, 210, 1970.
324. De Luca, L., Kleinman, H. K., Little, E. P., and Wolf, G., *Arch. Biochem. Biophys.*, 145, 332, 1971.
325. Tryfiates, G. P. and Krause, R. F., *Life Sci.*, 10, 1097, 1971.
326. Kaufman, D. G., Baker, M. S., Smith, J. M., Henderson, W. R., Harris, C. C., Sporn, M. B., and Saffiotti, U., *Science*, 177, 1105, 1072.
327. Sporn, M. B., Dunlop, N. M., and Yuspa, S. H., *Science*, 182, 722, 1973.
328. Tsai, C. H. and Chytil, F., *Life Science*, 23, 1461, 1978.
329. Blalock, J. E., and Gifford, G., *J. Gen. Virol.*, 32, 143, 1976.
330. Blalock, J. E., and Gifford, G. E., *Proc. Natl. Acad. Sci. U.S.A.*, 74, 5382, 1977.
331. Blalock, J. E., *Tex Rep. Biol. Med.*, 35, 69, 1977.
332. Sani, B. P. and Titus, B. C., *Cancer Res.*, 37, 4031, 1977.
333. Bashor, M. M., Toft, D. O., and Chytil, F., *Proc. Natl. Acad. Sci. U.S.A.*, 70, 3483, 1973.
334. Ong, D. E. and Chytil, F., *J. Biol. Chem.*, 250, 6113, 1975.
335. Wiggert, B., Russell, P., Lewis, M., and Chader, G., *Biochem. Biophys. Res. Commun.*, 79, 218, 1977.
336. Sani, B. P., *Biochem Biophys. Res. Commun.*, 75, 7, 1977.
337. Jetten, A. M. and Jetten, M. E. R., *Nature*, 278, 180, 1979.
338. Sani, B. P. and Donovan, M. K., *Cancer Res.*, 39, 2492, 1979.
339. Ong, D. E., Page, D. L., and Chytil, F., *Science*, 190, 60, 1975.
340. Sani, B. P. and Corbett, T. H., *Cancer Res.*, 37, 209, 1977.
341. Ong, D. E. and Chytil, F., *Proc. Natl. Acad. Sci. U.S.A.*, 73, 3976, 1976.
342. Abelev, G. I., *Adv. Cancer Res.*, 14, 295, 1971.
343. Schapira, F., *Adv. Cancer Res.*, 18, 77, 1973.
344. Ong, D. E. and Chytil, F., *Cancer Lett.*, 2, 25, 1976.
345. Ong, D. E., Market, C., and Chiu, J. F., *Cancer Res.*, 38, 4422, 1978.
346. Chytil, F. and Ong, D. E., *Nature*, 260, 49, 1976.
347. Takahashi, K., Nakamiya, Z., Kawakami, K., and Kitasato, T., *Sci. Pap. Inst. Phys. Chem. Res. Tokyo*, 3, 81, 1925.
348. Cohlan, S. Q., *Science*, 117, 535, 1953.
349. Kalter, H. and Warkany, J., *Am. J. Path.*, 38, 1, 1961.
350. Kochhar, D. M., *Acta Path. Microbiol. Scand.*, 70, 398, 1967.
351. Shenefelt, R. E., *Teratology*, 5, 103, 1972.
352. Kochhar, D. M., *Teratology*, 7, 289, 1973.
353. Fraser, B. A. and Travill, A. A., *J. Embryol. Exp. Morphol.*, 48, 23, 1978.
354. Bard, D. R. and Lasnitzki, I., *Br. J. Cancer*, 35, 115, 1977.
355. Bard, D. R. and Lasnitzki, I., *Br. J. Cancer*, 37, 475, 1978.
356. Meeks, R. G. and Chen, R. F., *Fed. Proc.*, 38, 540, 1979.
357. Goodman, D. W. S., Smith, J. E., Hembry, R. M., and Dingle, J. T., *J. Lipid Res.*, 15, 406, 1974.
358. Harrison, S. D., Jr., Hixson, E. J., Burdeshaw, J. A., and Denine, E. P., *Nature*, 269, 511, 1977.
359. Ziboh, V. A., Price, B., and Fulton, J., *J. Invest. Dermatol.*, 65, 370, 1975.
360. Flower, R. J. and Blackwell, G. J., *Biochem. Pharm.*, 25, 285, 1976.
361. Levine, L. and Ohuchi, K., *Nature*, 276, 274, 1978.
362. Hixson, J. and Denine, E. P., *Toxicol. Appl. Pharm.*, 43, 317, 1978.
363. Flower, R. J., *Pharm. Rev.*, 26, 33, 1974.
364. Polliack, A. and Levij, I. S., *Nature*, 216, 187, 1967.
365. Polliack, A. and Levij, I. S., *Cancer Res.*, 29, 327, 1969.
366. Levij, I. S. and Polliack, A., *Cancer*, 22, 300, 1968.
367. Levij, I. S., Rwomushama, J. W., and Polliack, A., *J. Invest. Dermatol.*, 53, 228, 1969.
368. Polliack, A. and Sasson, Z. B., *Nature*, 234, 547, 1971.
369. Polliack, A. and Sasson, Z. B., *J. Natl. Cancer Inst.*, 48, 407, 1972.
370. Smith, W. E., Yazdi, E., and Miller, L., *Environ. Res.*, 5, 152, 1972.
371. Epstein, J. H., *J. Invest. Dermatol.*, 64, 212, 1975.
372. Epstein, J. H. and Grekin, D. A., *J. Invest. Dermatol.*, 72, 272, 1979.
373. Schmähl, D. and Hobs, M., *Arzneim. Forsch.*, 28, 49, 1978.
374. Schmähl, D., Kruger, C., and Pressler, P., *Arzneim. Forsch.*, 22, 946, 1972.
375. Kuni, C. C., *Dermatologica*, 150, 47, 1975.
376. Wilson, E. L. and Reich, E., *Cell*, 15, 385, 1978.
377. Miskin, R., Easton, T. G., and Reich, E., *Cell*, 15, 1301, 1978.
378. Beers, W. H., Strickland, S., and Reich, E., *Cell*, 6, 387, 1975.
379. Wigler, M. and Weinstein, I. B., *Nature*, 259, 232, 1976.

Chapter 5

ANTITUMOR EFFECTS OF POLYENE ANTIBIOTICS

J. Brajtburg, G. Medoff, and G. S. Kobayashi

TABLE OF CONTENTS

I. INTRODUCTION

A variety of chemotherapeutic regimens have been employed in the treatment of cancer. They all attempt to utilize one or both of two basic concepts: a direct attack on the tumor cells, or an indirect effect which results when the host's defense mechanisms are enhanced. In general, because one agent does not achieve both effects, combinations of drugs are routinely used.

For the past few years, our group has been studying the polyene macrolide antibiotics and their possible use as antitumor agents. Although most of the previous studies with these drugs have concentrated on their antifungal properties, it is also known that they have potent effects on animal cells. Research with these compounds has high interest at the present time because they appear to possess both direct toxicity for tumor cells and an ability to enhance immune responses in the host. In this chapter, we will review the evidence for both of these effects and we will attempt to put the polyene antibiotics into perspective as to their potential for use in the treatment of human cancers.

II. GENERAL PROPERTIES OF THE POLYENE ANTIBIOTICS

More than 80 polyenes, produced by several different species of *Streptomyces,* have been described. Structurally, they are all large, lactone ring compounds (macrolides) containing variable stretches of unsaturated carbon atoms (hence "polyene"). The ring also possesses a hydrophilic portion due to the presence of hydroxyl groups. Differences in the chemical structures of polyenes include the number of conjugated double bonds, the size of the macrolide ring, and the absence or presence of a hexosamine sugar (mycosamine) or an aromatic moiety. Pandey and Rinehart have compiled a classification table which describes all of the known polyenes (Figure 1). Those with known structures are depicted above the lines and those with unknown structures are below.

The studies published before 1973 on the chemistry and pharmacology of polyenes have been reviewed by Hamilton-Miller.[1] Since that review, a great deal of work has been directed toward the elucidation of the structures of individual compounds,[2] and the development of accurate analytical methods for the measurement of polyene antibiotic levels in solutions and in biological fluids.[3]

The first observations on the biological action of polyenes closely followed their discovery.[4] In the early 1960s, Lampen and co-workers[5] and Kinsky[6] independently presented evidence indicating that polyenes bind to sterols located in the membranes of eukaryotic cells. The polyene-sterol interaction was later confirmed by physicochemical methods.[7,8] It was demonstrated that the interaction of polyenes with cellular sterols resulted in early changes in cell membrane permeability and lysis or cell death as the final events.[9]

The nature of the membrane lesion resulting from the polyene-sterol interaction leading to the increase in permeability and lysis has been studied extensively in natural[10] and artificial[11] membranes. Freeze-etch electron microscopy of erythrocytes and *Acholeplasma laidlawii* treated with polyenes has shown structural alterations such as pits, doughnut shaped defects, and protrusions in the membranes;[12] but it is not known how these structural changes are related to the enhanced permeability and lysis.

The study of polyene effects on membranes has led to the construction of several models aimed at the representation of assumed antibiotic-sterol interaction.[13] Although these models are being refined continuously,[14] it is fair to state that the mechanism of polyene action is complex and far from being elucidated.

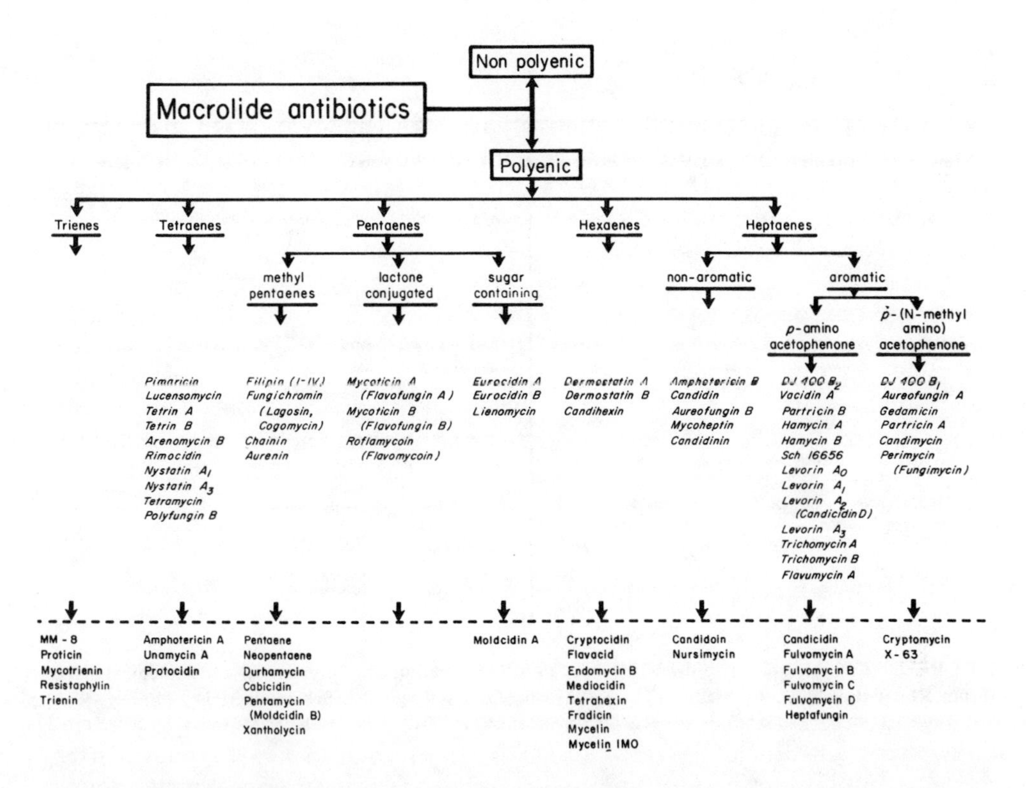

FIGURE 1. Classification of polyene macrolide antibiotics according to chemical structure. Those with known structures are above the broken line and those whose structures are not known are below the broken line. (Courtesy of Drs. Ramesh C. Pandey and Kenneth L. Rinehart, Jr.)

Work on the biologic effects induced by polyenes on yeast and animal cells[9,15] has shown that the changes in cell membranes are dependent upon the type and concentration of polyene used. Although polyenes vary greatly in potency, they can be categorized according to the type of biological action observed. Using the yeast, *Saccharomyces cerevisiae,* Lampen and Arnow[16] assayed the effects of several polyenes and were able to classify them into two groups. Our recent studies have[17] strengthened and extended this classification. We have compared 14 polyenes antibiotics and six of their semisynthetic derivatives for their abilities to induce lysis and potassium (K^+) leakage from red blood cells. On the basis of these investigations, we also concluded that polyenes could be divided into two groups (Figure 2). Group I included those causing little or no measurable change in membrane permeability to K^+ prior to lysis. This group of polyenes included those with less than seven double bonds in the macrolide ring and consisted of one triene (trienin), two tetraenes (pimaricin and etruscomycin), two pentaenes (filipin and chainin), and one hexaene (dermostatin). The polyenes classified in Group II were those which cause considerable K^+ leakage at lower concentrations (a reversible effect) and lysis (irreversible effects) at higher concentrations. Included in this group were the heptaenes (amphotericin B, candicidin, aureofungin A and B, hamycin A and B), nystatin (considered a degenerated heptaene), and six semisynthetic derivatives (amphotericin B methyl ester, N-acetyl-amphotericin B, hamycin A and hamycin B methyl ester, N-acetyl-candicidin, and N-acetyl-nystatin). The concentrations of the antibiotics required to produce the reversible and/or irreversible effects depended greatly on the experimental conditions of the assay such as temperature and length of incubation of cells with the antibiotic, antibiotic-to-cell ratio, and the composition of the medium. Figure 3 shows the structure of filipin as a represent-

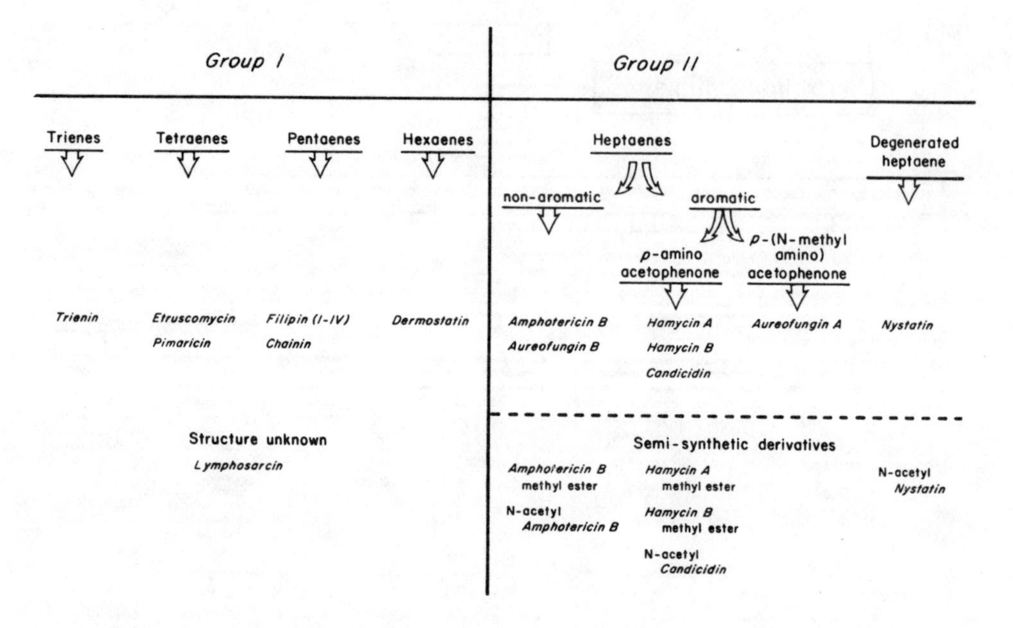

FIGURE 2. Classification of polyene macrolide antibiotics according to chemical and biological effects. (From Kotler-Brajtburg, J., Medoff, G., Kobayashi, G. S., Boggs, S., Schlessinger, D., Pandey, R. C., and Rinehart, K. L., Jr., *Antimicrob. Agents Chemother.*, 15, 716, 1979. With permission.)

ative of group I, amphotericin B as a representative of group II, and nystatin as the "degenerate" heptaene.

III. EFFECTS OF POLYENES ON ANIMAL CELLS

In the same way that fungi vary in sensitivity to amphotericin B,[15] animal cells also show different levels of susceptibility. This was demonstrated for comparable cells derived from different animals (e.g., rat erythrocytes were more sensitive to amphotericin B than were human erythrocytes[18]) and for cellular or subcellular cell membrane fractions taken from the same species (e.g., amphotericin B, which was active against rat erythrocytes was inactive against rat liver mitochondria[19]) Furthermore, diet and certain diseases appear to influence the susceptibility of some cells to polyenes, e.g., erythrocytes from rats fed orotic acid were more sensitive than cells taken from rats fed a normal diet.[20]

The variable susceptibility of cultured animal cells to polyenes also has been demonstrated. Amati and Lago[21] compared sensitivity of eight different cell lines to amphotericin B by measuring inhibitions of ^{3}H-uridine incorporation into acid-precipitable material. Some cell lines (3T3) were extremely sensitive to the antibiotic while others (BHK) were highly resistant. There was a positive correlation between levels of density-dependent inhibition of growth of cell lines and their sensitivity to amphotericin B because cells from lines with reduced membrane-mediated control of growth, measured by efficiency of plating in agar and saturation density, were less sensitive to amphotericin B than cells from lines with higher levels of contact inhibition of growth. This has led to the conclusion that the action of amphotericin B may be mediated through a membrane component involved in regulation of growth.

One determinant of sensitivity to polyene antibiotics is thought to be the level of membrane cholesterol which is known to be a target site for polyene antibiotics. The

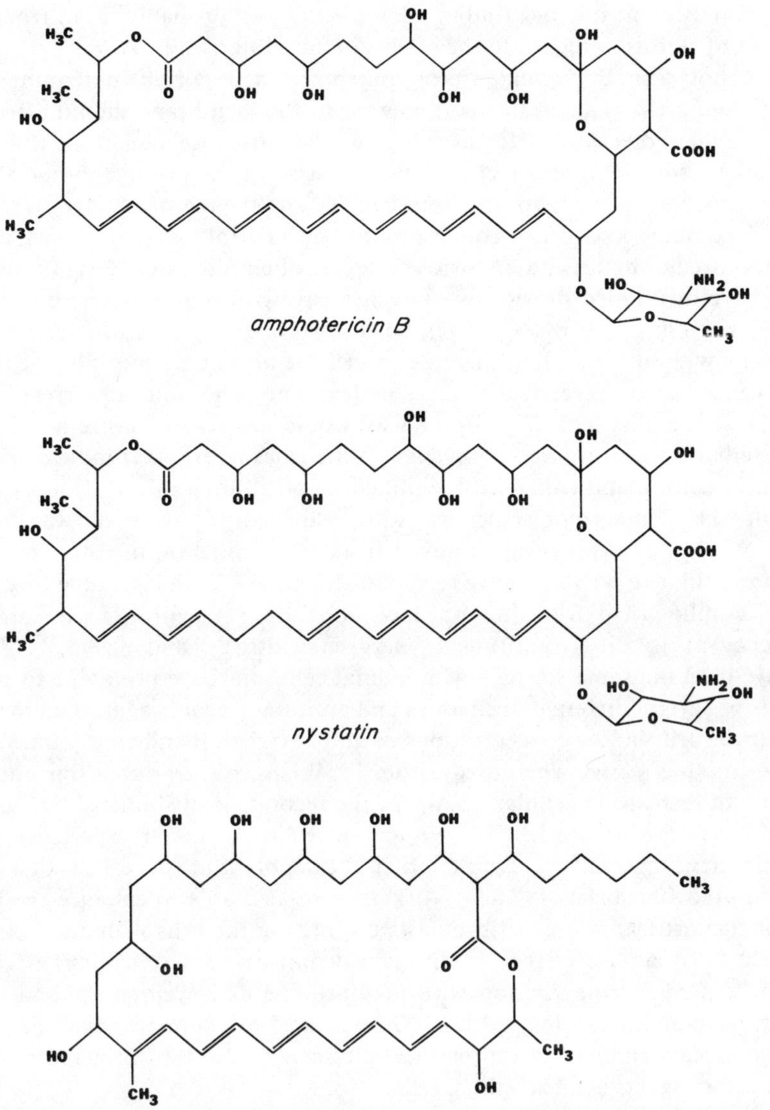

FIGURE 3. Chemical structures of amphotericin B, A heptaene; nystatin, a "degenerate heptaene"; and filipin, a pentaene.

evidence for this is that cells selected for resistance to amphotericin B had lower cholesterol content and also had about half the rate of cholesterol synthesis than sensitive cell lines.[22,23] Other results are not completely supportive of this concept and indicate that the absolute membrane content of free sterol does not correlate fully with sensitivity to amphotericin B.[24,25]

When we compared the sensitivity of four cultured cell lines to amphotericin B, we observed that cells with lower cholesterol-to-phospholipid ratios were more sensitive than cells with higher ratios.[26] Recently, using red blood cells with various cholesterol-to-phospholipid ratios, we have found that the regulation of cell sensitivity to polyenes was complex and involved several factors including the sterol content in relation to phospholipid levels in cell membranes, the concentration of amphotericin B used and the duration and temperature of incubation (Brajtburg, unpublished observation).

The seemingly contradictory findings described above probably result from a multifaceted role of sterols in regard to the effects of amphotericin B. As the primary receptor for amphotericin B binding, sterol content is an important determinant of cell sensitivity, and gross changes in sterol content of the membrane should affect its susceptibility to amphotericin B. However, sterols also influence membrane fluidity[24] and this and other factors[28] that affect the physical state of the cell membrane also determine the sensitivity of cells to amphotericin B. The degree of change in membrane cholesterol content, and other conditions of the environment, probably determine which effect predominates and, in turn, the degree of amphotericin B sensitivity.

As we have mentioned previously, polyenes have two major effects on cells. Polyenes categorized in group I (see above), when used at effective concentrations, caused lysis of cells without preceding increase in cell membrane permeability. Polyenes in group II caused early, reversible changes under some conditions and irreversible ones under others. It is this second group that we would like to emphasize in this chapter.

At nonlethal concentrations, the polyene antibiotics in group II induced K^+ leakage from cells in culture and enhanced the uptake of small molecules.[29] Under some conditions, uptake of macromolecules was also enhanced.[30] This effect was completely reversible when the polyene was removed from the incubation mixture. These observations are similar to what we have seen with fungi. We have been able to exploit the effects of amphotericin B on fungi at low, nonlethal concentrations in combination with other agents to achieve antifungal synergism in vitro[31,32] and in vivo.[33]

We have used the same strategy with animal cells and have been able to potentiate the effects of several different antibiotics and antitumor agents against animal cells by incubating the cells in low concentrations of amphotericin B and concentrations of the second agents, ineffective when used alone.[34,35] We have shown that this potentiation was due to an increase in cellular uptake of the second agents induced by the permeabilizing effects of amphotericin B.[36] Others have had a similar experience. Kuwano and Ikehara have reported that amphotericin B can potentiate the effects of α-amanitin against cultured fibroblasts.[37] These workers were also able to enhance the action of bleomycin against transformed fibroblasts by treating the cells with another polyene, pentamycin.[38] Recently, Peterson et al.[39] have demonstrated a synergistic effect of amphotericin B used in combination with prednisolone and cyclophosphamide against stimulated lymphocytes in culture. These studies support the notion that polyene-controlled changes in cell permeability can be utilized to increase the action of various other agents against animal cells.

This potentiation of the effects of second agents against cells can also be utilized to overcome certain kinds of drug resistance. One of the major mehanisms of cellular resistance developing during the course of treatment with anticancer agents is the selection or development of tumor cells which no longer take up the drugs. We used HeLa cells, selected for resistance to actinomycin D on the basis of decreased uptake of the antibiotic,[40] to test the possibility that the permeabilizing, nonlethal effects of amphotericin B and other group II agents could overcome the resistance. These cells were rendered sensitive to actinomycin D by nonlethal concentration of amphotericin B as measured by viable cell counts, morphologic changes, and measurements of RNA synthesis.[41]

We have already stated that different cell types vary in their sensitivity to polyene antibiotics. Several studies have also indicated that malignant or viral transformation can change the sensitivity of cells to these antibiotics. In 1972, Mondovi et al.[42] showed that oxygen consumption and DNA synthesis by Novikoff hepatoma cells and Ehrlich ascites tumor cells were inhibited by filipin and etruscomycin; whereas regenerating liver cells used as nonmalignant controls were much less sensitive. We have shown that

3T3 cells, transformed with mouse leukemia virus, were more sensitive to the potentiation effects of second agents by amphotericin B than were nontransformed cells.[43] In a study by Amati and Lago,[44] 3T3 cells transformed with SV40 were less sensitive to amphotericin B than nontransformed cells. The 3T3 cells, however, did not acquire resistance to amphotericin B if they were transformed by polyoma virus.[45] One interpretation of these results is that malignant transformation results in changes in cell membranes, which increase sensitivity to polyenes in some cases, and decrease it in others. Also, the effect produced is dependent on the agent inducing the transformation.

Up to now, we have described one type of effect of polyenes on animal cells. That is, cell membranes are made more permeable by polyenes and this increase in permeability leads to an increase in uptake of second drugs. There is also evidence that polyenes have effects on cell metabolism. Blum et al.[46] observed that increased K^+ flux induced in normal erythrocytes by amphotericin B was accompanied by an increased rate of K^+ pumping. As the cation pump increased in response to amphotericin B, the glycolytic pathway, utilizing glucose and producing ATP for the operation of the cation pump, increased ATP production. This was manifested by stability of the red cell ATP stores and an increase in the rate of glucose utilization. In more recent work, amphotericin B was shown to stimulate uridine transport and RNA and DNA synthesis in density-inhibited, nontransformed fibroblasts.[47] These changes may have important biologic effects when they occur in specific cell types. For example, stimulatory effects on cells of the immune system by polyenes has been demonstrated. Nystatin was found to increase in vitro antibody production to heterologous erythrocytes by mouse spleen cells.[48] It also stimulated DNA synthesis in thymus-independent (B-) cells but not in thymus-dependent (T-) cells and elicited polyclonal antibody synthesis in mouse spleen cell cultures and restored the antibody response of deficient spleen cells of congenitally athymic nude mice to heterologous erythrocytes.[49] In another study, Hammarström and Smith[50] found that nystatin and amphotericin B were polyclonal B cell activators of mouse lymphocytes. Candicidin was also capable of inducing DNA synthesis and polyclonal antibody production when added to normal spleen cells.

Another aspect of these studies relates to the varying susceptibility of different cell types to polyenes. Thymocytes were more sensitive to amphotericin B than lymph node and splenic lymphocytes.[51,52] This differential sensitivity supports the notion that polyene antibiotics can exert specifically directed toxicity or stimulation on cells of the immune system. In accord with these in vitro results, it was also shown that amphotericin B caused a reversible thymus involution in mice after administration of a single dose of 500 μg. After 48 hr, regeneration of the thymus occurred, and after 1 week the weight of the thymus exceeded those of control mice. Thymocytes from these amphotericin B-treated animals were much more responsible to lectin stimulation than those from control animals. The most recent evidence suggests that immunoadjuvant effects of amphotericin B may depend on a selective interaction of the polyene with a subset of T-cells and a resultant impairment of the normal suppressor regulation that limits the magnitude and duration of immune responses.[53]
effects of amphotericin B may depend on a selective interaction of the polyene with a subset of T-cells and a resultant impairment of the normal suppressor regulation that limits the magnitude and duration of immune responses.[53]

Studies of amphotericin B effects on the functional activities of single-cell types of the immune system have also included a study of murine macrophages and their precursors. Peritoneal cells harvested from mice that had been pretreated with amphotericin B showed higher phagocytic and antibacterial activities than those isolated from control mice.[54] The effect of polyenes on macrophage tumorcidal capability has been studied by Chapman and Hibbs.[55] Incubation of activated macrophages from *Bacillus*

Calmette-Guerin-infected mice with amphotericin B or its methyl ester enhanced the capability of activated macrophages to kill tumor cells.

Because the stimulatory effects of the polyenes were seen only when those from group II were used, but not by antibiotics from group I, these effects of the polyenes were probably related to changes in cell-membrane permeability. This notion was supported by our findings that the adjuvant property of amphotericin B methyl ester was lost in parallel with its loss of antifungal and erythrocyte membrane permeabilizing effects following inactivation by visible light irradiation.[56]

Our working hypothesis of the results we have described is that, as a result of amphotericin B binding to cell membrane sterols, membrane integrity is altered and ion flux occurs. This triggers a series of events which are not presently understood but can be exploited to increase the uptake of second agents into cells and enhance their effects against both sensitive and resistant cells. These membrane effects, either by the same or different mechanisms, lead to an activation of certain processes which, in cells of the immune system, lead to stimulation of T-cells, B-cells, and macrophage function. In addition, there appears to be specificity of the polyenes for certain cell types and, in some cases, tumor cells appear to be more sensitive than normal. Because the permeabilizing properties of the group II drugs are important for drug enhancement and adjuvanticity, a reasonable prediction would be that drugs from group II would be useful in potentiating the action of second agents and in stimulating host resistance to the tumors. Drugs from group I, on the other hand, might have direct tumorcidal capability but lack the permeabilizing and adjuvant effects of those in group II. Thus far, our animal experiments support this hypothesis.

IV. ANIMAL STUDIES

Our first in vivo test of the antitumor potential of polyene antibiotics was done using the AKR leukemia model system in mice.[57,58] In these experiments, AKR mice were injected with 10^6 syngeneic leukemic cells on day 0. Untreated mice and mice treated with AmB died on day 5. 1,3-6 bis (2-chloroethyl)-1-nitrosourea (BCNU), at the doses we used (0.2 mg per mouse), prolonged life 1 to 2 days. When amphotericin B (0.5 mg per mouse) was given on day 3 and BCNU on day 4, there was significant increase in the prolongation of life over that of BCNU alone and 40 to 80% of the mice survived indefinitely and were cured (Figure 4). These cured mice were immune to subsequent rechallenge with 10^6 leukemic cells, a dose that caused fatal leukemia in all untreated control mice. Therefore, it was apparent that the combination of amphotericin B and BCNU was more effective than either agent alone, and in some cases, was curative.

This experiment was repeated using the same treatment protocol, and groups of the treated mice were sacrificed at different times and a clonogenic assay of the number of leukemia cells in their femoral marrows was performed. Twenty-four hours after amphotericin B plus BCNU treatment, a log increase in killing of leukemic cells was seen when compared to animals treated with BCNU alone. Leukemia cells persisted for several days in these drug combination-treated animals and, finally, by days 12 to 16 they disappeared. These persisting leukemic cells could cause fatal leukemia when harvested from the treated animals and injected into nontreated animals.

Our interpretation of these experiments is that amphotericin B was affecting the AKR leukemia in these animals by two mechanisms. In one, it was increasing tumor cell membrane permeability and increasing the uptake and effects of BCNU. This led to the early one log decrement in cell number when compared to BCNU alone. The persistance of the leukemic cells without dissemination and multiplication and finally their elimination at 12 to 16 days (well after the effects of BCNU had dissipated) was

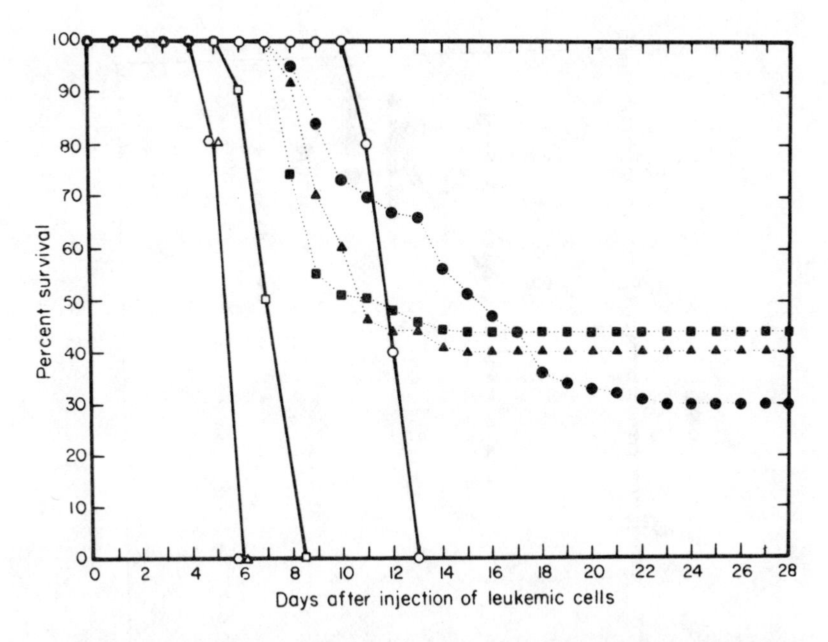

FIGURE 4. Percentage of survival of mice given i.v. injections on day 0 of 10 leukemic cells and then treated with different regimens. AMB was given i.p. on day 3 and BCNU was given i.p. on day 4. Each of the untreated and the single-drug-treated groups is made up of 10 animals receiving the following drugs: 0.5 mg AMB (⬡); 0.2 mg BCNU (□); 0.4 mg BCNU (○); no treatment (△). The groups treated with both drugs (closed symbols) consist of 100 mice each. (From Medoff, J., Medoff, G., Goldstein, M. N., Schlessinger, D., and Kobayashi, G. S., *Cancer Res.*, 35, 2548, 1975. With permission.)

probably secondary to an effect on the host. We believe this effect was due to a stimulation of the host-immune response. This was supported by the experiment which showed that the leukemic cells were viable and multiplied normally and were able to kill recipient mice. The combination of the pharmacological effects of amphotericin B on the tumor cells and the immunoadjuvant effects on the host led to a cure of the leukemia in a large percentage of animals. Subsequent studies which have attempted to dissect out both mechanisms by analyzing dose and schedule dependency of amphotericin B effects against tumor have supported this hypothesis.[59]

Clonogenic spleen assays of marrow cells taken from femurs of individual leukemic mice treated with amphotericin B and BCNU showed a characteristic kinetic response.[60] Figure 5 is taken from the paper of Valeriote et al.[60] and shows the survival data for leukemia colony-forming units (LCFU) in individual mice in a control group, and in individual mice given amphotericin B alone, BCNU alone, or amphotericin B + BCNU. The kinetics of response in the amphotericin B- + BCNU-treated animals showed proliferation of the residual leukemia cells for several days after treatment (as was seen in the previous study) and then a rapid rejection when the immune response is initiated.[61] This is exactly what was observed in rejection when a histoincompatible barrier is crossed[61] and confirmed our notion that an immune response was involved in the elimination of tumor cells.

Experiments with a variety of different antitumor agents have shown a consistent enhancement of cell killing by amphotericin B when used in combination with these drugs. It is interesting that those agents which showed the largest enhancement of killing of leukemic cells as measured by the spleen colony assay did not necessarily lead to cures of animals.[62,63] Those agents which showed this discrepant effect were

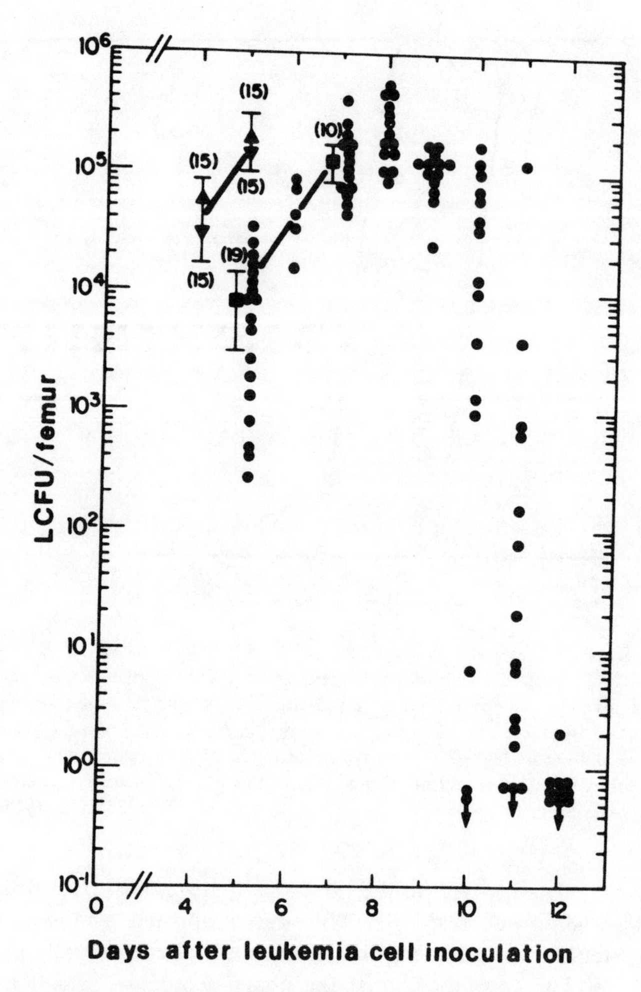

FIGURE 5. LCFU content of femoral marrow from individual leukemia-bearing mice. ▲, untreated; ▼, AmB only; ■, BCNU only, •, AmB + BCNU. Errors shown are ± 1 SD. Arrows indicate individual points that were below the limit of sensitivity of the assay. The number of animals pooled for each data point is in parenthesis. (From Valeriote, F., Lynch, R., Medoff, G., Tolen, S., and Dieckman, J., *J. Natl. Cancer Inst.*, 61, 399, 1978. With permission.)

highly immunosuppressive when compared to BCNU, and we think that they were suppressing the immunoadjuvant properties of amphotericin B required for cure.

Further evidence of the immunostimulant properties of amphotericin B in mice has been provided by direct measurements of immune responses induced by several different antigens in combination with amphotericin B.[64,65] Table 1 from the study by Blanke et al.[51] shows the immunostimulant effect of amphotericin B on the mouse secondary-immune response BALB/c mice received a single injection of antigen trinitrophenylated human serum albumin (TNP-HSA) plus amphotericin B, antigen alone, or amphotericin B as their primary stimulus. Then, after varying time intervals, all mice were given a single, secondary injection of TNP-HSA to induce the formation of cells producing IgG antibodies unless the primary immunization included both the immunogen and amphotericin B. Another study showed significant increase in the intensity of the graft-versus-host reactions induced by the transfer of parental thymo-

Table 1
EFFECT OF AmB ON THE SECONDARY RESPONSE IN BALB/c MICE

Primary immunization[a]	Secondary immunization[b]	Days after primary immunization	Days after secondary immunization	PFC/1 × 10^6 spleen cells[c]	
				IgM	IgG
AmB + TNP-HSA	TNP-HSA	20	7	25.1 ± 6.1	6.8 ± 12.0
TNP-HSA	TNP-HSA	20	7	18.0 ± 2.9	0
AmB	TNP-HSA	20	7	11.0 ± 4.3	0
AmB + TNP-HSA	TNP-HSA	27	7	30.2 ± 8.1	12.7 ± 8.8
TNP-HSA	TNP-HSA	27	7	21.6 ± 6.7	0
AmB	TNP-HSA	27	7	20.3 ± 6.0	0
AmB + TNP-HSA	TNP-HSA	34	7	37.4 ± 9.2	18.1 ± 11.6
TNP-HSA	TNP-HSA	34	7	20.6 ± 4.9	0
AmB	TNP-HSA	34	7	12.1 ± 4.3	0

[a] Primary immunization was performed as described in the legend to Figure 1, except that recipients of AmB received 300 μg each. All mice were 6- to 8-week-old males.

[b] Secondary immunization for all mice was 100 μg of TNP-HSA in 0.2 mℓ of PBS injected i.p.

[c] The PFC values are arithmetic means of six replicate assays ± 1 SD for spleen cell pools from three mice. A Student's *t* test was used to analyze six replicate experiments in BALB/c male mice assayed 34 days after primary and 6 to 7 days after secondary immunization. The derived *P* values for the difference in the pooled mean secondary IgG and secondary IgM PFC was <0.02 and <0.2, respectively, for recipients of antigen plus AmB compared to recipients of antigen alone.

From Blanke, T. J., Little, J. R., Shirley, S. F., and Lynch, R. G., *Cell. Immunol.*, 33, 180, 1977. With permission.

cytes from amphotericin B-treated mice to neonatal F_1 recipients.[66] More recent work has demonstrated that amphoreticin B, when it was given simultaneously with 2,4,dinitrofluorobenzene augmented contact sensitivity to this agent in mice.[66] We also have tested whether the immunoadjuvant properties of amphotericin B could prevent the development of spontaneous AKR lymphoma in AKR mice. Animals treated with amphotericin B every 2 weeks beginning at 6 months of age had a significant delay (12 weeks) in the onset of spontaneous lymphoma compared to untreated mice, supporting the role of the adjuvant property in the antitumor effect.[67]

We have tested several different polyenes from both groups of drugs and, in general, the in vivo data appeared to confirm the in vitro results. Drugs from group II were more potent immunoadjuvants and were more effective in potentiating the antitumor effects of several different drugs than the polyenes in group I (unpublished). In addition, we have compared the potentiation, by amphotericin B, of second agents against leukemia cells in normal bone marrow cells and it would appear that the leukemia cells were more sensitive. Thus, in this system, amphotericin B potentiation had relative specificity for tumor cells and this, combined with the specificity of the traditional antitumor agents, resulted in a high degree of lethality for leukemic cells without extreme toxicity for normal cells of the host.

We have begun to screen other tumors for responsiveness to amphotericin B combinations. Thus far, we have data on the L1210 and ASL1 leukemias, early transplants of a new AKR leukemia, the B16 melanoma, and the EMT 6 mammary carcinoma. In each case, we have been able to prolong survival by treating mice with amphotericin B and BCNU over that of the regimen of BCNU alone. We have also been able to overcome the resistance of leukemic cells to several different antitumor agents by adding amphotericin B to the treatment regimens (unpublished).

Other laboratories have reported similar effects with other tumors. Laurent et al., in 1976,[68] showed an increased rate of regression and cure of a subcutaneous mouse ependymoblastoma when amphotericin B was given 10 hr before 1-(2-chloroethyl)-3-cyclohexyl-1-nitrosourea (CCNU). In another study,[69] amphotericin B was shown to increase the antitumor effects of the thiophosphamide and cyclophosphane against a transplanted tumor of mice and rats. Muller and Tator, in 1976,[70] demonstrated that effects of the therapeutic action of CCNU against brain tumors in mice could be enhanced by amphotericin B. It is interesting that the enhancement of CCNU's tumorcidal activity occurred only after the amphotericin B was delivered directly to the tumor-bearing hemisphere. This suggests that, in this system, the antibiotic exerted its effect by acting directly on the tumor cells and not on cells of the host. As in other studies, the enhancing effects of amphotericin B depended on the schedule of administration of both the amphotericin B and the antitumor agent.

Thus, the studies in vivo by several different laboratories support the in vitro results that amphotericin B and other polyenes can enhance the effects of many different antitumor agents. In addition, amphotericin B appears to have immunoadjuvant properties, and these results also support the in vitro experiments.

V. HUMAN STUDIES

On the basis of the in vitro experiments and the work on tumor models in animals, it seemed timely to examine whether polyene antibiotics could also have therapeutic effects on cancer patients. The results are very preliminary but appear to support the notion that amphotericin B may be a useful addition to antitumor therapy.

Presant et al.[71] used amphotericin B plus BCNU in patients with advanced metastatic malignant disease. In this study, the dose limiting toxicity was myelosuppression.

Thrombocytopenia occurred more frequently and was more severe than granulocytopenia. Platelet and granulocyte count nadirs occurred on an average of 5 weeks after therapy and the patients were able to receive a subsequent course 8 weeks after the previous course. It appeared from this study that the toxicity was no greater with the amphotericin B and BNCU than that seen with BCNU alone. On the basis of this study, the recommended dose levels for phase two studies were established as 7.5 mg/m/day of amphotericin B on day 1, 15 mg/m/day on day 2, 30 mg/m/day on day 3, and this was to be followed on day 4 by 250 g/m of BCNU. Preliminary results of this phase one study suggested that the amphotericin B was potentiating the antitumor effects of BCNU. In a later study done by Presant et al.[72] patients who demonstrated clinical resistance to adriamycin appeared to have a response when amphotericin B was combined with adriamycin. In another study, Vogel et al.[73] used a combination of amphotericin B and CCNU in advanced lung cancer with promising results. However, Krutchik et al.[74] showed only limited effects of amphotericin B when used in combination with adriamycin in 19 patients resistant to an adriamycin-containing regime.

The notion that the permeabilizing effects of amphotericin B was selectively directed against cancer cells was supported by experiments with neocarzinostatin, an antitumor antibiotic of high molecular weight. Nakazawa et al.[75] showed that the treatment of gastric cancer patients with neocarzinostatin, in combination with small concentrations of amphotericin B, resulted in increased penetration of this drug into cancer tissue as compared with penetration into normal surrounding tissues. This is in contrast to the use of neocarzinostatin alone in which the concentration of drug in cancer and normal tissues was equal.

Immunologic responses in humans have been evaluated in only one study,[72] and the results were inconclusive because both stimulation and suppression were seen. This study is difficult to interpret because the patients placed on the protocols all had advanced cancer with concomitant depressed-immunologic responses. However, it was found that amphotericin B-BCNU therapy was more frequently associated with stimulation of immunological function than with depression and that the stimulation was most evident in tests of thymocyte function.

In summary, the results in human tumors are too preliminary to be properly evaluated. They appear encouraging, however, and tend to confirm the in vitro and animal results. Further studies are indicated and should include more polyenes and their semisynthetic derivatives. In a search for determinants of cell sensitivity, the polyene-induced changes in cell membranes should be investigated in detail. The animal studies, including investigation of immunoadjuvant effects and effects directed against cancer cells, should be expanded to other tumor models. Comparison of the action of polyenes from group I and group II may help to understand the mechanism of polyene action. A great deal more work has to be done, but we think that it is indicated because polyene antibiotics represent a class of drugs with unique properties and are of high interest as potential antitumor agents.

ACKNOWLEDGMENTS

Studies from this laboratory were supported by U.S. Public Health Service grants AI06213 and AI10622 and training grants AI00459 and AI07015; by the John A. Hartford Foundation, Inc.; and by the Research Corporation (Brown-Hazen Fund).

REFERENCES

1. **Hamilton-Miller, J. M. T.,** *Bacteriol. Rev.*, 37, 166, 1973.
2. **Pandey, R. C. and Rinehart, K. L.,** *J. Antibiot.*, 29, 1035, 1976.
3. **Thomas, A. H.,** *Analyst (London)*, 101, 321, 1976.
4. **Hazen, E. L. and Brown, R.,** *Science*, 112, 423, 1950.
5. **Lampen, J. O., Arnow, P. M., Borowska, Z., and Laskin, A. I.,** *J. Bacteriol.*, 84, 1152, 1962.
6. **Kinsky, S. C.,** *Proc. Natl. Acad. Sci.*, 48, 1049, 1962.
7. **Norman, A. W., Spielvogel, A. M., and Wong, R. G.,** Polyene antibiotic-sterol interaction, in *Advances in Lipid Research*, Vol. 2, Paoletti, R. and Kritchevsky, D., Eds., Academic Press, New York, 1976, 127.
8. **Bittman, R.,** *Lipids*, 13, 686, 1978.
9. **Hammond, S. M.,** *Prog. Med. Chem.*, 14, 105, 1977.
10. **Kitajima, Y., Sekiya, T., and Nozawa, Y.,** *Biochim. Biophys. Acta*, 445, 452, 1976.
11. **Gent, M. P. N. and Prestgard, J. H.,** *Biochim. Biophys. Acta*, 426, 17, 1976.
12. **Verkleij, A. J., De Kruijff, B., Gerritsen, W. F., Demel, R. A., Van Deenen, L. L. M., and Verver-gaert, P. H. J.,** *Biochim. Biophys. Acta*, 291, 577, 1973.
13. **Gomperts, B. D.,** Models for structure and function, in *The Plasma Membrane*, Academic Press, London, 1977, chap. 3.
14. **Van Hoogevest, P. and De Kruijff, B.,** *Biochim. Biophys. Acta*, 511, 397, 1978.
15. **Kobayashi, G. S. and Medoff, G.,** *Annu. Rev. Microbiol.*, 31, 291, 1977.
16. **Lampen, J. O. and Arnow, P.,** *Bull. Res. Counc. Isr. Sect. A.*, 114, 286, 1963.
17. **Kotler-Brajtburg, J., Medoff, G., Kobayashi, G. S., Boggs, S., Schlessinger, D., Pandey, R. C., and Rinehart, K. L., Jr.,** *Antimicrob. Agents Chemother.*, 15, 716, 1979.
18. **Kinsky, S. C.,** *Arch. Biochem. Biophys.*, 102, 180, 1963.
19. **Pressman, B. C.,** *Proc. Natl. Acad. Sci. U.S.A.* 53, 1076, 1965.
20. **McBride, J. A. and Jacob, H. S.,** *Br. J. Haematol.*, 18, 383, 1970.
21. **Amati, P. and Lago, C.,** *Nature*, 247, 466, 1974.
22. **Hidaka, K., Endo, H., Akiyama, S., and Kuwano, M.,** *Cell*, 14, 415, 1978.
23. **Hidaka, K., Matsui, K., Endo, H., Akiyama, S. I., and Kuwano, M.,** *Cancer Res.*, 38, 4650, 1978.
24. **Archer, D. B.,** *Biochim. Biophys. Acta*, 436, 68, 1976.
25. **Singer, M. A.,** *Can. J. Physiol. Pharmacol.*, 53, 1072, 1975.
26. **Kotler-Brajtburg, J., Medoff, G., Kobayashi, G. S., Schlessinger, D., and Atallah, A.,** *Biochem. Pharmacol.*, 26, 705, 1977.
27. **Cooper, R. A., Leslie, M. H., Fischkoff, S., Shinitzky, M., and Shattil, S. J.,** *Biochemistry*, 17, 327, 1978.
28. **Demel, R. A. and DeKruyff, B.,** *Biochim. Biophys. Acta*, 457, 109, 1976.
29. **Kotler-Brajtburg, J., Medoff, G., Schlessinger, D., and Kobayashi, G. S.,** *Antimicrob. Agents Chemother*, 11, 803, 1977.
30. **Kumar, V. J., Medoff, G., Kobayashi, G. S., and Schlessinger, D.,** *Nature*, 250, 323, 1974.
31. **Medoff, G., Comfort, M., and Kobayashi, G. S.,** *Proc. Soc. Exp. Biol. Med.*, 138, 571, 1971.
32. **Medoff, G., Kobayashi, G. S., Kwan, C. N., Schlessinger, D., and Venkov, P.,** *Proc. Natl. Acad. Sci. U.S.A.*, 69, 196, 1972.
33. **Kitahara, M., Kobayashi, G. S., and Medoff, G.,** *J. Infect. Dis.*, 133, 663, 1976.
34. **Medoff, G., Kwan, C. N., Schlessinger, D., and Kobayashi, G. S.,** *Antimicrob. Agents Chemother.*, 3, 441, 1973.
35. **Medoff, G., Schlessinger, D., and Kobayashi, G. S.,** *J. Natl. Cancer Inst.*, 50, 1047, 1973.
36. **Medoff, G., Kwan, C. N., Schlessinger, D., and Kobayashi, G. S.,** *Cancer Res.*, 33, 1146, 1973.
37. **Kuwano, M. and Ikehara, Y.,** *Exp. Cell Res.*, 82, 454, 1973.
38. **Nakashima, T., Kuwano, M., Matsui, K., Komiyama, S., Hiroto, I., and Endo, H.,** *Cancer Res.*, 34, 3258, 1974.
39. **Peterson, L. R., Christianson, L. D., Sarosi, G. A., and Costas-Martinez, C.,** *Current Ther. Res. Clin. Exp.*, 24, 858, 1978.
40. **Goldstein, M. N., Hamm, K., and Amrod, E.,** *Science*, 151, 1555, 1966.
41. **Medoff, J., Medoff, G., Goldstein, M. N., Schlessinger, D., and Kobayashi, G. S.,** *Cancer Res.*, 35, 2548, 1975.
42. **Mondovi, B., Strom, R., Agro, A. R., Caiafa, P., DeSole, P., Bozzi, A., Rotilio, G., and Fanelli, A. R.,** *Cancer Res.*, 31, 505, 1971.
43. **Medoff, G., Schlessinger, D., and Kobayashi, G. S.,** *Cancer Res.*, 34, 823, 1974.
44. **Amati, P. and Lago, C.,** *Cold Spring Harbor Symp. Quant. Biol.*, Part I, 39, 371, 1975.

43. Medoff, G., Schlessinger, D., and Kobayashi, G. S., *Cancer Res.*, 34, 823, 1974.
44. Amati, P. and Lago, C., *Cold Spring Harbor Symp. Quant. Biol.*,Part I, 39, 371, 1975.
45. Lago, C., Sartorius, B., Tramontano, D., and Amati, P., *J. Cell Physiol.*, 92, 265, 1977.
46. Blum, S. F., Shohet, S. B., Nathan, D. G., and Gardner, F. H., *J. Lab. Clin. Med.*, 73, 980, 1969.
47. Kitagawa, T. and Andoh, T., *Exp. Cell Res.*, 115, 37, 1978.
48. Ishikawa, H., Narimatsu, H., and Saito, K., *Cell. Immunol.*, 17, 300, 1975.
49. Ishikawa, H., Narimatsu, H., and Saito, K., *Microbiol. Immunol.*, 21, 137, 1977.
50. Hammarström, L. and Smith, C. I. E., *Acta Pathol. Microbiol. Scand.*, 85, 277, 1977.
51. Blanke, T. J., Little, J. R., Shirley, S. F., and Lynch, R. G., *Cell. Immunol.*, 33, 180, 1977.
52. Stein, S. H., Plut, E. J., Shine, T. E., and Little, J. R., *Cell. Immunol.*, 40, 211, 1978.
53. Shirley, S. F. and Little, J. R., *J. Immunol.*, in press.
54. Lin, S. H., Medoff, G., and Kobayashi, G. S., *Antimicrob. Agents Chemother.*, 11, 154, 1977.
55. Chapman, H. A., Jr. and Hibbs, J. B., Jr., *Proc. Natl. Acad. Sci. U.S.A.*, 75, 4349, 1978.
56. Little, J. R., Plut, E. J., Kotler-Brajtburg, J., Medoff, G., and Kobayashi, G. S., *Immunochemistry*, 15, 219, 1978.
57. Medoff, G., Schlessinger, D., and Kobayashi, G. S., *J. Natl. Cancer Inst.*, 50, 1047, 1973.
58. Medoff, G., Valeriote, F., Lynch, R. G., Schlessinger, D., and Kobayashi, G. S., *Cancer Res.*, 34, 974, 1974.
59. Medoff, G., Valeriote, F., Ryan, J., and Tolen, S., *J. Natl. Cancer Inst.*, 58, 949, 1977.
60. Valeriote, F., Lynch, R., Medoff, G., Tolen, S., and Dieckman, J., *J. Natl. Cancer. Inst.*, 61, 399, 1978.
61. Ducos, R. and Valeriote, F., *Transplantation*, 21, 279, 1976.
62. Valeriote, F., Medoff, G., and Dieckman, J., *Cancer Res.*, 39, 3051, 1979.
63. Valeriote, F. and Medoff, G., *Proc. Am. Assoc. Cancer Res.*, 20, 7, 1979.
64. Blanke, T., Lynch, R., and Little, J. R., *Surg. Forum*, 26, 126, 1975.
65. Little, J. R., Blanke, T. J., Valeriote, F., and Medoff, G., Immunoadjuvant and antitumor properties of amphotericin B, in *Immune Modulation and Control of Neoplasia by Adjuvant Therapy*, Chirigos, M. A., Raven Press, New York, 1978, 381.
66. Shirley, S. F. and Little, J. R., *J. Immunol.*, in press.
67. Valeriote, F., Lynch, R., Medoff, G., and Kumar, B. V., *J. Natl. Cancer Inst.*, 56, 557, 1976.
68. Laurent, G., Atassi, G., and Hildebrand, J., *Cancer Res.*, 36, 4069, 1976.
69. Iaremenko, K. V., Eremeeva, A. A., and Tereshin, I. M., *Antibiotiki*, 22, 153, 1977.
70. Muller, P. J. and Tator, C. H., *J. Neurosurg.*, 49, 579, 1978.
71. Presant, C. A., Klahr, C., and Santala, R., *Ann. Intern. Med.*, 86, 47, 1977.
72. Presant, C. A., Klahr, C., Olander, J., and Gatewood, D., *Cancer*, 38, 1917, 1976.
73. Vogl, S. E., *Proc. Am. Assoc. Cancer Res.*, 18, 169, 1977.
74. Krutchik, A. N., Buzdar, A. U., Blumenschein, G. R., and Sinkovics, J. G., *Cancer Treat. Rep.*, 62, 1565, 1978.
75. Nakazawa, I., Ouchi, E., Ishida, N., Ouchi, K., and Wagai, K., *Tohoku J. Exp. Med.*, 124, 97, 1978.

Chapter 6

CYTOTOXIC AND ANTITUMOR TERPENOIDS

Renuka Misra and Ramesh C. Pandey

TABLE OF CONTENTS

I. INTRODUCTION

The study of terpenoids[1] has attracted considerable attention in recent years because of our increasing knowledge of their structural diversity[2] and biological activities.[3-10] This is partly the result of a systematic plant-screening program,[11] developed by the National Cancer Institute around 1960, aiming to isolate and identify a broad spectrum of chemical compounds that might be useful as antitumor agents. The early investigations involved collection of the plant products with proper identification,[12] extraction with suitable solvents,[9,13,14] and testing of the crude extracts for anticancer activity.[15] Folklore literature was also examined for information on possible sources of antitumor activity.[16] Many groups of scientists are involved in this giant task, and a great number of extraction[9] and assay[15] procedures have been reported in the literature. As the program developed, many of the procedures were improved, both for extraction and biological assay. Today, cell culture screen (KB or 9KB)* and lymphocytic leukemia P-388 (PS, 3PS, P-388)** systems are used for in vitro and in vivo studies,[17] respectively. Further fractionation, purification, and isolation of pure material by procedures such as column chromatography, thin layer chromatography, countercurrent distribution, and high-pressure liquid chromatography, guided by bioassay have yielded a number of interesting anticancer compounds.[3-10,14,18]

The current Drug Research and Development Criteria for acceptable minimal activity[17] are: KB$\leqslant$ 4 μg/mℓ and PS (% T/C) $\geqslant$ 130. Once a compound has been found active in these systems, particularly in PS, it is tested in vivo against one or more animal tumors,[19] viz., lymphoid leukemia L-1210, Walker carcinosarcoma 256 (W-256), melanotic melanoma B16, Lewis lung carcinoma, mouse colon, human colon xenograft, mouse breast, human breast xenograft, mouse lung, and human lung xenograft. Because of the low isolation yield of the active constituents, in vivo work often cannot be done; therefore, only cytotoxicity (KB:ED_{50}) data are reported in the literature.[3-10]

This program has led to the isolation and identification of various groups of biologically active compounds, which include terpenoids, steroids, alkaloids, cardenolides, lignans, ansamacrolides, saponins, tannins, and phytosterols. The last two groups of compounds are invariably present in higher plants and are not regarded as potentially useful anticancer agents.[6] A number of reviews,[3-6,8] monographs,[10] and books,[7,9] have appeared covering different aspects of these natural products, including terpenoids, but there is no review which deals exclusively with the active terpenoids. Since terpen-

* KB is expressed in ED_{50}, ID_{50}, or IC_{50} and is the dose level in μg/mℓ at which 50% inhibition of growth in vitro is noted vs. untreated controls.

** Activity in PS is expressed by a number which is the ratio of average survival of treated animals to that of control, in days (T/C) × 100 or increase in life span (ILS) + 100.

oids from marine sources and vitamin A and other related tetraterpenoids have been dealt with separately in this book, these will not be discussed here. This discussion will be restricted to cytotoxic and antitumor terpenoids isolated from higher plants and ferns, i.e., spermatophytes (phanerogams) and pteridophytes. Active terpenoids isolated from lower forms, e.g., algae, fungi, mosses, and bacteria, will not be included.

Various groups which have made significant contributions to our understanding of the chemistry and antineoplastic activities of terpenoids have been led by (the late) S. M. Kupchan, W. Herz, M. E. Wall, and M. C. Wani, G. R. Pettit, K. H. Lee, J. L. Hartwell, G. A. Cordell, and N. R. Farnsworth.

The available vast amount of data could be organized on the basis of either the taxonomy of the plant source, or on the activity of the compound against test systems, or structure/activity relationship. The latter system would appear to be quite attractive and various speculations[6,20-22] have been made. However, not all the compounds can be classified on this basis. Therefore, we have adopted the simplest approach based on structural classification. The active terpenoids have been divided into five major classes, viz., monoterpenoids (C_{10}), sesquiterpenoids (C_{15}), diterpenoids (C_{20}), quassinoids (C_{20}), and triterpenoids (C_{30}). Each class is further divided into biogenetically related structural types. Some structure activity relations could then be drawn for the compounds within each related structural type.

Because of the limited space, only significant structural features are discussed. The biological activities are discussed and/or are shown with the structures, sometimes in a tabular form. In all figures, ED_{50}, ID_{50}, or IC_{50} always represents the dose level in $\mu g/m\ell$ at which 50% inhibition of growth in vitro is noted vs. untreated control. Wherever in vivo results are known, these are shown in abbreviated form just after ED_{50} results, and the values in the brackets indicate the percentage of the ratio of average survival of treated animals to that of control, in days (% T/C), or as ILS + 100. If these values are not known, only the test systems are reported.

II. MONOTERPENOIDS

Four naturally occurring monoterpenoids have been shown to have cytotoxic and/or in vivo activity. All of these (Figure 1), allamandin[23] (**1**), plumericin[23] (**2**), isoplumericin[23] (**3**), and penstemide[24] (**4**) are iridoids and contain an α, β-unsaturated-γ-lactone ring in their structure, which may be the active site.

III. SESQUITERPENOIDS

This is one of the largest groups of terpenoids which possess significant activity. It is noteworthy that, with only a few exceptions, all contain an α-methylene-γ-lactone ring in their structures. It has been suggested that this structural element is essential for biological activity.[5,22] Additional α, β-unsaturated carbonyls, or an increase in the lipophilic character, enhances the biological activity.[22] In order to compare the activity, we have classified the active sesquiterpenoids according to their carbon skeletons into germacranolides, guaianolides, pseudoguaianolides, elemanolides, eudesmanolides, and seco-eudesmanolides, and miscellaneous compounds. We will discuss each class separately.

A. Germacranolides

The characteristic features of this group, besides the α-methylene-γ-lactone ring, is a ten-membered carbocyclic ring oxygenated at various centers. In order to avoid confusion in identifying these, Rogers et al.[25,26] have recently suggested a classification for germacranolides based on the nature of Δ^4 and $\Delta^{1(10)}$ actual or masked double

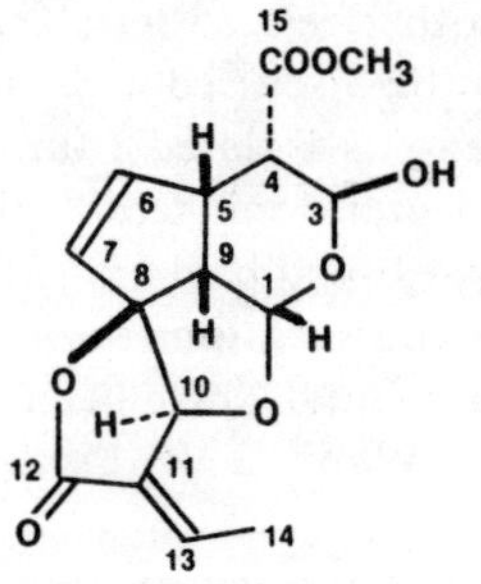

1 Allamandin
ED_{50} 2.1, PS(145)

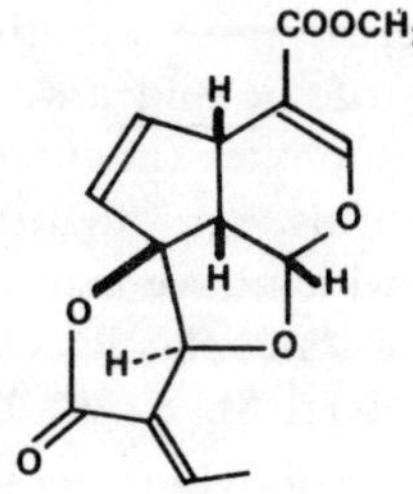

2 Plumericin
ED_{50} 2.7

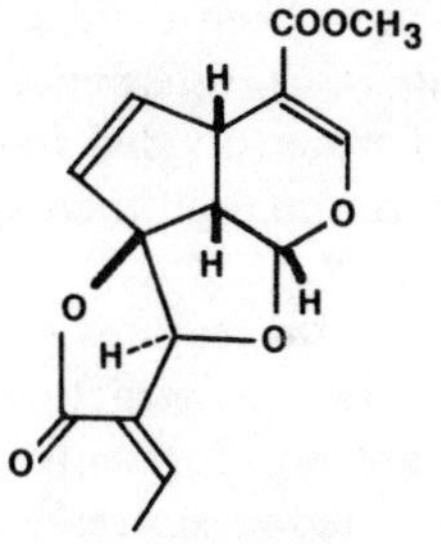

3 Isoplumericin
ED_{50} 2.6, PS(145)

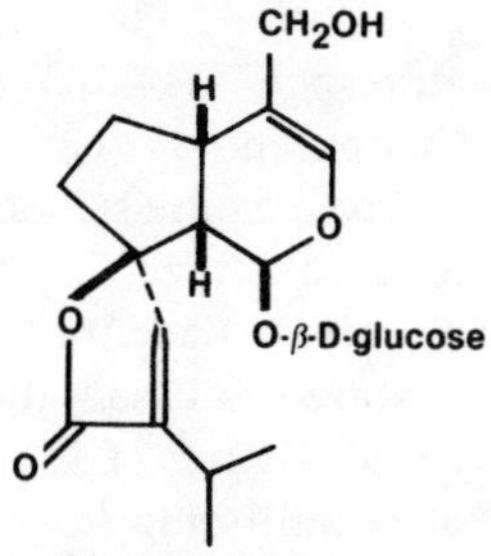

4 Penstemide
PS(184)

FIGURE 1. Allamandin,[23] plumericin,[23] isoplumericin,[23] and penstemide.[24]

bonds into germacrolides (Δ^4 and $\Delta^{1(10)}$ *trans, trans*) melampolides (Δ^4 and $\Delta^{1(10)}$ *trans, cis*), and heliangolides (Δ^4 and $\Delta^{1(10)}$ *cis,trans*) and also suggested a fourth group of Δ^4, $\Delta^{1(10)}$ *cis, cis* germacranolides. We here, propose the name phantomolides for this group after the biologically active member phantomolin **(53)**. All the active germacranolides here thus have been divided into germacrolides, heliangolides and phantomolides.

1. Germacrolides

At least 29 germacrolides have been found to have cytotoxic and/or antitumor activity. The orientation of the α-methylene-γ-lactone ring in this group has been either to C-6 (Figures 2 to 6 and 9) or to C-8 (Figures 7 and 8). Costunolide[20,27,28] **(5)** isolated from *Saussurea lappa* Clarke and *Liriodendron tulipifera* L. is the simplest member of this group. It has C-6 α-oriented lactone ring and shows inhibitory activity against the KB cell culture of a human carcinoma of nasopharynx (ED_{50} 0.26). Tamaulipins[22,29] A and B **(6)** and **(7)** isolated from *Ambrosia confertiflora* DC, eupatolide[30] **(8)** isolated from *Eupatorium formosanum* HAY, tulipinolide[20,28,31] **(9)** and epitulipinolide[31] **(10)** isolated from *L. tulipifera* L., eupaserrin[32] **(11)**, and deacetyleupaserrin[32] **(12)** isolated from *E. semiserratum* DC, all could be considered derivatives of costunolide and are probably formed by oxidation at C-2, C-3, or C-8. The structures of these with their activity are shown in Figure 2. Only eupaserrin **(11)** and deacetyleupasernin **(12)** show significant in vivo antileukemic activity (P-388) in mice at 30 and 18 mg/kg dose levels, respectively. This activity could be attributed to the presence of acetylsarracinyl and sarracinyl groups at C-8 in **(11)** and **(12)**, respectively.

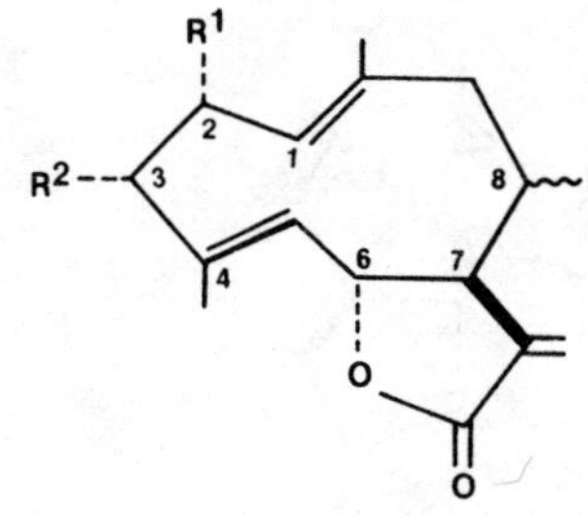

	R^1	R^2	R^3	ED_{50}	
5 Costunolide	H	H	H	0.26	
6 Tamaulipin A	OH	H	H	1.26	
7 Tamaulipin B	H	OH	H	2.60	
8 Eupatolide	H	H	β-OH	0.034	
9 Tulipinolide	H	H	α-OAc	0.46	
10 epi-Tulipinolide	H	H	β-OAc	2.1	
11 Eupaserrin	OH	H	$O.COC(CH_2OCOCH_3){=}CHCH_3$	0.23 ,	PS
12 Deacetyleupaserrin	OH	H	$O.COC(CH_2OH){=}CHCH_3$	0.29,	PS

FIGURE 2. Costunolide,[20,27,28] tamaulipin A,[22,29] tamaulipin B,[22,29] eupatolide,[30] tulipinolide,[20,28,31] epitulipinolide,[31] eupaserrin,[32] and deacetyleupaserrin.[32]

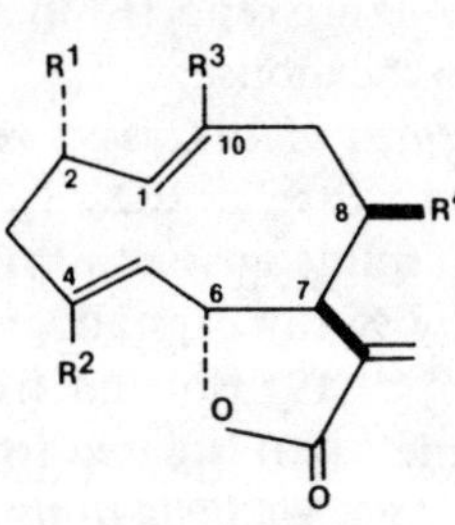

	R^1	R^2	R^3	R^4	ED_{50}
13 Eriofertopin	OH	CH_3	CH_2OH	$O.CO.C(CH_3){=}CH_2$	1.2 , PS(167)
14 2-Acetyleriofertopin	OAc	CH_3	CH_2OH	$O.CO.C(CH_3){=}CH_2$	1.75, PS(130)
15 Ovatifolin	H	CH_3	CH_2OAc	OH	KB, PS(143)
16 Alatolide	H	CH_2OH	CH_2OH	$O.CO.CH(CH_3)_2$	0.8~1.8
17 Euserotin	H	COOH	CH_3	$O.CO.CH{=}C(CH_3){-}C_2H_5$	
18 Cnicin	H	CH_2OH	CH_3	$O.CO.C({=}CH_2){-}C(CH_2CH_2OH){-}CHOH$	0.1 (Hela Cells)

FIGURE 3. Eriofertopin,[33] 2-acetyleriofertopin,[33] ovatifolin,[34] alatolide,[35] euserotin,[36] and cnicin.[37]

Figure 3 shows germacrolides in which, besides oxygenation in the ring, the methyl groups at C-4 and/or C-10 are also oxygenated. Among these, eriofertopin[33] (13) and

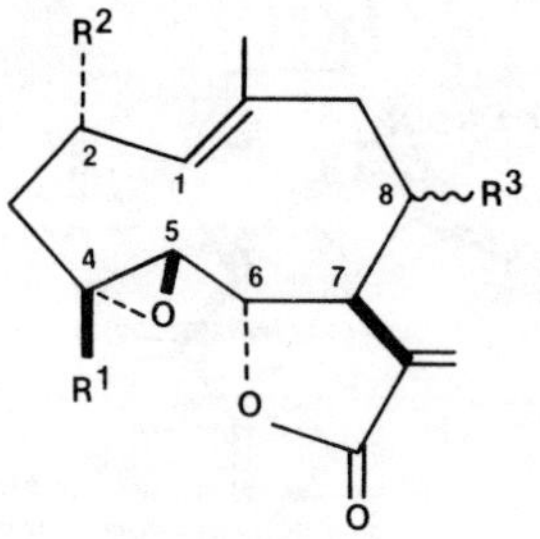

	R^1	R^2	R^3	ED_{50}	
19 Parthenolide	CH_3	H	H	0.45	
20 Lipiferolide	CH_3	H	β-OAc	0.16,0.18	
21 Dehydrolanuginolide	CH_3	H	α-OAc		
22 Eupahyssopin	CH_2OH	H	β-$OCOC(CH_3)=CHCH_2OH$		PS(132), WA(330), LL(147)

FIGURE 4. Parthenolide,[38-40] lipiferolide,[41] dehydrolanuginolide,[42] and eupahyssopin.[43,44]

2-acetyleriofertopin[33] **(14)** isolated from *Eriophyllum confertiflorum* Gray and ovatifolin[34] **(15)** from *Podanthus ovatifolius* are reported to have significant in vivo activity against lymphocytic leukemia (PS) in the mouse.

Alatolide[35] **(16)** isolated from *Jurinea alata* Cass is reported to have significant cytotoxic activity against the HeLa and KB cells (ED_{50} 1.3 and 0.8 μg/mℓ, respectively). The structure of the unusual germacrolide euserotin[36] **(17)** isolated from *Eupatorium serotinum* Michx. was established by X-ray crystallography. The germacrolide monoepoxides (Figure 4), parthenolide[38-40] **(19)** isolated from *Michelia Champaca*[38] and *Magnolia grandiflora* L.,[39] lipiferolide[41] **(20)** isolated from *Liriodendron tulipifera* L., dehydrolanuginolide[42] **(21)** isolated from *Michelia doltsopa* (Buch.-Ham.) ex DC., and eupahyssopin[43,44] **(22)** isolated from *Eupatorium hyssopifolium* L. all have been reported to have significant in vitro or in vivo activity. As expected, because of the α-β-unsaturated side chain at C-8, eupahyssopin **(22)** in particular is reported to be active against Walker 256 carcinosarcoma Sprague-Dawley rat tumor system at 2.5 mg/kg/day dose level (T/C 330), lymphocytic leukemia (PS) screen in DBA/2 mice at 25 mg/kg/day dose level (T/C 132), and in Lewis lung screen in $C_{57}Bl_6$ mice at 25 mg/kg/day dose level (T/C 147). It caused 93% inhibition of Ehrlich ascites cell growth in CF_1 mice at 25 mg/kg/day dose level.

Two diepoxides (Figure 5), epitulipinolide diepoxide[41] **(23)** isolated from *L. tulipifera* L. and michelenolide[40] **(24)** isolated from *Michelia compressa* (Maxim.) Sarg. are reported to be cytotoxic (ED_{50} 0.34 and 1.0 μg/mℓ, respectively). No in vivo work has been reported on these compounds. Elephantin[22,45] **(25)**, elephantopin[22,45] **(26)** and deoxyelephantopin[44,46] **(27)** (Figure 6) isolated from *Elephantopus elatus* Bertol. are dilactones with a C-4, C-5 epoxide or unsaturation and an α-β-unsaturated ester side chain. These show both in vitro and in vivo activity against a number of test systems. The dilactone diepoxide containing germacrolide mikanolide[20,47] **(28)** (Figure 7) isolated from *Mikania scandens* (L.) Willd.[47] has been shown cytotoxic (ED_{50} $<$ 1μg/mℓ).[20] In this compound, the α-methylene-γ-lactone ring is closed to C-8 instead to C-6 as in the earlier discussed germacrolides. Two other members in which the orientation

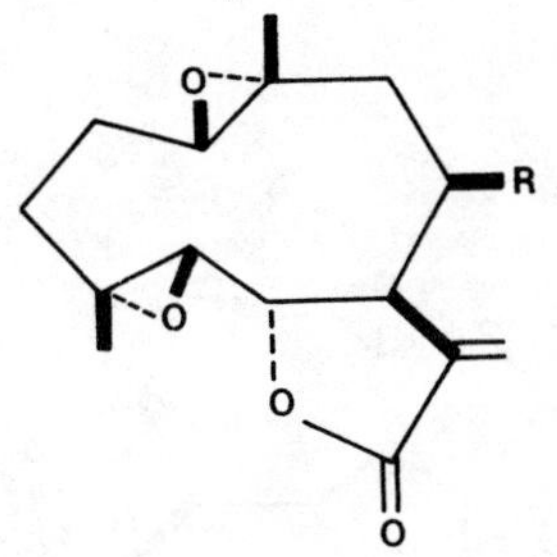

		R	ED_{50}
23	Epitulipinolide diepoxide	OAc	0.34
24	Michelenolide	H	1.0

FIGURE 5. Epitulipinolide diepoxide[41] and michelenolide.[40]

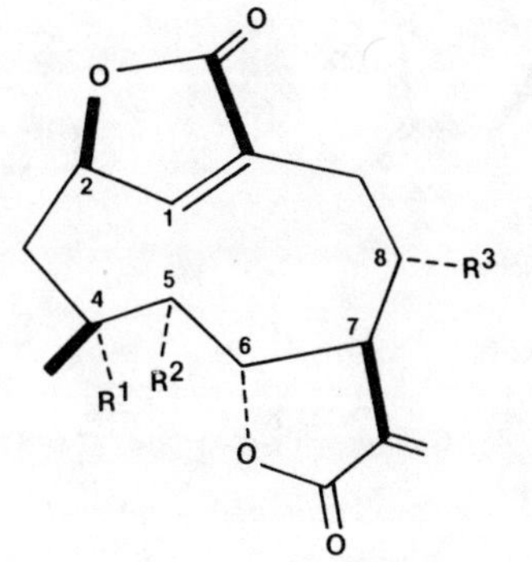

		R^1	R^2	R^3	ED_{50}	
25	Elephantin	—O—		O.CO.C = C(CH$_3$)$_2$	0.28,	WA(12)
26	Elephantopin	—O—		O.CO.C(CH$_3$) = CH$_2$	0.32,	WA(22), PS(140)
27	Deoxyelephantopin	—Δ—		O.CO.C(CH$_3$) = CH$_2$		WM(226) PS(123)

FIGURE 6. Elephantin,[22,45] elephantopin,[22,45] and deoxyelephantopin.[44,46]

of α-methylene-γ-lactone ring has changed from C-6 to C-8 are chamissonin diacetate[22,48] **(29)** and baileyin[49,50] **(30)**. These were isolated from *Baileya multiradiata* Harv. and Gray (Figure 8), and are cytotoxic (ED_{50} 2.13 and 2.9 μg/mℓ, respectively). Very recently, peroxycostunolide[51] **(31)**, peroxyparthenolide[51] **(32)** isolated from *Magnolia grandiflora* L. and peroxyferolide[52,53] **(33)** isolated from *Liriodendron tulipifera* L. (Figure 9) have been shown to be cytotoxic (ED_{50} 2.7 μg/mℓ, 2.8 μg/mℓ and 0.29 μg/mℓ, respectively).

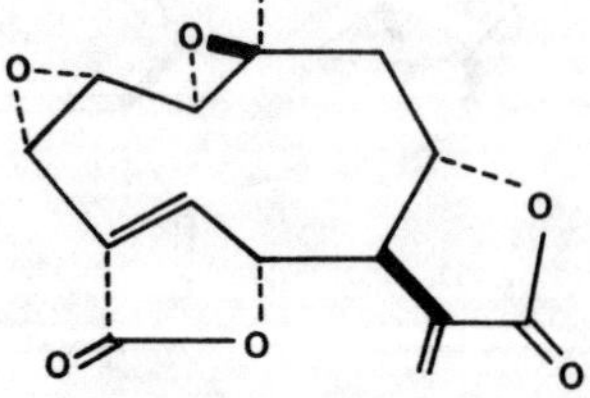

28 Mikanolide
ED_{50} <1.0

FIGURE 7. Mikanolide.[20,47]

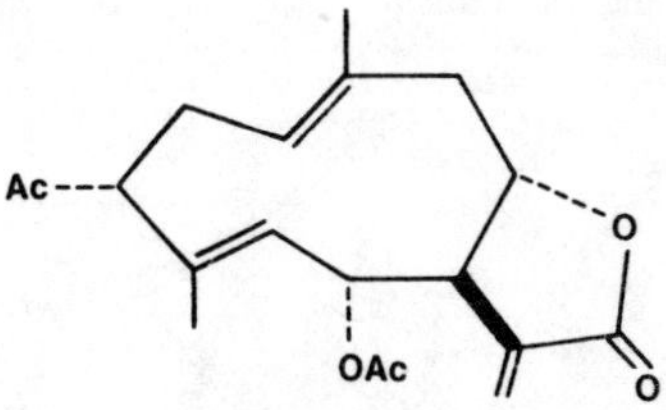

29 Chamissonin diacetate
ED_{50} 2.13

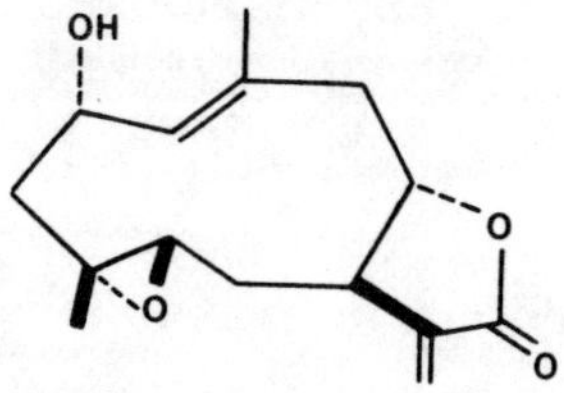

30 Baileyin
ED_{50} 2.9(P-388)
PS(109)

FIGURE 8. Chamissonin diacetate[22,48] and baileyin.[49,50]

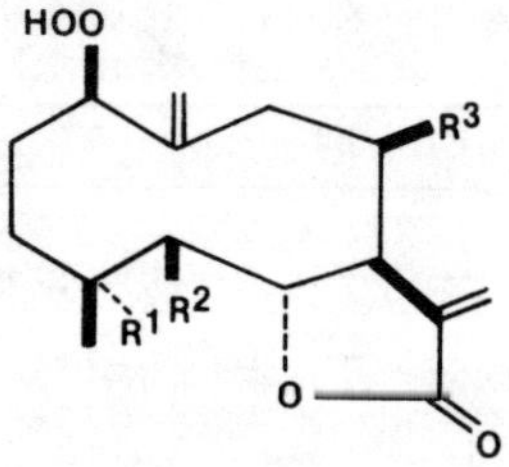

	R^1	R^2	R^3	ED_{50}
31 Peroxycostunolide	—Δ—		H	2.7
32 Peroxyparthenolide	—O—		H	2.8
33 Peroxyferolide	—O—		OAc	0.29

FIGURE 9. Peroxycostunolide,[51] peroxyparthenolide,[51] and peroxyferolide.[52,53]

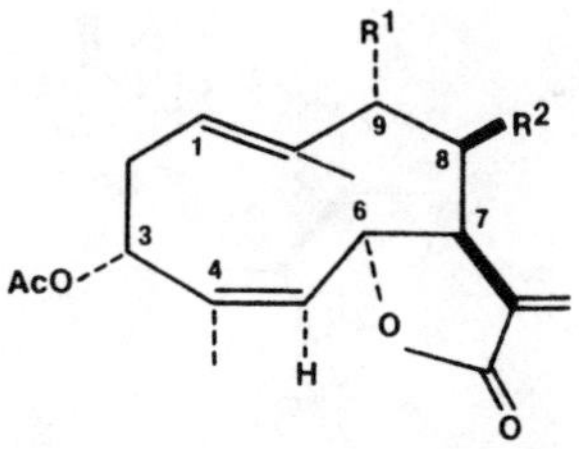

	R^1	R^2	ED$_{50}$	
34 Eupaformin	H	OH	Active (H.Ep.-2)	
35 Provincialin	H	O.CO.C(CH$_2$O.C(CH$_2$OH)=CHCH$_3$)=CHCH$_2$OH	3.5	
36 Eupatocunin	OH	O.CO.C(CH$_3$)=CHCH$_3$	0.11,	PS(135)
37 Eupatocunoxin	O.CO.C(CH$_3$)—(O)—CH.CH$_3$	OH	1.7	
38 Eupaformosanin	H	O.CO.C(CH$_2$OH)=CHCH$_2$OH		WA(471) PS(147)

FIGURE 10. Eupaformin,[54,55] provincialin,[56] eupatocunin,[57,58] eupatocunoxin,[58] and eupaformosanin.[44,59]

2. Heliangolides

About 16 active heliangolides have been reported (Figures 10 to 16). As discussed earlier, the characteristic features of this group of compounds are, in contrast with the germacrolides, that they contain a Δ^4 *cis* double bond and a $\Delta^{1(10)}$ *trans* or masked double bond and that all have C-6-oriented α-methylene-γ-lactone ring (which, though, is not a criteria to be a heliangolide).

Eupaformin[54,55] **(34)** isolated from *Eupatorium formosanum* HAY is the simplest member of this group and its structure (Figure 10) was established by a single crystal X-ray analysis.[54] It is reported to show significant in vitro activity against human epidermoid carcinoma of larynx (H. Ep.-2).[55] The other members of this group which have Δ^4 *cis* and $\Delta^{1(10)}$ *trans* double bonds, provincialin[56] **(35)** isolated from *Liatris provincialis* Godfrey, eupatocunin[57,58] **(36)** isolated from *Eupatorium cuneifolium* (Tourn.) L., eupatocunoxin[58] **(37)** isolated from *E. cuneifolium* Willd., and eupaformosanin[44,59] **(38)** isolated from *E. formosanum* Hay, all show significant activity. Their structures and activities are reported in Figure 10. Provincialin **(35)** possesses an unusual C_{10} ester side chain comprising of two C_5 acyl units. Erioflorin[34,60] **(39)** isolated from *Eriophyllum confertiflorum* Gray and *Podanthus ovatifolius* Lag., erioflorin acetate[34] **(40)** and erioflorin methacrylate[34] **(41)** isolated from *P. ovatifolius* Lag. all contain a C-1 (10)-epoxide group and have a β-oxygenation at position C-3. Their structures and activities are shown in Figure 11. The structure of the cytotoxic heliangolide diepoxide eleganin[61,62] **(42)** isolated from *Liatris elegans* (Walt.) Michx. and *L. scabra* (Greene) K. Schum. is shown in Figure 12.

Lee et al. have recently reported[63-65] on two novel cytotoxic germacranolides, molephantinin[44,63] **(43)** and molephantin[44,63-65] **(44)** isolated from *Elephantopus mollis* H.S. K. They contain a C-2 carbonyl group and differ in the configuration of the C-5 hydroxyl group. The structure of molephantin **(44)** was established by physicochemical

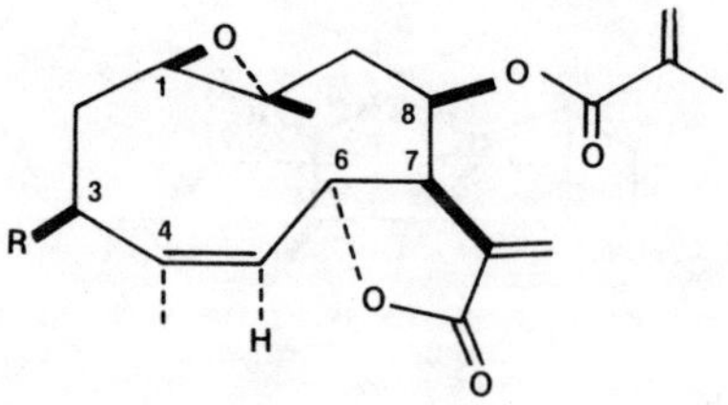

	R	ED_{50}	
39 Erioflorin	OH		
40 Erioflorin acetate	OAc	KB	PS(131)
41 Erioflorin methacrylate	$OCOC(CH_3)=CH_2$		PS

FIGURE 11. Erioflorin,[34,60] erioflorin acetate,[34] and erioflorin methacrylate.[34]

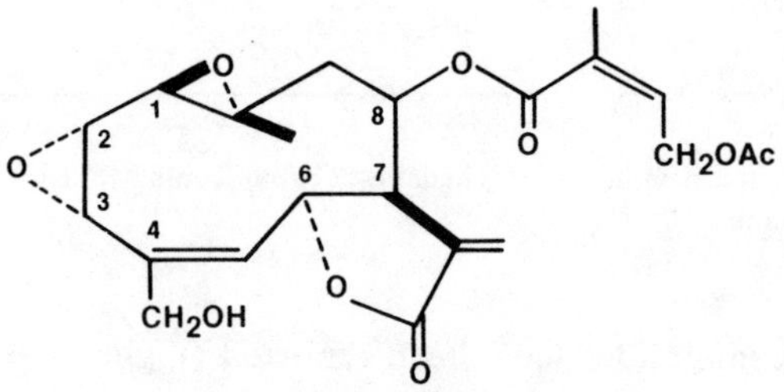

42 Eleganin
KB active

FIGURE 12. Eleganin.[61,62]

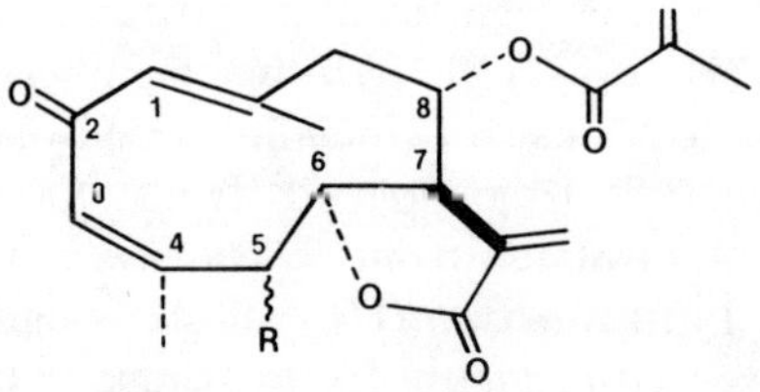

	R	ED_{50}	
43 Molephantinin	β-OH		WA(397),LL(123), PS(146)
44 Molephantin	α-OH	0.33 (H.Ep.-2)	WA(149),LL(80),PS(118)

FIGURE 13. Molephantinin[44,63] and molephantin.[44,63-65]

data and X-ray crystallography.[65] Their structures and biological activities are shown in Figure 13.

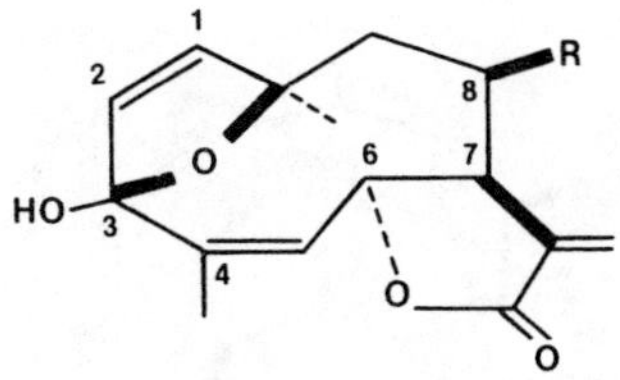

	R	ED_{50}	
45 Liatrin	$O.CO.C(CH_2OAc)=CHCH_3$	1.5,	PS
46 Tagitinin F	$O.CO.CH\ (CH_3)_2$		PS(161)

FIGURE 14. Liatrin[66,67] and tagitinin F.[68,69]

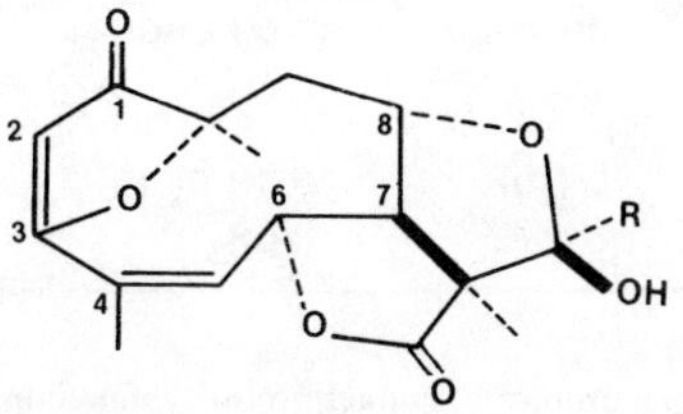

	R	ED_{50}
47 Ermantholide A	$-CH\ (CH_3)_2$	2.0
48 Ermantholide B	$-CH(CH_3)-C_2H_5$	2.0

FIGURE 15. Ermantholide A[70,71] and ermantholide B.[70,71]

Kupchan et al. reported[66] in 1971 that the chloroform extract of *Liatris chapmanii* shows significant in vitro activity. Fractionation of this extract led to the isolation of an antileukemic (PS) sesquiterpenoid lactone, liatrin[66,67] **(45)** (Figure 14), whose structure was determined by X-ray crystallography.[66,67] Another member of this group tagitinin, F[68,69] **(46)** (Figure 14) was soon isolated from *Tithonia tagitiflora* Desf. It also showed significant activity against P-388 lymphocytic leukemia. (T/C 161, 155, and 161 at dose levels of 50, 25, and 12.5 mg/kg, respectively).[68]

The eremantholides[70,71] A and B **(47)** and **(48)** (Figure 15) isolated from *Eremanthus elaeagnus* (Schultz-Bip.) do not contain the α,β-unsaturated-γ-lactone, which grouping has been regarded as the site of nucleophilic attack by enzymes and/or nucleic acids. Reactions with propane-1-thiol have suggested a Michael-type addition to the 4,5-double bond. It is suggested that eremantholides are novel modified germacranolide derivatives, rather than modified diterpenoids.[71] The structure of eremantholide A was determined by X-ray crystallography.[70,71]

Eupacunin[22,57] **(49)**, eupacunoxin[57] **(50)** isolated from *Eupatorium cuneifolium* (Tourn.) L., eupacunolin[58] **(51)** isolated from *E. cuneifolium* Willd. and eurecurvin[72,73] **(52)** isolated from *E. recurvans* Small., *E. anomalum* Nash., and *E. mohrii* Greene

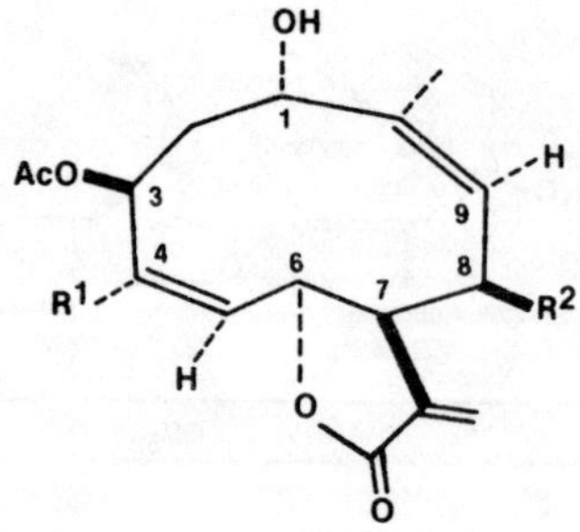

	R^1	R^2	ED_{50}	
49 Eupacunin	CH_3	O.CO.C(CH$_3$)=CHCH$_3$	2.1 0.84	PS(135) WA(39)
50 Eupacunoxin	CH_3	O.CO.C(CH$_3$)(–O–)CHCH$_3$	2.1	
51 Eupacunolin	CH_2OH	O.CO.C(CH$_3$)=CHCH$_3$	3.7	
52 Eurecurvin	CH_2OH	O.CO.CH(CH$_3$) C_2H_5		

FIGURE 16. Eupacunin,[22,57] eupacunoxin,[57] eupacunolin,[58] and eurecurvin.[72,73]

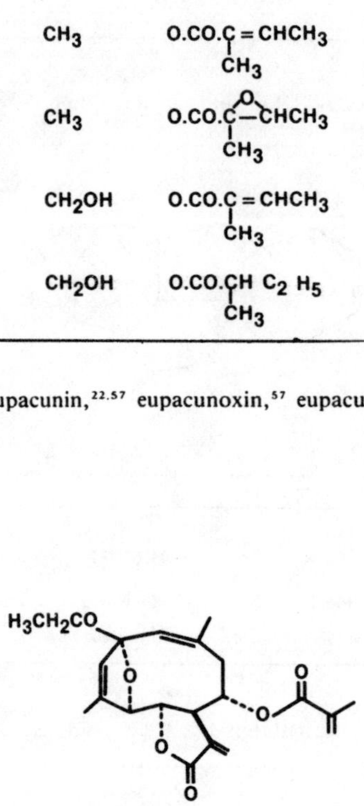

53 Phantomolin
ED_{50} 0.66 (H.Ep.-2)
WM-256(378), LL(123), PS(113)

FIGURE 17. Phantomolin.[64]

are all Δ^4 *cis*, Δ^9 *cis*, germacranolides (Figure 16). They could be considered to contain $\Delta^{1(10)}$ masked *trans* double bond, therefore regarded as heliangolides. Their structures and biological activities are shown in Figure 16.

3. Phantomolides

As discussed earlier, this group is characterized by the Δ^4 *cis*, $\Delta^{1(10)}$ *cis* double bonds in the germacranolide structure. The only biologically active member of this group phantomolin[64] **(53)** (Figure 17) isolated from *Elephantopus mollis* and having $\Delta^{3,4}$ *cis*, $\Delta^{1(10)}$ *cis* double bonds is considered here to have a masked Δ^4 *cis* bond and is, there-

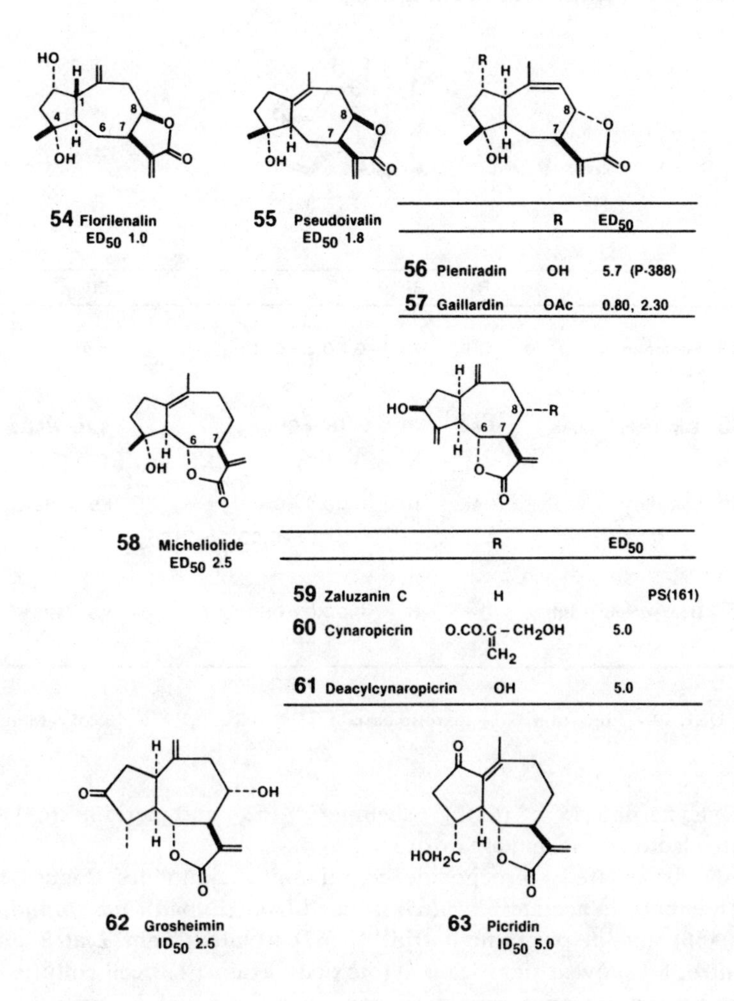

	R	ED_{50}
56 Pleniradin	OH	5.7 (P-388)
57 Gaillardin	OAc	0.80, 2.30

	R	ED_{50}
59 Zaluzanin C	H	PS(161)
60 Cynaropicrin	O.CO.C(=CH$_2$)–CH$_2$OH	5.0
61 Deacylcynaropicrin	OH	5.0

FIGURE 18. Florilenalin,[74] pseudoivalin,[20,75] pleniradin,[76] gaillardin,[20,22,77-79] micheliolide,[40] zaluzanin C,[80,81] cynaropicrin,[37,82-84] deacylcynaropicrin,[37,84] grosheimin,[37,83] and picridin.[37]

fore, placed in this group. It shows significant in vitro activity against human epidermoid carcinoma of larynx (H. Ep.-2) at 0.66 μg/mℓ dose level.

B. Guaianolides

Guaianolides are bi-cyclic sesquiterpenoids with a five-membered and a seven-membered ring, and an α-methylene-γ-lactone ring closed to either C-6 or C-8. The bi-cyclic ring can be oxygenated at various centers leading to different active components. These compounds have been grouped based on the similarities in their structures as shown in Figures 18 to 23 with the biological activities.

Figure 18 shows the simplest members of this class. Florilenalin[74] **(54)** isolated from *Helenium autumnale* L. and pseudoivalin[20,75] **(55)** isolated from *Iva microcephala* Nutt. both contain a *cis*-fused α,β-unsaturated-γ-lactone ring closed at C-8 and possess significant in vitro activity (ED_{50} 1.0 and 1.8 μg/mℓ, respectively). The structure of florilenalin **(54)** was proved by X-ray analysis.[74] Pleniradin[76] **(56)** isolated from *Baileya multiradiata* and gaillardin[20,22,77-79] **(57)** isolated from *Gaillardia pulchella* Foug. have a *trans*-fused lactone closed at C-8. Gaillardin **(57)**, which is the acetyl derivative of pleniradin **(56)** also possesses significant activity in vitro (ED_{50} 0.80 μg/mℓ).[20] The other members of this group micheliolide[40] **(58)**, zaluzanin C[80,81] **(59)**, cynaropicrin[37,82-

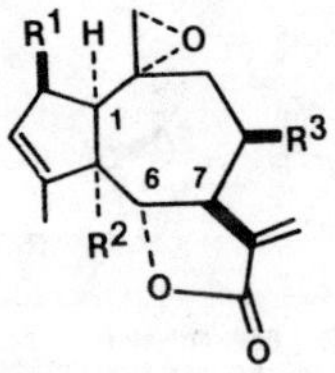

	R_1	R_2	R_3	ED_{50}
64 Euparotin	OH	OH	O.CO.C = CHCH$_3$ CH$_3$	0.21
65 Euparotin acetate	OAc	OH	OCOC = CHCH$_3$ CH$_3$	0.21, WA(23)
66 Spicatin	OAc	H	O.CO.C = CHCH$_3$ CH$_2$O.CO.C = CHCH$_3$ CH$_2$OH	KB , PS
67 Deoxygraminiliatrin	OH	H	O.CO.C = CHCH$_2$OAc CH$_3$	KB , PS

FIGURE 19. Euparotin,[20,85] euparotin acetate,[20,85,86] spicatin,[62,87] and deoxygraminiliatrin.[62,87]

[84] **(60)**, deacylcynaropicrin[37,84] **(61)**, grosheimin[37,83] **(62)**, and picridin[37] **(63)** all contain a C-6-oriented lactone ring and are cytotoxic.

The highly oxygenated spiroepoxide containing guaianolides (Figure 19) euparotin[20,85] **(64)**, euparotin acetate[20,85,86] **(65)** isolated from *Eupatorium rotundifolium* L., spicatin[62,87] **(66)** and deoxygramniliatrin[62,87] **(67)** isolated from *Liatris graminifolia* (Walt.) Kuntze. all showed significant cytotoxicity against KB cell culture. Euparotin acetate **(65)** has also been reported to have significant inhibitory activity against Walker carcinosarcoma 256 in rats at 75 mg/kg.[86] Further epoxidation of the spiroepoxides at C-3, C-4 seems to reduce the in vitro cytotoxicity of these compounds, as is observed in the case of eupatoroxin[85] **(68)**, graminiliatrin[62,87] **(69)** and 10-epi-eupatoroxin[85] **(70)** (Figure 20).

The naturally occurring chlorosesquiterpenoids eupachlorin[85] **(71)**, acetyleupachlorin[85,88] **(72)**, isolated from *E. rotundifolium* L. and centaurepensin[89,90] **(73)** isolated from *Centaurea solstitialis* L. (Figure 21) all show significant cytotoxic activity (ED_{50} 0.21, 0.18 and 1.2 μg/mℓ dose level, respectively). Again, epoxidation of Δ^3 double bond seems to reduce the activity, as is observed in eupachloroxin **(75)** (Figure 22). Hydroxy isopatchoulenone[91] **(77)** (Figure 23) isolated from *Pleocarpus revolutus* has recently been shown to have moderate in vitro activity. Its structure, **(77)**, has an α,β-unsaturated cyclopentenone moiety and does not contain any lactone ring.

C. Pseudoguaianolides

In this group of compounds, the methyl group is on C-4 instead of C-5, as in the guaianolides. Based on the orientation of the C-10 methyl group, these are further divided into ambrosanolides (C-10 methyl is β-oriented) and helenanolides, the characteristic feature of which is an α-oriented C-10 methyl group.

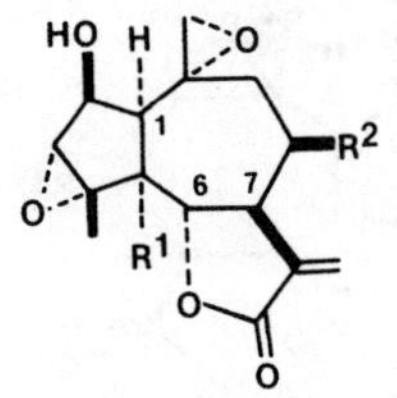

	R_1	R_2	ED_{50}
68 Eupatoroxin	OH	O.CO.C(CH$_3$)=CHCH$_3$	2.8
69 Graminiliatrin	H	O.CO.C(CH$_3$)=CHCH$_2$OAc	KB, PS

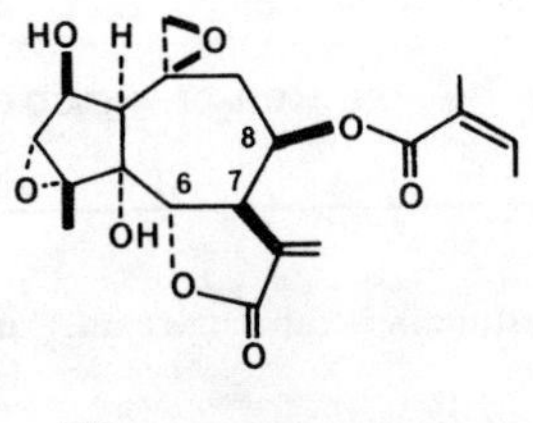

70 10-epi-Eupatoroxin
ED_{50} 2.6

FIGURE 20. Eupatoroxin,[85] graminiliatrin,[62,87] and 10-epi-eupatoroxin.[85]

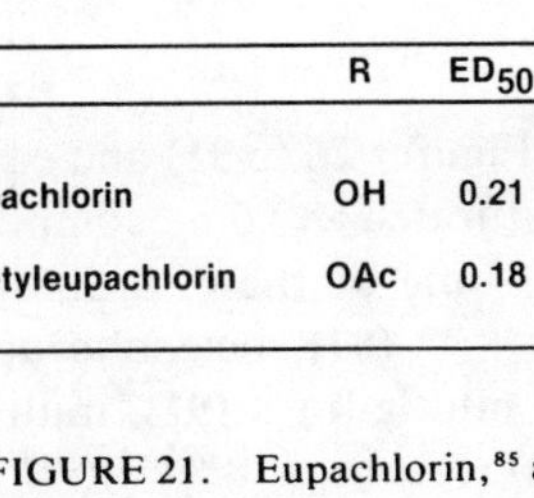

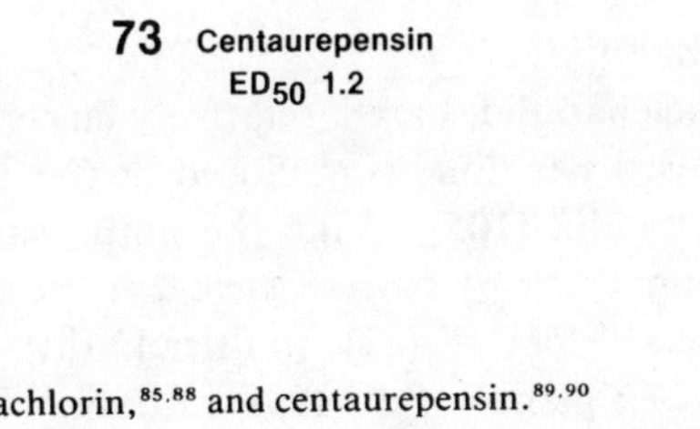

	R	ED_{50}
71 Eupachlorin	OH	0.21
72 Acetyleupachlorin	OAc	0.18

73 Centaurepensin
ED_{50} 1.2

FIGURE 21. Eupachlorin,[85] acetyleupachlorin,[85,88] and centaurepensin.[89,90]

1. Ambrosanolides

This is a small group of compounds which contain an α,β-unsaturated-γ-lactone and a cyclopentanone or a cyclopentenone moiety in their structures, similar to several guaianolides. These are shown in Figures 24 and 25.

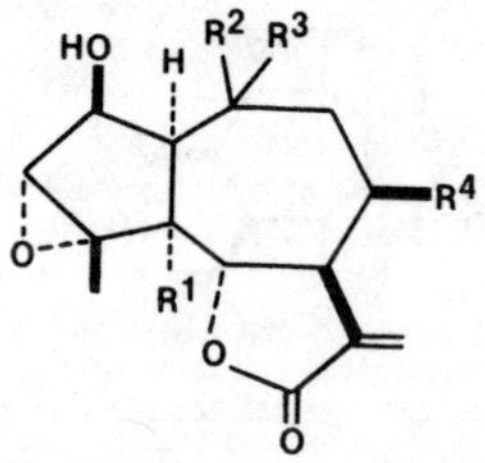

	R^1	R^2	R^3	R^4	ED_{50}
74 Eupatundin	OH	—CH_2—		O.CO.C(CH$_3$)=CHCH$_3$	0.39
75 Eupachloroxin	OH	α-OH	β-CH_2Cl	O.CO.C(CH$_3$)=CHCH$_3$	3.6
76 Graminichlorin	H	α-OH	β-CH_2Cl	O.CO.C(CH$_3$)=CHCH$_2$OH	

FIGURE 22. Eupatundin,[85] eupachloroxin,[85] and graminichlorin.[87]

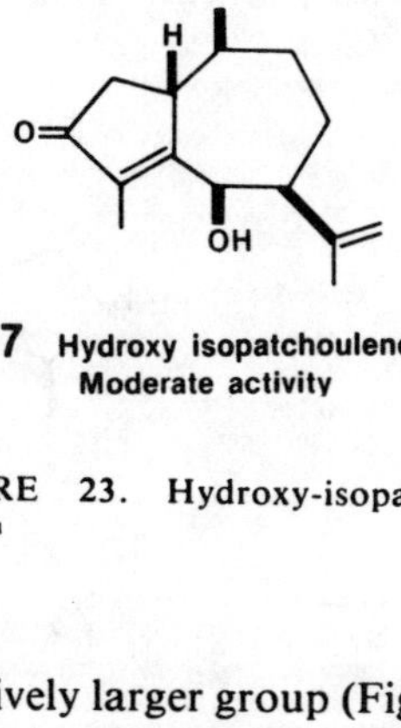

77 Hydroxy isopatchoulenone
Moderate activity

FIGURE 23. Hydroxy-isopatchoulenone.[91]

2. *Helenanolides*

The helenanolides are a relatively larger group (Figures 26 to 31) and contain a *trans-* or *cis-*fused γ-lactone ring closed at C-8 and is saturated in some compounds [Figure 28 and tenulin **(105)**]. Like the ambrosanolides, some of these, aromaticin[22,99] **(85)**, mexicanin I[22,100,101] **(86)** (Figure 26); helenalin[99,102-105] **(87)**, helenalin acetate[106] **(88)**, fastigilin C[20,49,107,108] **(89)**, multiradiatin[49,108] **(90)**, multigilin[108] **(91)**, multistatin[108] **(92)** (Figure 27); plenolin[44,109] **(93)**, radiatin[49,108,110] **(94)**, fastigilin B[20,49,107,108] **(95)**, hymenoflorin[111] **(96)**, microhelenin B[112] **(97)**, microhelenin C[112] **(98)**, fastigilin A[107,108] **(99)** (Figure 28); isohelenol[113] **(100)** (Figure 29), tenulin[44,118-120] **(105)** (Figure 30), and microlenin[122,123] **(107)** (Figure 31), have an α,β-unsaturated cyclopentenone moiety in their structure. Most of these are oxygenated at C-6 and are α,β-unsaturated ester derivatives at that position.

The most active component of this group, helenalin (87) isolated from the common sneezeweed *Helenium autumnale* L. var. *montanum* (Nutt.) Fern. has been shown to have significant antitumor activity against Walker 256 ascites carcinosarcoma

	R^1	R^2	R^3	ED_{50}
78 Damsin	H	H	H	0.58 ,0.032
79 3-Hydroxydamsin	OH	H	H	2.65
80 Confertiflorin	H	H	OAc	
81 Desacetylconfertiflorin	H	H	OH	2.30
82 Coronopilin	H	OH	H	1.45

FIGURE 24. Damsin,[20,92,93] 3-hydroxydamsin,[22,94] confertiflorin,[95] desacetylconfertiflorin,[22,95] and coronopilin.[20,22,96]

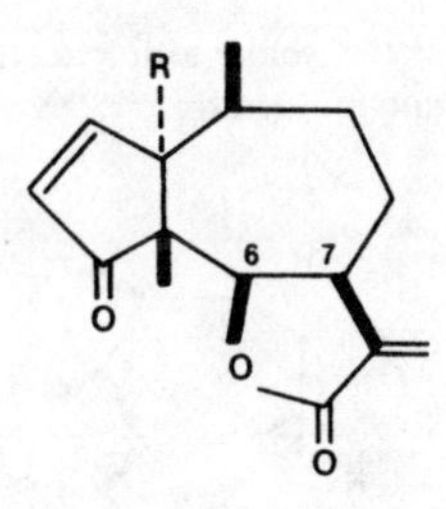

	R	ED_{50}
83 Ambrosin	H	0.04 , PS(180)
84 Parthenin	OH	0.025

FIGURE 25. Ambrosin[20,22,97-99] and parthenin.[20,22,98]

	R	ED_{50}
85 Aromaticin	H	0.34
86 Mexicanin I	OH	0.33

FIGURE 26. Aromaticin[22,99] and mexicanin I.[22,100,101]

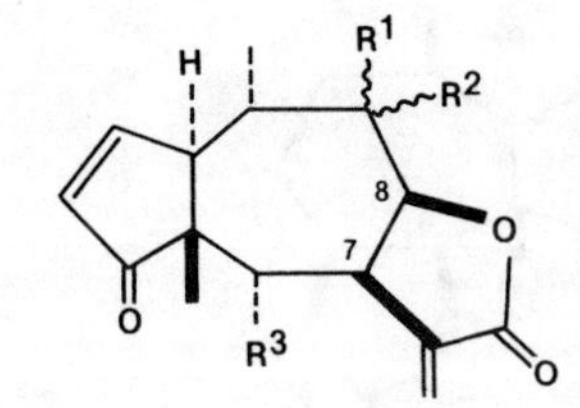

		R^1	R^2	R^3	ED_{50}	
87	Helenalin	H	H	OH	0.19,	PS(220),WA(316),LL(142)
88	Helenalin acetate	H	H	OAc		PS(165)
89	Fastigilin C	β-OH	α-H	O.CO.CH = C(CH$_3$)$_2$	0.34,	PS(153)
90	Multiradiatin	—O—		O.CO.CH = C(CH$_3$)$_2$	0.12	
91	Multigilin	β-OH	α-H	O.CO.C = CHCH$_3$ CH$_3$		PS(164)
92	Multistatin	—O—		O.CO.C = CHCH$_3$ CH$_3$	0.37,	PS(131)

FIGURE 27. Helenalin,[99,102-105] helenalin acetate,[106] fastigilin C,[20,49,107,108] multiradiatin,[49,108] multigilin,[108] and multistatin.[108]

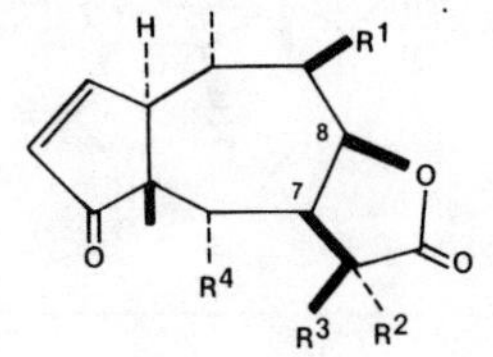

		R^1	R^2	R^3	R^4	ED_{50}	
93	Plenolin	H	H	CH$_3$	OH	0.814	WA(207),PS(138)
94	Radiatin	OH	H	CH$_3$	O.CO.C = CH$_2$ CH$_3$	0.39	PS(161),LE(103) LL(113).
95	Fastigilin B	OH	H	CH$_3$	O.CO.CH = C(CH$_3$)$_2$	0.078,(P-388)	PS(141)
96	Hymenoflorin	H	CH$_2$OH	OH	H		L1210
97	Microhelenin B	H	H	CH$_3$	O.CO.CH-C$_2$H$_5$ CH$_3$		WA(≥125)
98	Microhelenin C	H	H	CH$_3$	O.CO.C = CHCH$_3$ CH$_3$		WA(≥125)
99	Fastigilin A	OH	H	CH$_3$	O.CO.C = CHCH$_3$ CH$_3$	3.9,2.1 (P-388)	

FIGURE 28. Plenolin,[44,109] radiatin,[49,108,110] fastigilin B,[20,49,107,108] hymenoflorin,[111] microhelenin B,[112] microhelenin C,[112] and fastigilin A.[107,108]

(Sprague-Dawley rats, 2.5 mg/kg/day dose level, T/C 316%), Ehrlich ascites (CF$_1$ mice, 33.3 mg/kg/day dose level, 99% inhibition, P-388 lymphocytic leukemia (BDA-2 mice, 25 mg/kg/day dose level, T/C 127%), and Lewis lung carcinoma (C57BLG mice at 25 mg/kg/day dose level, T/C 142%).[102]

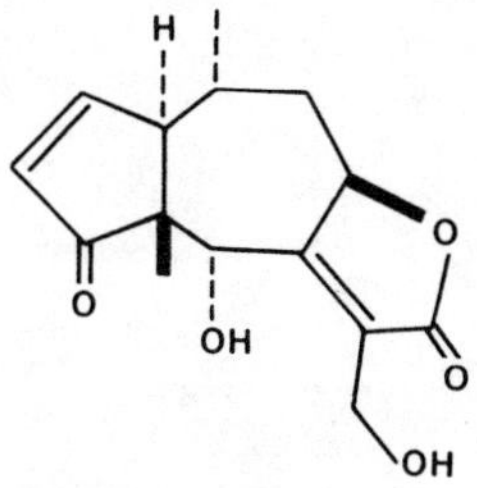

100 Isohelenol
PS(133)

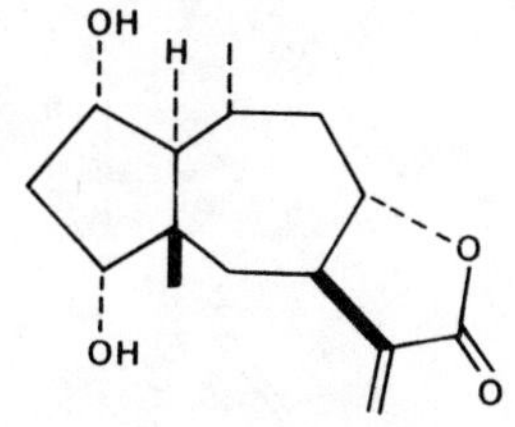

101 Pulchellin
ED_{50} 1.8

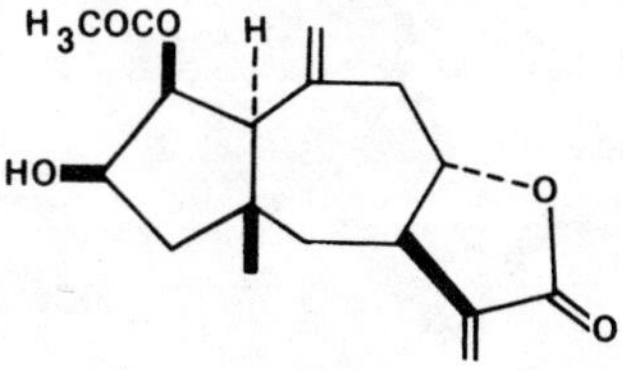

102 Pulchellin E
ED_{50} 1.0

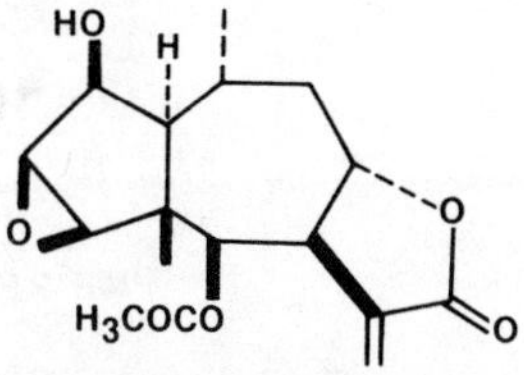

103 Gaillardilin
ED_{50} 2.2

FIGURE 29. Isohelenol,[113] pulchellin,[20,114,115] pulchellin E,[20,116] and gaillardilin.[20]

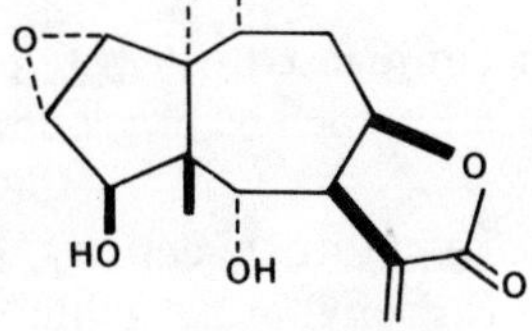

104 Autumnolide
ED_{50} 3.1

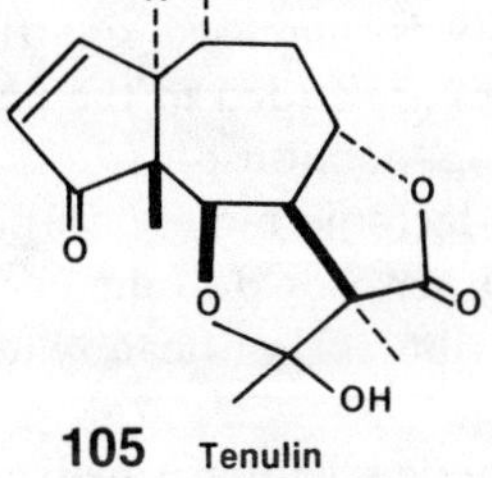

105 Tenulin
WA(266),LL(126)
PS(131)

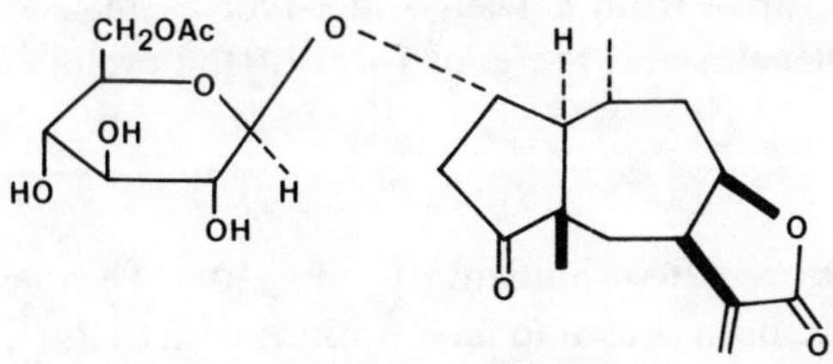

106 Paucin
PS active

FIGURE 30. Autumnolide,[103,117] tenulin,[44,118-120] and paucin.[111,121]

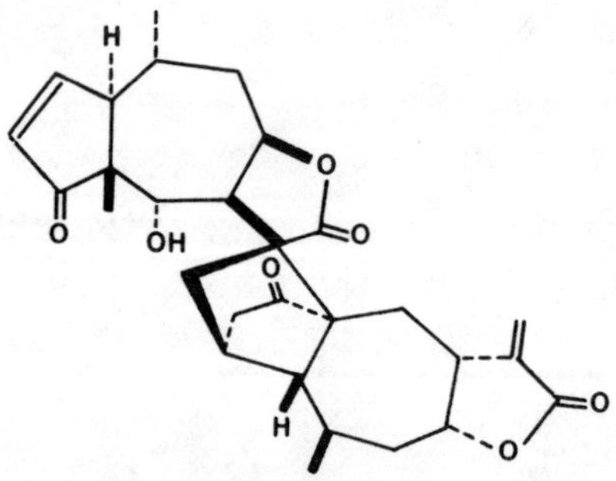

107 Microlenin
WA(173)

FIGURE 31. Microlenin.[122,123]

Lee et al.[102] and Pettit et al.[103] have studied several helenalin derivatives to determine the structure-activity relationship in this group of lactones and the following conclusions were made:

1. The α,β-unsaturated cyclopentenone ring contributed significantly to the in vitro and in vivo activity, whereas the α-methylene-γ-lactone system was less important for the same activity.
2. Cytotoxicity is increased significantly when conjugated esters (e.g., cinnamate ester group) are added to the parent compound, which ester grouping might act as alkylating functions.
3. α-Epoxy-γ-lactonic moiety do not contribute to cytotoxic activity, but an epoxy function either α or β to the cyclopentanone carbonyl in helenalin-related derivatives contributes significantly to cytotoxicity.

The novel dimeric sesquiterpene lactone, microlenin[122,123] **(107)** isolated from *Helenium microcephalum*, and whose structure was established by X-ray analysis (Figure 31), has been reported to show significant activity against Walker 256 carcinosarcoma in Sprague-Dawley rats (T/C 173) at 2.5 mg/kg dose level. The formation of microlenin **(107)** is suggested to arise from a Diels-Alder-type condensation involving the $\Delta^{11(13)}$ double bond of helenalin and the enol form of the cyclopentenone ring of a norpseudoguaianolide.

D. Elemanolides

Three elemanolides, (Figure 32) vernomenin[20,124,125] **(108)**, vernolepin[20,124-127] **(109)**, and vernodalin[128] **(110)** have been shown to have antitumor activity.

E. Eudesmanolides and Seco-Eudesmanolides

Five eudesmanolides (Figure 33), ivalin[20,129] **(111)**, invasperin[20,130] **(112)**, asperilin[20,130] **(113)**, alantolactone[20,131] **(114)**, and pinnatifidin[20,132] **(115)**, and two seco-eudesmanolides (Figure 34) eriolanin[133] **(116)**, and eriolangin[133] **(117)**, have been shown to have cytotoxic activity.

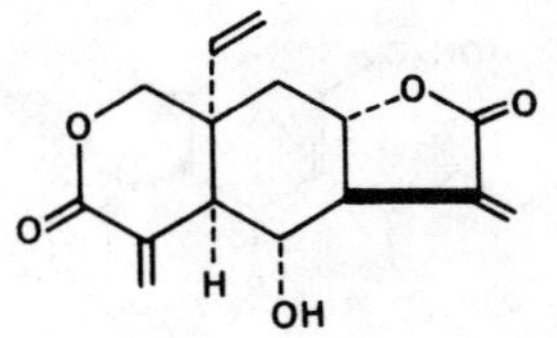

108 Vernomenin
ED_{50} 20.0,WA(64)

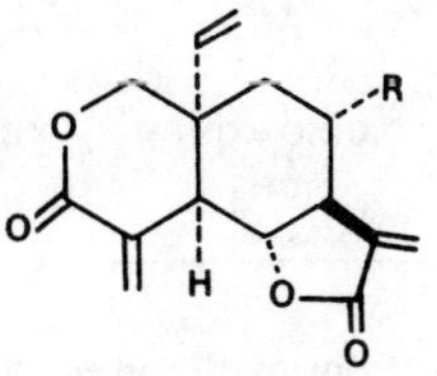

		R	ED_{50}	
109	Vernolepin	OH	2.0,	WA(32)
110	Vernodalin	$O.CO.C(=CH_2)CH_2OH$	1.8	

FIGURE 32. Vernomenin,[20,124,125] vernolepin,[20,124-127] and vernodalin.[128]

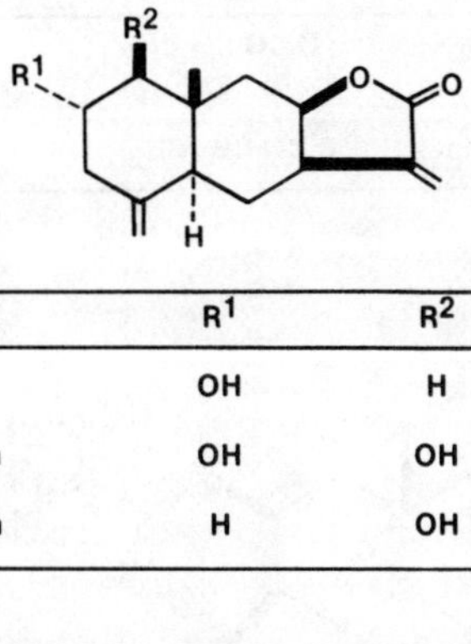

		R^1	R^2	ED_{50}
111	Ivalin	OH	H	0.72
112	Ivasperin	OH	OH	1.6
113	Asperilin	H	OH	1.0

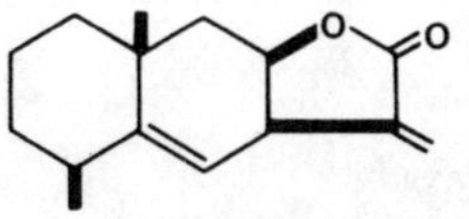

114 Alantolactone
ED_{50} 1.4

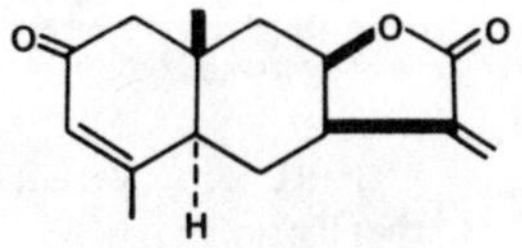

115 Pinnatifidin
ED_{50} 1.7

FIGURE 33. Ivalin,[20,129] ivasperin,[20,130] asperilin,[20,130] alantolactone,[20,131] and pinnatifidin.[20,132]

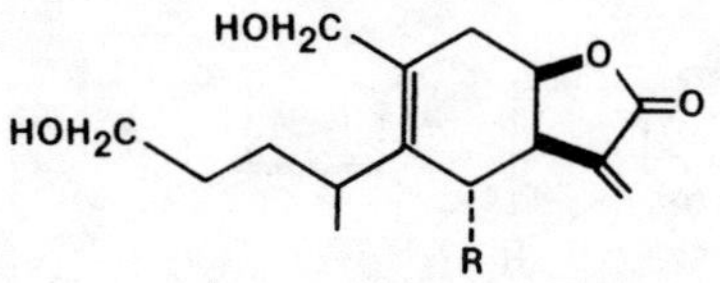

	R	ED_{50}
116 Eriolanin	O.CO.C(CH$_3$)=CH$_2$	P388
117 Eriolangin	O.CO.C(CH$_3$)=CHCH$_3$	KB PS

FIGURE 34. Eriolanin[133] and eriolangin.[133]

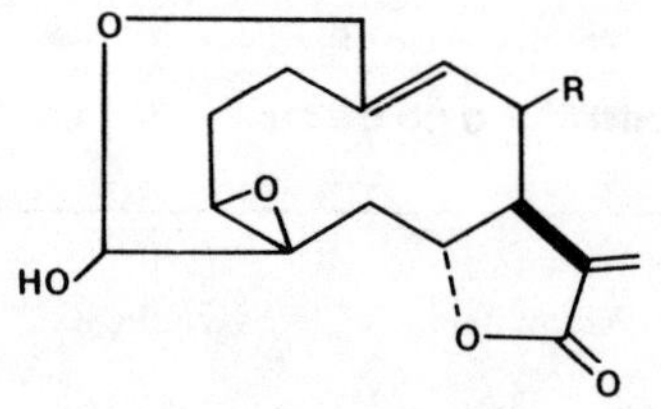

	R	ED_{50}
118 Vernolide	O.CO.C(CH$_3$)=CH$_2$	2.0
119 Vernomygdin	O.CO.CH (CH$_3$)$_2$	1.5

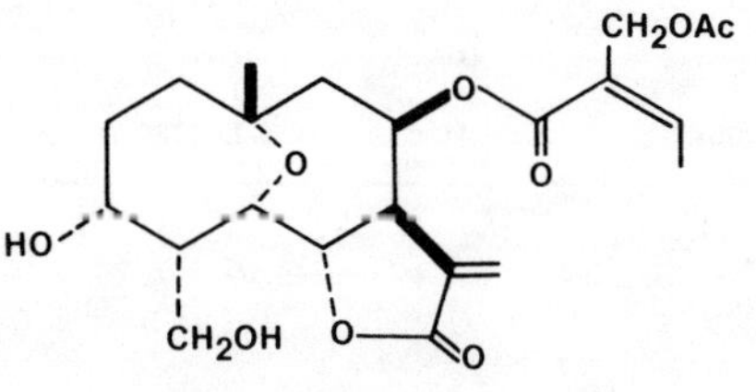

120 Chapliatrin
PS active

FIGURE 35. Vernolide,[134] vernomygdin,[128] and chapliatrin.[62]

F. Miscellaneous

This group (Figures 35 to 37) comprises a number of unusual sesquiterpenoids which do not fit in the foregoing list, and which have cytotoxic and/or antitumor activity.

IV. DITERPENOIDS

Based on the carbon skeleton and functionalities, these have been further classified

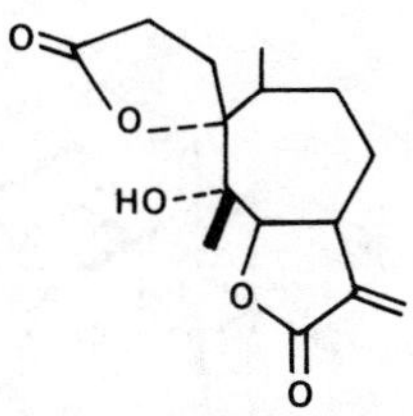

121 Psilostachyin A
ED_{50} 5.4

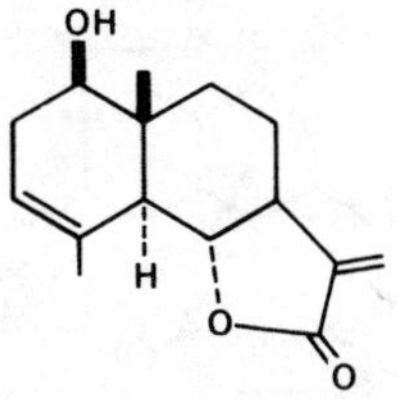

122 Santamarine
ED_{50} 1.1

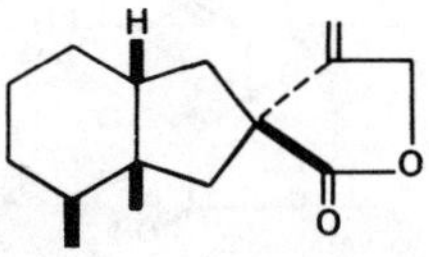

123 Bakkenolide A
ED_{50} 1.0

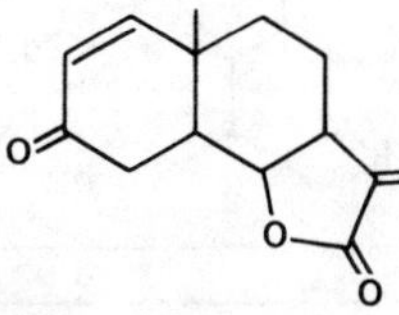

124 Tuberiferin
ED_{50} 1.0

FIGURE 36. Psilostachyin A,[135] santamarine,[40] bakkenolide A,[136] and tuberiferin.[37,137]

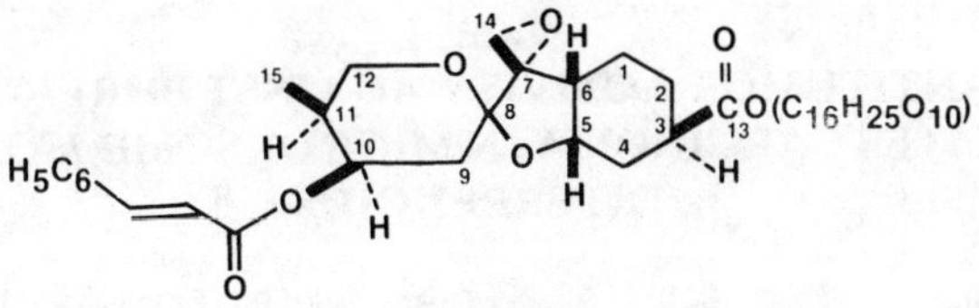

125 Phyllanthoside
ED_{50} 0.01, PS(153)

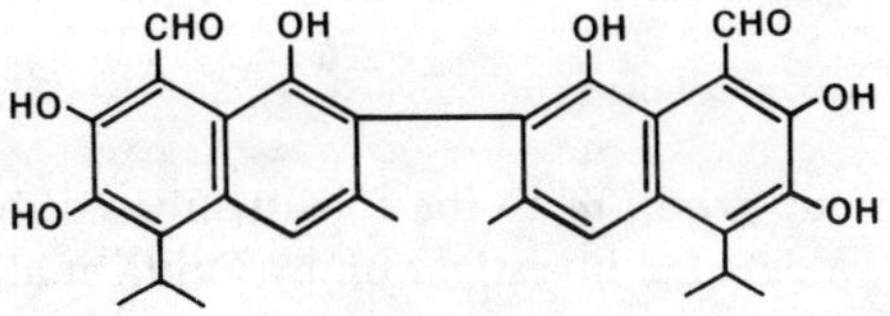

126 Gossypol
PS(150)

FIGURE 37. Phyllanthoside[138] and gossypol.[139]

into kauranes and seco-kauranes, daphenotoxanes, tiglianes, ingenanes, norditerpenoid dilactones, quinone methides and quinones, alkaloids, and a miscellaneous group.

A. Kauranes and Seco-Kauranes

In 1963, Arai et al.[140] indicated that enmein **(127)** (Figure 38) and its diacetate had antitumor activity while the dihydro derivative was devoid of activity. They concluded

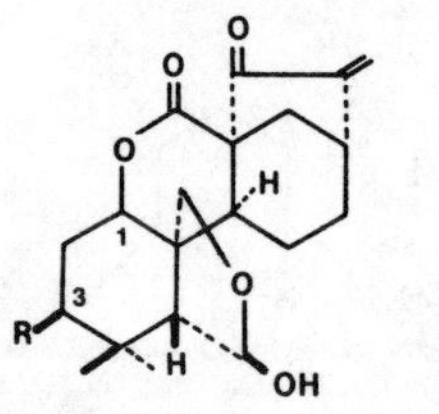

	R	ED_{50}
127 Enmein	OH	
128 Enmein-3-acetate	OAc	

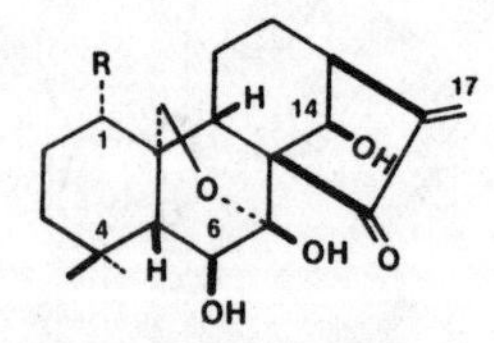

	R	ED_{50}
129 Oridonin	OH	2.8
130 Lasiokaurin	OAc	

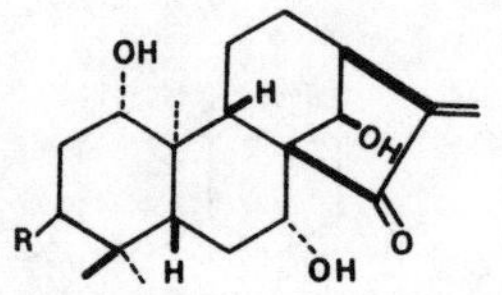

	R	ED_{50}
131 Kamebanin	H	
132 Isodomedin	β-OAc	4.0

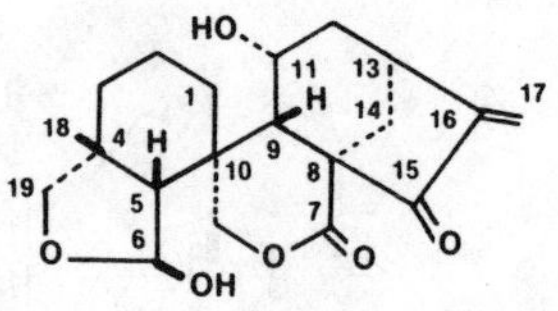

133 Shikodonin
ED_{50} 4.2

FIGURE 38. Enmein,[141,142] enmein-3-acetate,[141] oridonin,[141,142,144] lasiokaurin,[141,142] kamebanin,[143] isodomedin,[143] and shikodonin.[144]

Table 1
ANTITUMOR ACTIVITY AGAINST EHRLICH ASCITES CARCINOMA IN MICE OF SOME *ISODON* DITERPENOIDS

Structure no.	Compound	Dose (mg/kg)	PS (LD_{50}) (mg/kg)	I.L.S.[a] (%)
127	Enmein	25	>80	66
128	Enmein-3-acetate	40		56
129	Oridonin	10	37.5	115
130	Lasiokaurin	10	>80	117

[a] I. L. S. — increase of life span.

that the exocyclic methylene group α- to the five-membered ring ketone was essential to the activity. Recently, a detailed investigation by Fujita et al.[141] on *Isodon* diterpenoids has confirmed their findings and has shown that oridonin **(129)** and lasiokaurin **(130)** (Figure 38) are also antitumor agents. Based on a comparison of the activity of various derivatives,[142] they concluded that the activity of oridonin and lasiokaurin can be attributed to

1. The presence of an α-methylene cyclopentanone system in the molecule.
2. Some hydroxyl groups located in a suitable position for contact with, and binding to, a special enzyme carrying a nucleophile in a tumor cell.
3. Hydrogen bonding which increases the electro-philicity of the C-17 atom.

Table 1 shows the antitumor activity against Ehrlich ascites carcinoma in mice[141] of some of the *Isodon* diterpenoids.

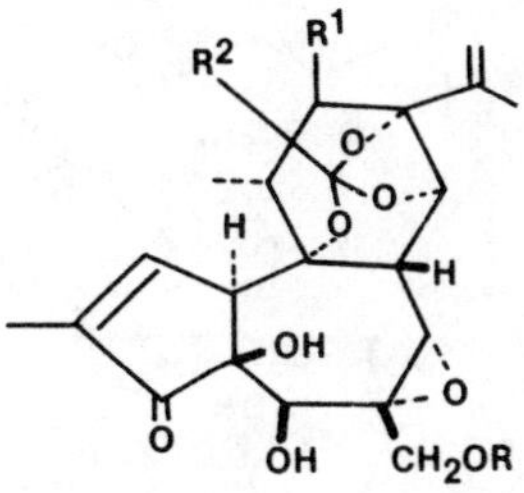

		R	R^1	R^2	ED_{50}
134	Gnidicin	H	$O.OC.CH = CH.C_6H_5$	C_6H_5	PS
135	Gnididin	H	$O.CO.(CH = CH)_2(CH_2)_4CH_3$	C_6H_5	PS
136	Gniditrin	H	$O.CO.(CH = CH)_3(CH_2)_2CH_3$	C_6H_5	PS
137	Mezerein	H	$O.OC.(CH = CH)_2.C_6H_5$	C_6H_5	PS,LE
138	Gnidilatin	H	$O.CO.C_6H_5$	$(CH_2)_8CH_3$	PS(140)
139	Gnidilatin-20-Palmilate	$CO.C_{15}H_{31}$	$O.CO.C_6H_5$	$(CH_2)_8CH_3$	PS(~170)
140	Gnidilatidin-20-Palmitate	$CO.C_{15}H_{31}$	$O.CO.C_6H_5$	$(CH = CH)_2(CH_2)_4CH_3$	PS(~170)
141	Huratoxin	H	H	$(CH = CH)_2(CH_2)_8CH_3$	PS
142	Montanin	H	H	$(CH_2)_{10}CH_3$	PS

FIGURE 39. Gnidicin,[145] gnididin,[145] gniditrin,[145] mezerein,[146] gnidilatin,[147] gnidilatin-20-palmitate,[147] gnidilatidin-20-palmitate,[147] huratoxin,[148,149] and montanin.[148]

Three more compounds of this class, kamebanin[143] **(131)**, isodomedin **(132)** (KB LD_{50} 4.0 μg/mℓ),[143] and shikodonin **(133)** (KB LD_{50} 4.2 μg/mℓ)[144] have been reported recently. Structural studies have suggested that isodomedin is a β-acetoxy kamebanin.[143] The absolute structure of shikodonin was established by X-ray diffraction methods performed on its derivatives,[144] and it was found to have a unique spiro skeleton **(133)** and is the first *ent*-kaurene type diterpene oxygenated at C-19.

B. Daphnetoxanes

The first members of this class gnidicin **(134)**, gnididin **(135)** and gniditrin **(136)** (Figure 39) were isolated from *Gnidia lamprantha* Gilg. (Thymelaeaceae) by Kupchan et al.[145] in 1975 and were shown to have potent antileukemic activity against P-388 leukemia in the mouse at 20 to 100 μg/kg dose level. Since then, Kupchans group has reported further[146,147] on mezerein **(137)** (from *Daphne mezerenum* L.) gnidilatin **(138)**, gnidilatin-20-palmitate **(139)**, and gnidilatidin-20-palmitate **(140)** (from *Gnidia* species); and Ogura et al.[148] have reported on huratoxin[149] **(141)** and montanin **(142)** (from *Baliospermum montanum* roots); all of the same class (Figure 39). These are highly oxygenated diterpenoids and contain an α,β-unsaturated cyclopentenone. Kupchan et al. suggested that the ester attached to C-12 may be responsible for the compounds antitumor activity, and it may be acting as a carrier moiety in cell penetration or selective molecular complex formation. The significance of the C-6, C-7 epoxide, the cyclopentenone, the ortho ester and other structural features has not yet been established.

From another species *Gnidia subcordata* (Messn.) Engl., Kupchan et al.[150] have recently reported on the isolation and structure of the most potent antileukemic agents

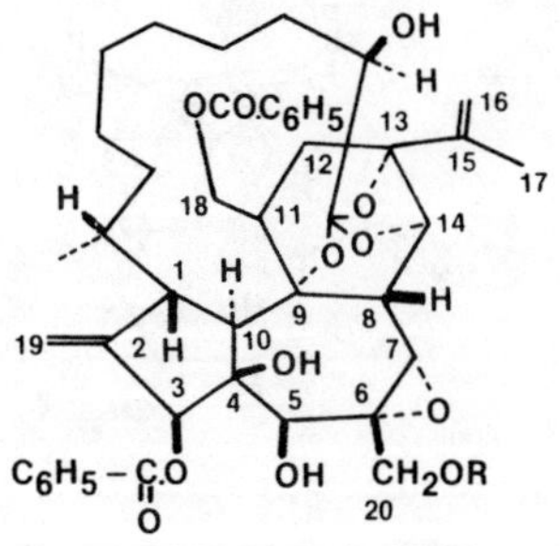

	R	PS
143 Gnidimacrin	H	180 at 12-16 μg/kg
144 Gnidimacrin 20- Palmitate	$CO.C_{15}H_{31}$	190 at 30-50 μg/kg

FIGURE 40. Gnidimacrin[150] and gnidimacrin-20-palmitate.[150]

gnidimacrin **(143)** and gnidimacrin 20-palmitate **(144)** (Figure 40), the first diterpenoids which contain a novel macrocyclic ring with one terminus at the ortho ester carbon. The absolute stereostructure of gnidimacrin was established by X-ray analysis of the 20-p-iodobenzoate. Gnidimacrin showed a T/C of about 180 at 12 to 16 μg/kg dose level and gnidimacrin 20-palmitate a T/C of about 190 at 30 to 50 μg/kg dose level.

C. Tiglianes

Although tigliane esters are best known for their toxic and tumor-promoting activity,[151] Kupchan et al.[152] for the first time showed that phorbol 12-tiglate 13-decanoate **(145)**, (Figure 41) isolated from the oil of *Croton tiglium,* possess antileukemic activity. Since then, five more phorbol esters baliospermin **(146)**, mancinellin **(147)**, 12-deoxyphorbol 13-palmitate **(148)**, 12-deoxy-5β-hydroxy phorbol 13-myrsitate **(149)**, and 12-deoxy-16-hydroxy phorbol 13-palmitate **(150)**, (Figure 41) isolated from *B. montanum* roots, have been shown to possess antitumor activity.[148] The close structural similarity between these antitumor tigliane (phorbol) esters and the toxic tumor-promoting tigliane esters have led to the suggestion[153] that the difference between the tumor-promoting action and antitumor action must lie in their detailed structural chemistry.

D. Ingenanes

This unusual group of diterpenes are structurally related to tiglianes, a group discussed earlier. Although a member of ingenane derivatives have been isolated from the genus *Euphorbia,* only ingenol dibenzoate **(151)** (Figure 42) isolated from *Euphorbia esula* L., has been shown to have PS, LL, and WA activity at a very low dose of 130 to 360 μg/kg dose level.

E. Norditerpenoid Dilactones

A number of biologically active nor-diterpenoid dilactones have been isolated from various *Podocarpus* species.[154-156] Their growth-inhibiting activity is well recognized.[156] Podolide **(152)**, (Figure 43) from *Podocarpus gracilor* Pilg (Taxaceae), was the first compound of this class, shown to have significant in vivo antitumor activity against P-388 leukemia in mice, and in vitro cytotoxicity against human carcinoma of the nasopharynx and P-388 murine leukemia.[154]

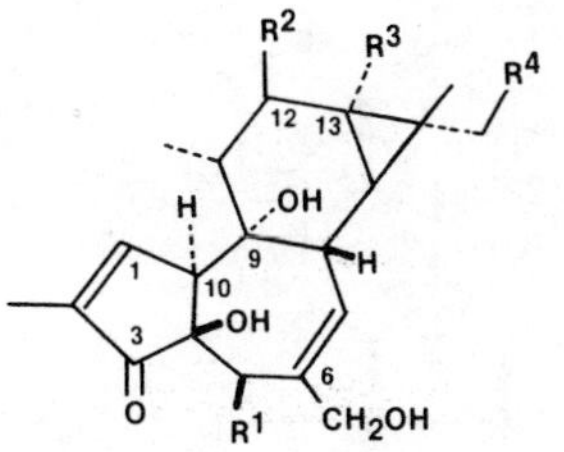

	R^1	R^2	R^3	R^4	ED_{50}
145 Phorbol-12-tigliate-13-decanoate	H	H	$O.CO.(CH_2)_8CH_3$	$O.CO.C(CH_3) = CH.CH_3$	
146 Baliospermin	OH	H	$O.CO(CH_2)_{10}CH_3$	H	
147 Mancinellin	OH	H	$O.CO(CH = CH)_3 (CH_2)_8CH_3$	H	
148 12-Deoxyphorbol-13-palmilate	H	H	$O.CO.(CH_2)_{14}CH_3$	H	
149 12-Deoxy-5β-hydroxy-phorbol-13-myristate	OH	H	$O.CO(CH_2)_{12}.CH_3$	H	
150 12-Deoxy-16-hydroxyphorbol-13-palmitate	H	H	$O.CO.(CH_2)_{14}CH_3$	H	

FIGURE 41. Phorbol-12-tigliate-13-decanoate,[152] baliospermin,[148] mancinellin,[148] 12-deoxyphorbol-13-palmitate,[148] 12-deoxy-5β-hydroxyphorbol-13-myristate,[148] and 12-deoxy-16-hydroxyphorbol-13-palmitate.[148]

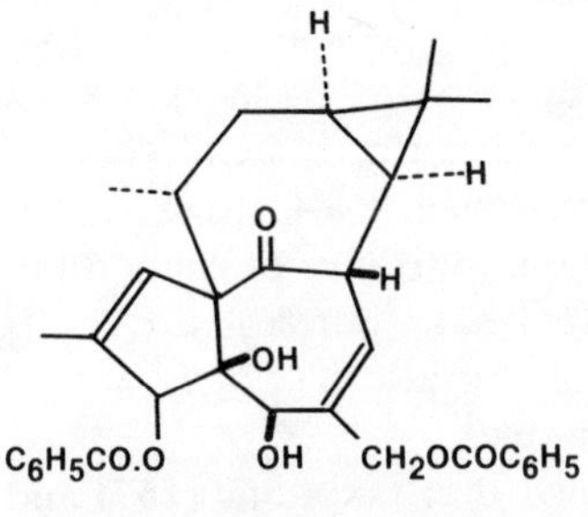

151 Ingenol dibenzoate
PS, LL, WA

FIGURE 42. Ingenol dibenzoate.[152]

Since then, a number of other nor-diterpenoid dilactones, nagilactones E, B, C, and D (**153, 154, 155,** and **156,** respectively) (Figures 43 and 44) have been shown to be strongly cytotoxic and to have antileukemic activity.[155,156] These compounds all exhibited an IC_{50} value of 10^{-3} to 10^{-4} mM against Yoshida sarcoma, nagilactones D and E being stronger. Nagilactones C and E were also effective against P-388 murine leukemia with a T/C of 145 and 125 at a dose level of 4 mg/kg per day, and 20 mg/kg per day.

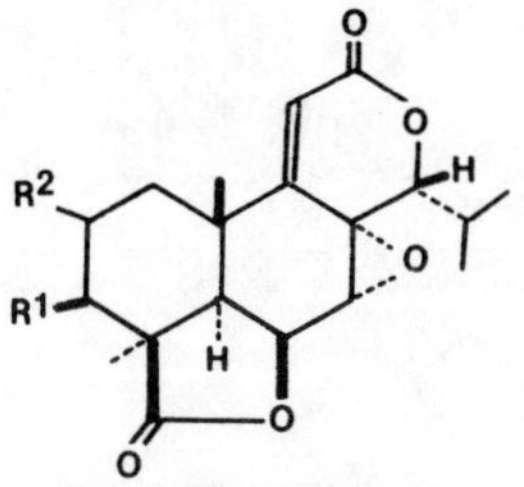

	R^1	R^2	IC_{50} (mM)	
152 Podolide	—Δ—		KB	PS
153 Nagilactone-E	OH	H	3.36×10^{-4}	PS(125)

FIGURE 43. Podolide[154] and nagilactone E.[155,156]

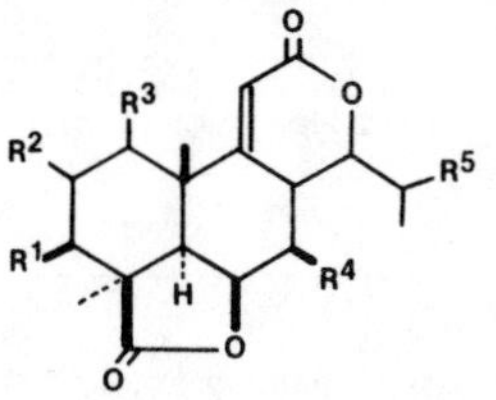

	R^1	R^2	R^3	R^4	R^5	IC_{50}	
154 Nagilactone B	H	β-OH	β-OH	OH	CH_3	1.72×10^{-3}	
155 Nagilactone C	OH	" —O—		OH	CH_3	2.25×10^{-3}	PS(145)
156 Nagilactone D	OH	α —O—		H	H	3.32×10^{-4}	

FIGURE 44. Nagilactone B,[155,156] nagilactone C,[155,156] and nagilactone D.[155,156]

These compounds all have a basic ring system consisting of a degraded totarol, and contain an α,β-unsaturated δ-lactone, a γ-lactone, and olefinic and/or epoxy groups in their structures. The relative importance of the α,β-unsaturated δ-lactone, epoxide, and other functions, has not yet been reported.

F. Quinone Methides and Quinones

Kupchan et al.[157] have shown that taxodone **(157)** and taxodione **(158)** (Figure 45) isolated from *Taxodium distichum* Rich (Taxodiaceae), demonstrated a significant inhibitory activity against Walker carcinoma 256 in rats at 40 and 25 mg/kg dose level, respectively. The structures of these two unusual diterpenoids were established by spectroscopic and chemical methods.[157]

Recently, Marletti et al.[158] reported on the isolation and structure of 14-methoxytaxodione **(159)** (Figure 45) from *Horminum pyrenaicum* and indicated that it had antitumor activity. Barbatusin **(160)**, (Figure 45) isolated from *Coleus barbatus*, Benthan (Labiatae), possesses significant tumor-inhibiting activity against Lewis lung carcinoma and lymphocytic leukemia P-388 in mice at dose levels of 200 and 400 mg/kg respectively.[159] It is the first example of a diterpene of the abietane group containing a spirocyclopropane ring in lieu of an open side-chain.

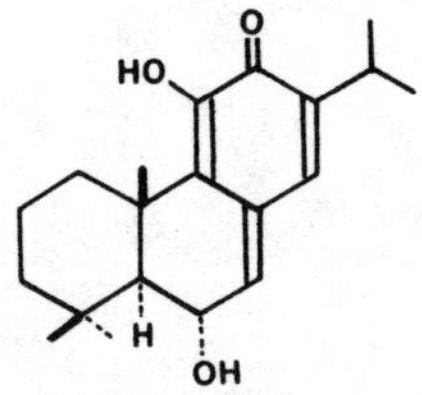

157 Taxodone

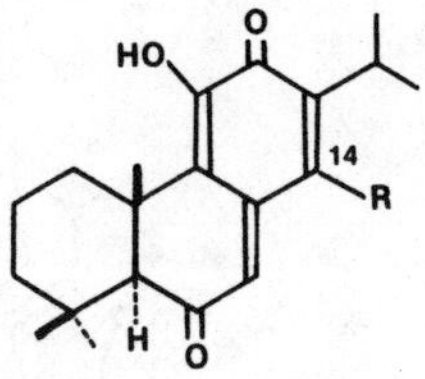

		R	ED_{50}
158	Taxodione	H	
159	14-Methoxytaxodione	OCH_3	

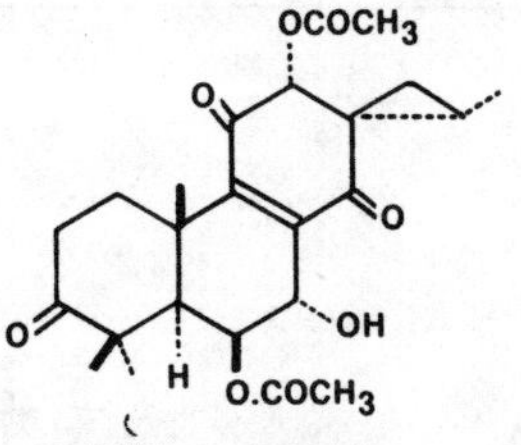

160 Barbatusin

FIGURE 45. Taxodone,[157] taxodione,[157] 14-methoxytaxodione,[158] and barbatusin.[159]

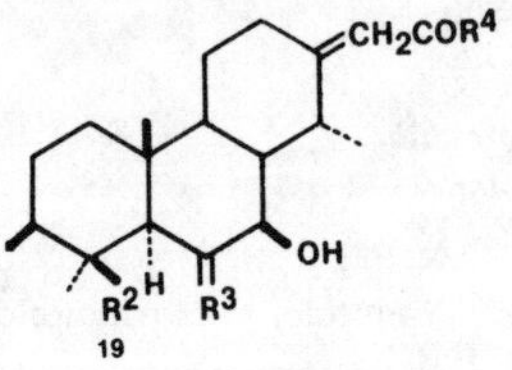

		R^1	R^2	R^3	R^4	ED_{50}
161	Norcassamidine	H	$COOCH_3$	H,H	$O.CH_2CH_2NHCH_3$	
162	Norerythrostachamine	OH	$COOCH_3$	H,H	$O.CH_2CH_2NHCH_3$	
163	3β-Acetoxynorerythrosuamine	OAc	$COOCH_3$	O	$O.CH_2CH_2NHCH_3$	0.0003
164	Norerythrostachaldine	OH	CHO	H,H	$O.CH_2CH_2NHCH_3$	0.029*
165	3β-Acetoxynorerythrostachaldine	OAc	CHO	H,H	$O.CH_2CH_2NHCH_3$	0.0073*

*values for the hydrochloride

FIGURE 46. Norcassamidine,[160] norerythrostachamine,[160] 3-β-acetoxynorerythrosuamine,[161] norerythrostachaldine,[162] and 3β-acetoxynorerythrostachaldine.[162]

G. Alkaloids

The diterpene alkaloids from *Erythrophleum chlorostachys* (F. Muell) Bail. norcassamidine[160] **(161)**, norerythrostachamine[160] **(162)**, 3β-acetoxynorerythrosuamine[161] **(163)**, norerythrostachaldine[162] **(164)**, and 3β-acetoxynorerythrostachaldine[162] **(165)** (Figure 46) have been shown to exert significant cytotoxic activity against a human carcinoma of the nasopharynx in cell culture. It was observed that the 3β-acetates were more cytotoxic than 3β-hydroxy alkaloids, and the corresponding amides were not significantly active.

	R	ED_{50}
166 Triptolide	H	0.0017, L-1210, P-388
167 Tripdiolide	OH	0.0042

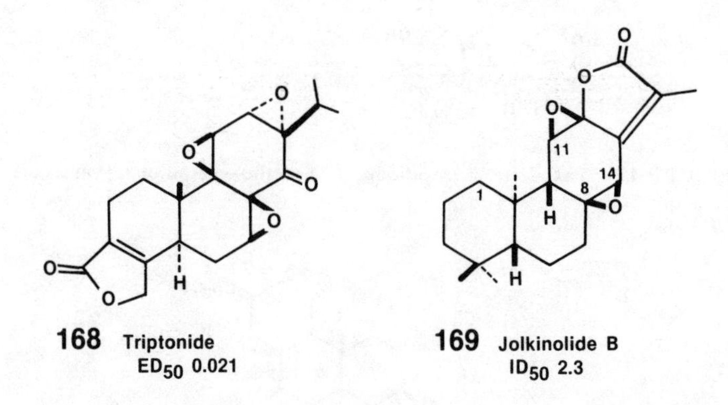

FIGURE 47. Triptolide,[163-165] tripdiolide,[163-165] triptonide,[163,164] and jolkinolide B.[166,167]

H. Miscellaneous

There are eight compounds in this class, two of which, triptolide **(166)** (Figure 47) and taxol **(170)** (Figure 48), are in the list of chemical structures currently of interest to the Division of Cancer Treatment of the National Cancer Institute.

Kupchan et al.[163,164] reported the isolation of triptolide **(166)**, tripdiolide **(167)** and triptonide **(168)**, three antileukemic compounds (Figure 47) from the roots of *Tripterygium wilfordii* Hook F., as the first natural products containing the 18 (4 → 3) abeo-abietane skeleton and the first diterpenoids triepoxides. The structure of triptolide was determined by direct X-ray crystallographic analysis.[163] Triptolide **(166)** and tripdiolide **(167)** have been shown to exert significant life-prolonging effects in mice afflicted with the L-1210 lymphoid leukemia ($T/C \geqslant 230$ at 0.1 mg/kg).[164] Jolkinolide B **(169)**, a crystalline lactone diepoxide of abietane class, (Figure 47) isolated from *Euphorbia jolkini* Boiss, has been shown active against Hela cells (ID_{50} 2.3 μg/mℓ).[166,167]

The stem bark of the western yew, *Taxus brevifolia*, has yielded a highly potent antileukemic and tumor-inhibiting compound, taxol **(170)**, (Figure 48) which has been shown to possess activity against L-1210, P-388, and P-1534 leukemia.[168] It is also highly active against WM-256 carcinosarcoma and 9KB assay, ED_{50} 5.5×10^{-5} μg/mℓ. Impure taxol has been active in Sarcoma 180 and Lewis lung tumors.[168]

Very recently, another compound of this class, cephalomannine **(171)**, (Figure 48) an alkaloidal taxane derivative isolated from *Cephalotaxus mannii*, has been shown

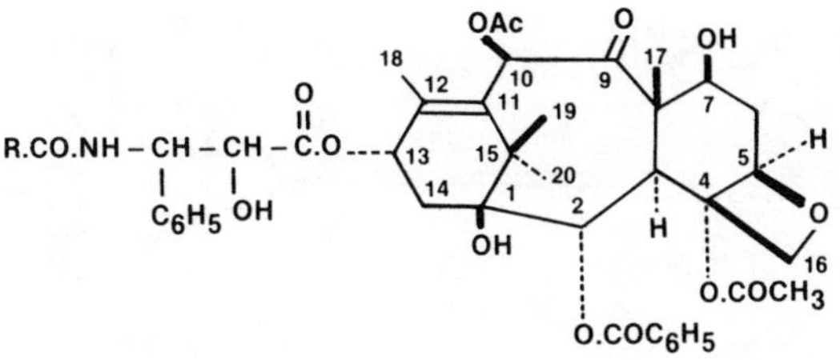

	R	ED50
170 Taxol	C_6H_5	5.5×10^{-5}
171 Cephalomannine	$C(CH_3)=CHCH_3$	3.8×10^{-3}

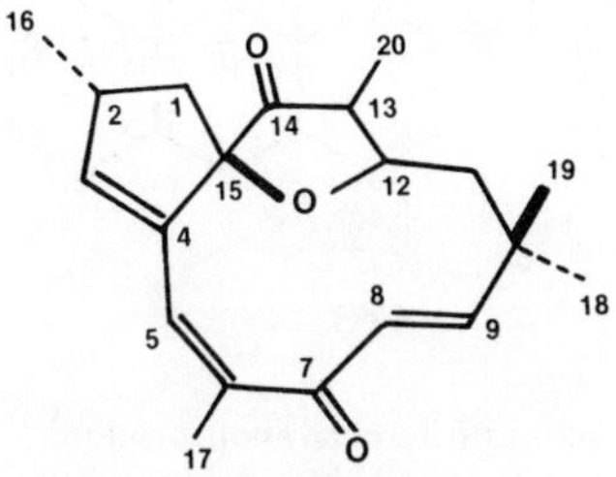

172 Jatrophone
ED_{50} 0.17

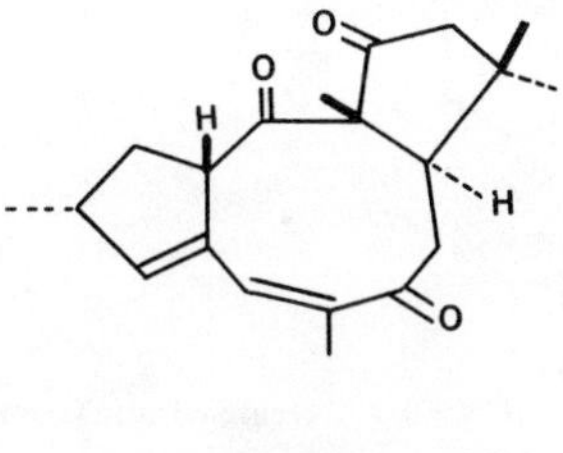

173 Jatrophatrione
PS(141)

FIGURE 48. Taxol,[168] cephalomannine,[169] jatrophone,[170] and jatrophatrione.[171]

to be cytotoxic in KB cell culture (LD_{50} 3.8×10^{-3} $\mu g/m\ell$), and a potent inhibitor of PS leukemia in mice.[169]

The unusual macrocyclic diterpene, jatrophone **(172)**, (Figure 48) isolated from *Jatropha gossypiifolia* L. (Euphorbiaceae), has been shown to exert significant inhibiting activity against S-180, Lewis Lung carcinoma, P-388 lymphocytic leukemia in the mouse at 27 mg and 12 mg/kg dose level, and Walker 256 intramuscular carcinoma in the rats.[170] From the chloroform extract of the roots of *J. macrorhiza* Benth. (Euphorbiaceae) Torrance et al.[171] have isolated jatrophatrione **(173)** (Figure 48) which shows significant activity against P-388 lymphocytic leukemia (T/C 130 and 141 at 1.0 and 0.5 mg/kg dose level, respectively). Jatrophatrione **(173)** is thus active, even though it lacks the Δ^8 double bond to which thiols add nucleophilically in jatrophone **(172)**. It is suggested[171] that the activity of jatrophatrione occurs because of its ability to undergo reverse Michael addition to give **(174)**, which can undergo addition similar to that of **(172)** (Figure 49).

V. QUASSINOIDS

Quassinoids (simaroubolides) bitter principles attracted great attention after the report of the antileukemic activity of holacanthone.[172,173] The structure of a related compound, simarubolide **(176)**, suggests that quassinoids arise from prototriterpene apotirucallol **(175)** by oxidative degradation[174] (Figure 50).

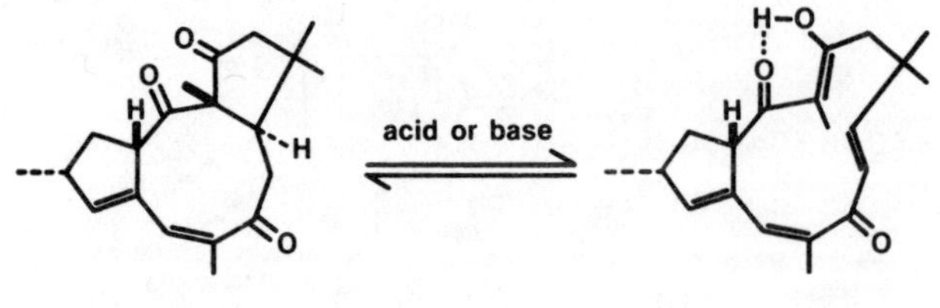

FIGURE 49. Probable active form of jatrophatrione.

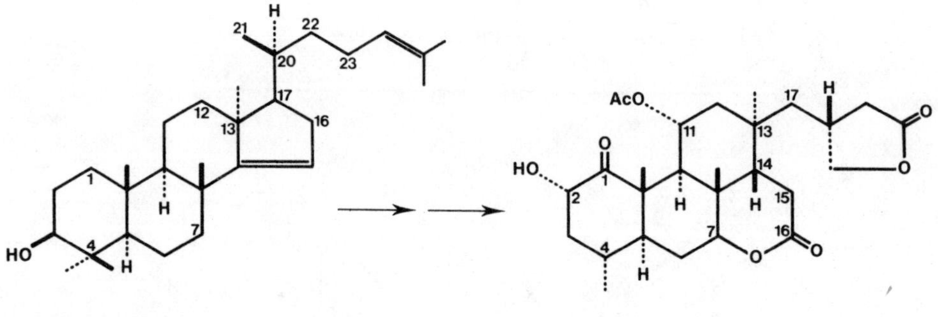

FIGURE 50. Orgin of quassinoids from prototriterpene apotirucallol.[174]

In general, quassinoids are highly oxygenated, structurally complex molecules. The antitumor quassinoids can be divided into bruceolides and glaucarubolones based on structural similarities. The structures of these compounds have been elucidated either by physicochemical methods or by chemical correlation. The common features of both groups are the presence of an α,β-unsaturated six-membered ketone and a δ-lactone.

A. Bruceolides

Bruceolides are shown in Figure 51 and contain a methyl or a carboxymethyl at C-13, a side-chain ester at C-15, and an oxo-methylene bridge between C-8 and C-13 carbon atoms.

A comparison of bruceolides antileukemic activity (PS) in the mouse indicated that samaderine E[175] **(177)** and bruceantarin[176] **(178)** had moderate activity and the acetate ester brucein B[177] **(179)** had marginal antileukemic activity. The parent quassinoid bruceolide[176,177] **(180)** had very low activity or was inactive. The most potent of the bruceolides were brusatol[178-180] **(181)**, bruceantin[174,176,178,181] **(182)** and bruceantinol[174,176] **(183)**, all of which contain α,β-unsaturated ester side-chains. When the side chain was reduced, compounds with diminished activity were obtained, as was the case with simalikalactone D[182] **(184)** and quassimarin[182] **(185)**, both of which have a saturated ester side chain.

Bruceantin **(182)** has been selected for toxicological investigation in preparation for clinical trials by the National Cancer Institute and is thus in the list of chemical structures currently of interest to the Division of Cancer Treatment.

B. Glaucarubolones

These compounds contain an exocyclic methylene or an α-methyl group at C-13, a side chain ester either at C-15 or C-6, and a hemiketal linkage between C-8 and C-11 carbon atoms. These are shown in Figure 52. Chaparrinone[183,184] **(186)** shows only in

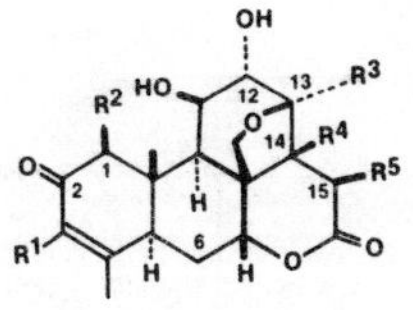

	R^1	R^2	R^3	R^4	R^5	ED_{50}	
177 Samaderine E	H	OH	CH_3	OH	H		PS(moderate)
178 Bruceantarin	OH	H	$COOCH_3$	H	$O-C(=O)-C_6H_5$	0.01-0.001	PS(moderate)
179 Bruceine B	OH	H	$COOCH_3$	H	OAC		PS(marginal)
180 Bruceolide	OH	H	$COOCH_3$	H	OH		PS(low)
181 Brusatol	OH	H	$COOCH_3$	H	$O-C(=O)-CH=C(CH_3)_2$		PS(158)
182 Bruceantin	OH	H	$COOCH_3$	H	$O-C(=O)-CH=C(CH_3)-CH(CH_3)_2$	0.01-0.001	B-16(178), LL,PS(197)
183 Bruceantinol	OH	H	$COOCH_3$	H	$O-C(=O)-CH=C(CH)-C(OH)(CH_3)_2$	0.01-0.001	PS(potent)
184 Simalikalactone D	H	OH	CH_3	H	$O-C(=O)-CH(CH_3)-CH_2-CH_3$	0.01-0.001	PS(165-175)
185 Quassimarin	H	OH	CH_3	H	$O-C(=O)-C(CH_3)(OAC)-CH_2-CH_3$	0.01-0.001	PS(165-175)

FIGURE 51. Samaderine E,[175] bruceantarin,[176,177] bruceine B,[177] bruceolide,[176,177] brusatol,[178-180] bruceantin,[174,176,178,181] bruceantinol,[174,176] simaliklactone D,[182] an quassimarin.[182]

vitro activity, but holacanthone[173] **(187)**, ailanthinone[185] **(188)**, dehydroailanthinone[186] **(189)**, glaucarubinone[176,181] **(190)**, 13,18-dehydroglaucarubinone[187] **(191)**, 6α-senecioyloxy-chaparrinone[173,188] **(192)**, 6α-tigloyloxychaparrinone[185] **(193)**, and undulatone[189] **(194)** show significant in vitro and in vivo activities.

Kupchan et al.[176] have demonstrated the significance of an ester side-chain for the biological activity from the test results of glaucarubinone (190) and glaucarubolone [**(190)**, $R^3 = OH$]. Both compounds were equally active in KB system; however, glaucarubolone was inactive towards P-388 leukemia in mice. It was also shown that variation of the ester side-chain had little effect on in vivo activity, and thus the ester may be involved primarily in transport. Furthermore, reduction of ring A significantly decreases the cytotoxicity of the compound. This portion of the molecule, containing a conjugated ketone group was therefore thought to play an important role with biological nucleophiles in the mechanism of their activity. Michael-type additions of model biological nucleophiles to highly electrophillic conjugated systems, especially methylene lactones and α,β-unsaturated esters, are known in sesquiterpenoids.[3,4,21] Methylation of C-1 alcohol seemed to result in a diminution of cytotoxicity and of in vivo antileukemic activity.[186] It was therefore suggested that the hydroxyl group may enhance the reactivity of the conjugated ketone through intramolecular hydrogen bonding (Figure 53).

More recently, the isolation and high antitumor activity of 6α-senecioyloxychaparrinone[173,188] **(192)**, 6α-tigloyloxychaparrinone[185] **(193)** and undulatone[189] **(194)** (Figure 52) against several leukemia systems and solid tumor have demonstrated that C-15 ester function is not necessary for antitumor activity in quassinoids.

Wani et al.[188] have reached certain conclusions from the in vivo tests of several quassinoids. The parent compound chaparrinone **(186)**, which lacks the ester function-

	R1	R2	R3	ED_{50}	
186 Chaparrinone	H	α-CH_3,β-H	H	0.1	
187 Holacanthone	H	α-CH_3,β-H	OAc		WA
188 Ailanthinone	H	α-CH_3,β-H	O.C(=O).CH(CH_3)—CH_2CH_3	0.01-0.001	PS(148)
189 Dehydroailanthinone	H	CH_2	O.C(=O).CH(CH_3)—CH_2CH_3	0.01-0.001	PS
190 Glaucarubinone	H	α-CH_3,β-H	O.C(=O)—C(CH_3)(OH)—CH_2CH_3	0.01-0.001 0.08,0.34	PS(161)
191 13,18-Dehydro-glaucarubinone	H	CH_2	O.C(=O)—C(CH_3)(OH)—CH_2CH_3	0.95	PS
192 6α-Senecioyloxy-chaparrinone	O.C(=O).CH = C(CH_3)$_2$	α-CH_3,β-H	H	0.007	PS(198),L-1210(151), B-16(149)
193 6α-Tigloyloxy-chaparrinone	O.C(=O).C(CH_3) = CHCH$_3$	α-CH_3,β-H	H	0.015	PS(163)
194 Undulatone	O.C(=O).C(CH_3) = CHCH$_3$	α-CH_3,β-H	OAc	0.1-0.001	PS(163)

FIGURE 52. Chaparrinone,[183,184] holacanthone,[173] ailanthinone,[185] dehydroailanthinone,[186] glaucarubinone,[176,186] 13, 18-dehydroglaucarubinone,[187] 6-α-senecioyloxychaparrinone,[173,188] 6α-tigloyloxychaparrinone,[184,185] and undulatone.[189]

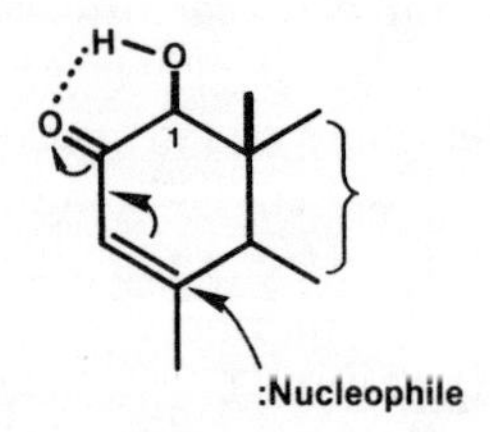

FIGURE 53. Intramolecular hydrogen bonding of C-1 hydroxyl in quassinoids.

ality either at C-6 or C-15, was weakly active in both 9KB and P-388 systems. The 6α-senecioyl ester **(192)** is the most active compound in the series, showing a good therapeutic index in P-388 system (maximum T/C 198 at 1 mg/kg dose level). As was also observed by Kupchan et al.,[186] the methylatin of C-1 hydroxyl resulted in marked diminution of P-388 activity (maximal T/C 142 at 4 mg/kg dose level) but the cytotoxicity was of the same order. A reduction in activity was also observed on partial acetylation at C-1. Recent X-ray studies[189] indicate that the acetylation or methylation destroys the intramolecular hydrogen bonding between the C-1 β and C-11 α-hydrogen groups which may be a requirement for maximal antitumor activity. It was concluded, there-

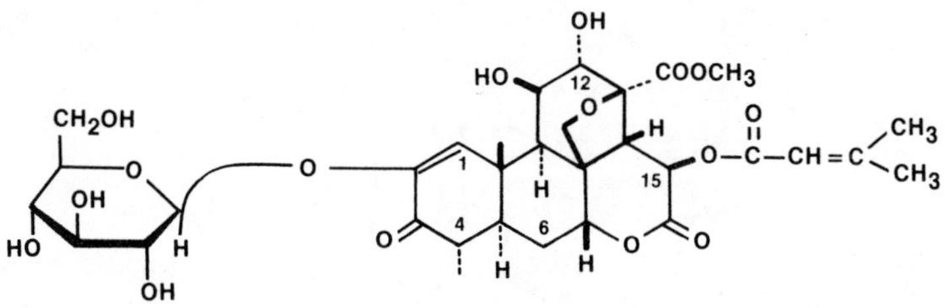

195 Bruceoside A
PS(156)

FIGURE 54. Bruceoside A.[178,179,190]

fore, that an ester group either at C-6 or C-15 is required for antitumor activity. The epoxy methano bridge is essential for activity in both series.

Lee et al.[190] have recently reported on the isolation and structure of a novel and potent antileukemic quassinoid bruceoside A **(195)** (Figure 54). It is the first quassinoid glycoside which has been shown to have such activity.[176,178] Bruceoside A showed significant antileukemic activity in P-388 leukemia (T/C 156 at 6 mg/kg/day dose level). More recently, Hall et al.,[179] while studying the mechanism of action of bruceoside A and brusatol on nucleic acid metabolism of P-388 lymphocytic leukemia cell, indicated that suppression of RNA and protein synthesis is not the critical event resulting in tumor cell death. DNA synthesis inhibition appears to correlate more directly with antitumor activity.

VI. TRITERPENOIDS

A number of triterpenoids are known to possess cytotoxic and/or antitumor activity.[6] These could be divided[2] into cucurbitacins, dammaranes, lanostanes, limonoids, lupanes, ursanes, oleananes, taraxeranes, and quinonoid nortriterpenoids.

A. Cucurbitacins

Cucurbitacins are highly toxic, closely related bitter principles, present in several medicinal plants of the family Cucurbitaceae, Scrophulariaceae, Begoniaceae, and Cruciferae, used in primitive medicines. These occur either free or as glucosides which are easily hydrolysed under the influence of glucosidase elaterase. Based on the functionalities, these have been placed in several groups.

Figure 55 shows cucurbitacin D[191-193] **(196)**, cucurbitacin B[191,193-195] **(197)**, fabacein[196] **(198)**, cucurbitacin A[197] **(199)**, and datiscoside[192] **(200)**. The common characteristics of these compounds are an α-hydroxy ketone in ring A, carbonyls at C-11 and C-22, the latter being conjugated to $\Delta^{23,24}$ double bond. These also all contain a $\Delta^{5,6}$ double bond in their structure. The X-ray structure of datiscoside di-p-iodobenzoate[192] established the configuration of the cucurbitacins at C-20 and C-2 for the first time. Figure 55 compares the differences in their structures with their activities. Cucurbitacins D, B, fabacein and datiscoside have been shown to exhibit antitumor activity in vivo. The second group of cucurbitacins cucurbitacin I[191-193] **(201)**, cucurbitacin E[191,193,194] **(202)**, and datiscacin[198] **(203)** differ from the first in having a diosphenol system in ring A and are shown in Figure 56 with their activities. Datiscacin was the first cucurbitacin 20-acetate isolated from *Datisca glomerata*.[198] The third group, comprised of isocucurbitacin B[194] **(204)**, isocucurbitacin D[199] **(205)**, and 3-epi-isocucurbitacin D[199] **(206)** (Figure 57), has an α-hydroxyketone in ring A, but the positions are changed compared

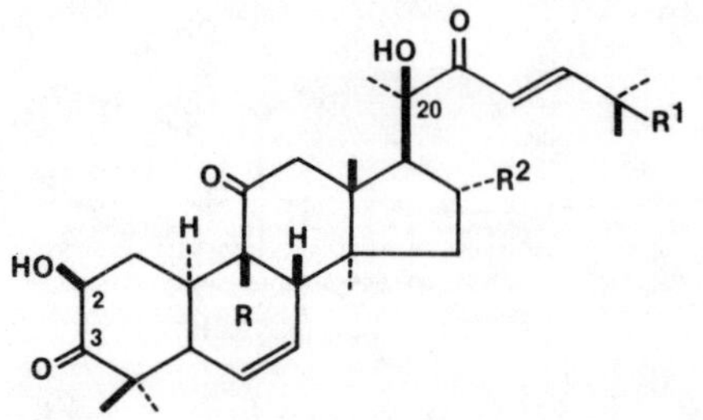

		R	R^1	R^2	ED_{50}	
196	Cucurbitacin D	CH_3	OH	OH	0.005 ~ 0.01	SA-180
197	Cucurbitacin B	CH_3	OAc	OH	1.0	SA-180
198	Fabacein	CH_3	OAc	OAc	2.5 x 10^{-6}	WA,LL
199	Cucurbitacin A	CH_2OH	OAc	OH	0.0014	
200	Datiscoside	CH_3	OH	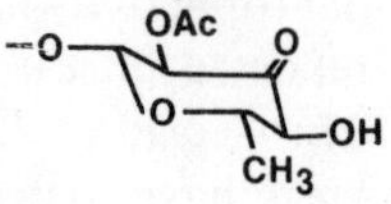 —O— OAc O O OH CH_3	0.16	PS,WM

FIGURE 55. Cucurbitacin D,[191-193] cucurbitacin B,[191,193-195] fabacein,[196] cucurbitacin A,[197] and datiscoside.[192]

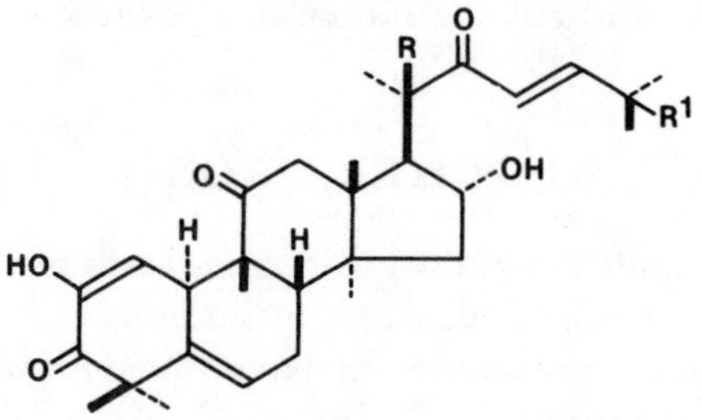

		R	R^1	ED_{50}	
201	Curcurbitacin I	OH	OH	0.005 ~ 0.01	SA-180
202	Cucurbitacin E	OH	OAc	4.5 x 10^{-7}	SA-180
203	Datiscacin	OAc	OH	Sig. Act.	

FIGURE 56. Cucurbitacin I,[191,193] cucurbitacin E,[191,193,194] and datiscacin.[198]

with the first group. The fourth group of cucurbitacins cucurbitacin C[200] **(207)**, cucurbitacin O[201] **(208)**, cucurbitacin Q[201] **(209)**, and cucurbitacin F[202] **(210)** (Figure 58) does not contain a carbonyl in ring A. The fifth group of cucurbitacins is similar to the second group in having a diosphenol system in ring A, but differs from those at C-23, C-24, which is reduced and/or hydroxylated. This includes cucurbitacin L[191,203] **(211)**, cucurbitacin J[191,203] **(212)**, and cucurbitacin K[191,203] **(213)** as shown in Figure 59. Cucurbitacins J and K differ from each other only in the configuration of the hydroxyl group at C-24. Dihydrocucurbitacin B[194] **(214)** and cucurbitacin P[201] **(215)** (Figure 60) have also been found to be cytotoxic.

Konopa et al.[193] have recently indicated, after examining the cytotoxic and antitumor activity of eight cucurbitacins, that the antitumor activity in vivo depends on the exist-

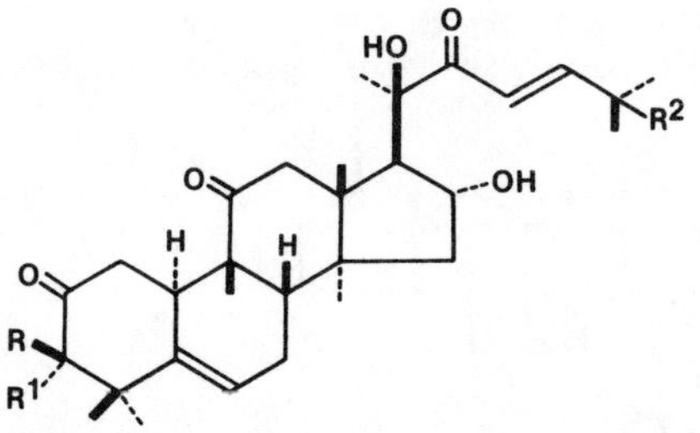

		R	R1	R2	ED_{50}
204	Isocucurbitacin B	H	OH	OAc	0.4
205	Isocucurbitacin D	H	OH	OH	0.024
206	3-epi-Isocucurbitacin D	OH	H	OH	0.24

FIGURE 57. Isocucurbitacin B,[194] isocucurbitacin D,[199] and 3-epi-isocucurbitacin D.[199]

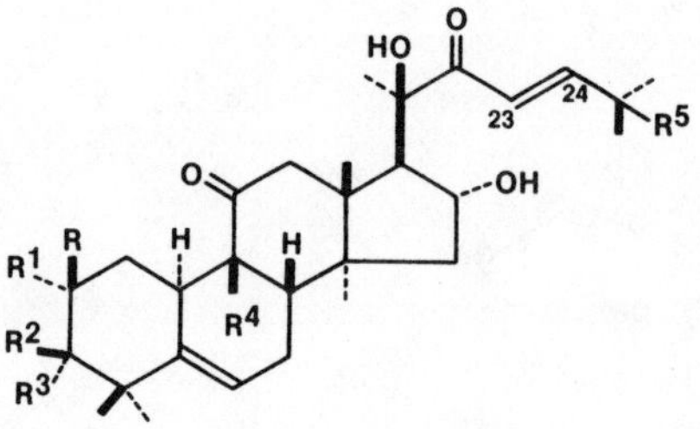

		R	R1	R2	R3	R4	R5	ED_{50}
207	Cucurbitacin C	H	H	OH	H	CH_2OH	OAc	0.001
208	Cucurbitacin O	H	OH	H	OH	CH_3	OH	
209	Cucurbitacin Q	H	OH	H	OH	CH_3	OAc	0.032
210	Cucurbitacin F	OH	H	OH	H	CH_3	OH	

FIGURE 58. Cucurbitacin C,[200] cucurbitacin O,[201] cucurbitacin Q,[201] and cucurbitacin F.[202]

ence of a double bond in the side-chain of these compounds. Further, though they exhibit exceptionally high cytotoxicity in tissue culture test on KB and HeLa cells (ED_{50} 1.0 or 0.005 $\mu g/m\ell$), they do not have any practical value as potential antitumor drugs because of their high toxicity and low in vivo antitumor activity.

B. Dammaranes

Anisimov et al.[204] have shown that the aglycons of panaxoside A and dihydropanaxoside A, and the diglucoside of panaxadiol[205] possess high cytotoxic effects at concentrations of 10 and 25 $\mu g/m\ell$ on developing eggs of the sea urchin.

C. Lanostanes

Two compounds of this class, sapelin A[206,207] **(216)** and sapelin B[206,207] **(217)** (Figure 61), isolated from *Entandrophragma cylindricum* Sprague (sapele)[206] and *Brusera Klugii* Macbr (Burseraceae)[207] have been shown to be active against the P-388 lymphocytic

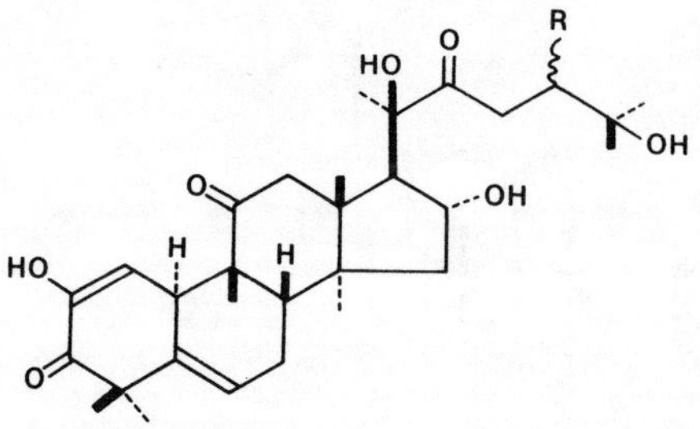

	R	ED_{50}
211 Cucurbitacin L	H	0.01 ~0.1
212 Cucurbitacin J	OH	0.1 ~1.0
213 Cucurbitacin K	OH	0.1 ~1.0

FIGURE 59. Cucurbitacin L,[191,203] cucurbitacin J,[191,203] and cucurbitacin K.[191,203]

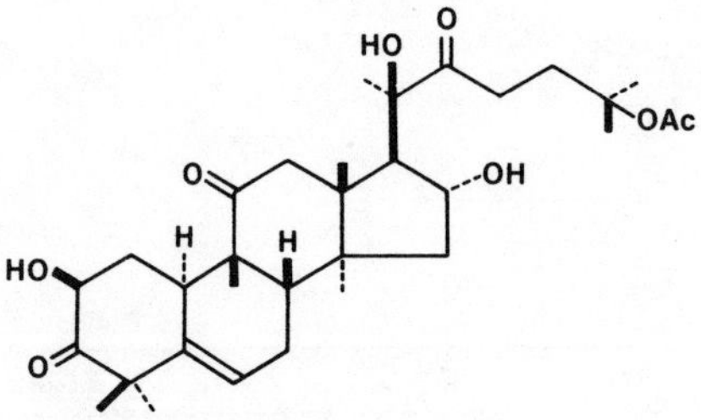

214 Dihydrocucurbitacin B
ED_{50} 0.0017

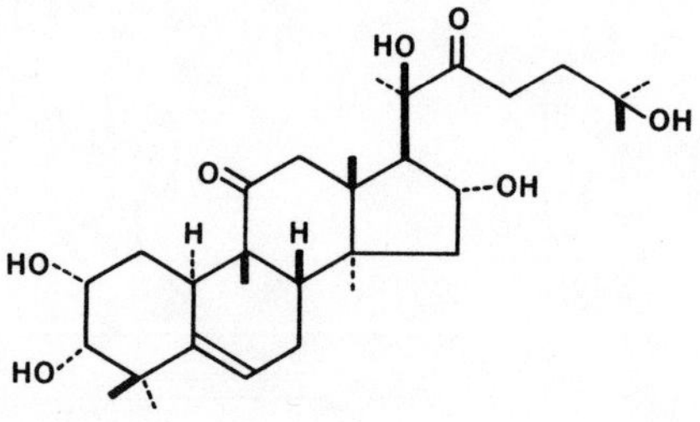

215 Cucurbitacin P
ED_{50} 0.61

FIGURE 60. Dihydrocucurbitacin B[194] and cucurbitacin P.[201]

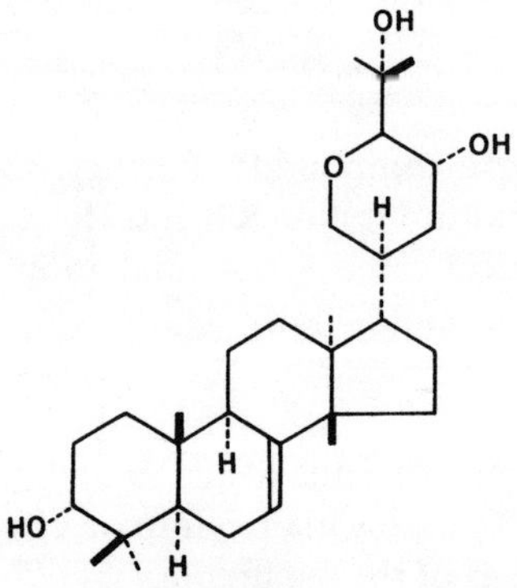

216 Sapelin A
PS (136)

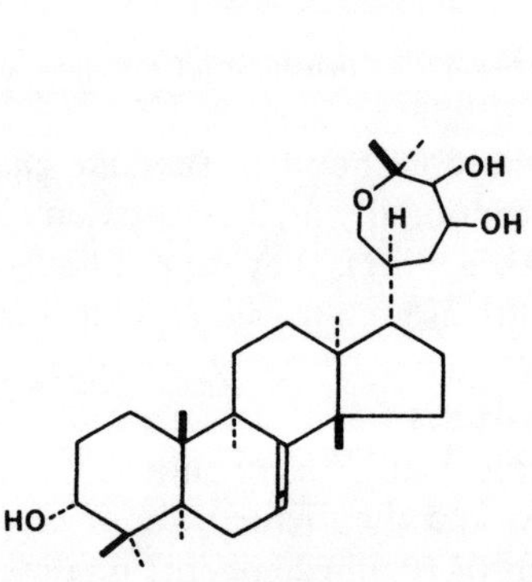

217 Sapelin B
PS (138)

FIGURE 61. Sapelin A[206,207] and sapelin B.[206,207]

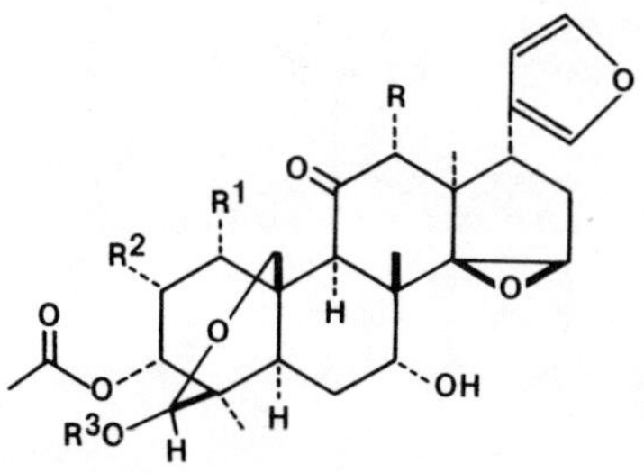

		R	R¹	R²	R³	ED_{50}
218	Aphanastatin	OH	OAc	OH	$CO.CH(CH_3)\text{-}C_2H_5$	0.065
219	Amoorastatin	H	OH	H	H	<0.001

FIGURE 62. Aphanastatin[208] and amoorastatin.[209]

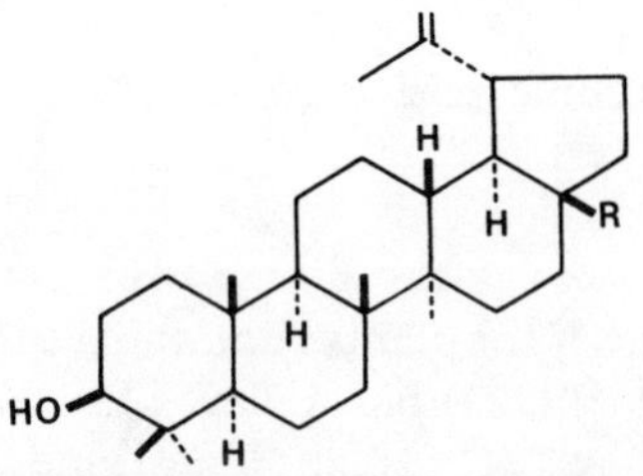

		R	ED_{50}
220	Lupeol	CH_3	WA
221	Betulin	CH_2OH	WA
222	Betulinic acid	COOH	WA,PS

FIGURE 63. Lupeol,[210] betulin,[210] and betulinic acid.[211]

leukemia (3PS) and the human epidermoid carcinoma of the nasopharynx (9KB). Sapelin A had a T/C of 130, 136, 127, and 130, and sapelin B had T/C of 136, 136, 130, and 138 at 10.0, 5.0, 2.5, and 1.25 mg/kg dose levels, respectively. The structures of these compounds were established by spectroscopic and chemical methods.

D. Limonoids

Polonski et al.[208,209] have shown recently that aphanastatin[208] **(218)** and amoorastatin[209] **(219)** (Figure 62) isolated from *Aphanamixis grandifolia* B1 seeds possess high cytotoxicity against P-388 cell lines (ED_{50} 0.65 and <0.001 μg/mℓ, respectively). The structures of these two highly oxygenated limonoids were established by X-ray analysis. A comparison of their structures suggested that the 1 α-acetoxy, 2 α- and 12 α-dihydroxy, and 28 α-methyl butyryl groups of aphanastatin are unnecessary and reduce the antineoplastic activity.[209]

E. Lupanes

Recently, it has been shown that lupeol[210] **(220)**, betulin[210] **(221)**, and betulinic acid[211] **(222)** (Figure 63) are active against Walker carcinoma 256 (intramuscular) tumor system. Lupeol shows a T/C of 39 at 200 mg/kg dose level,[210] betulin has a T/C of 13 at

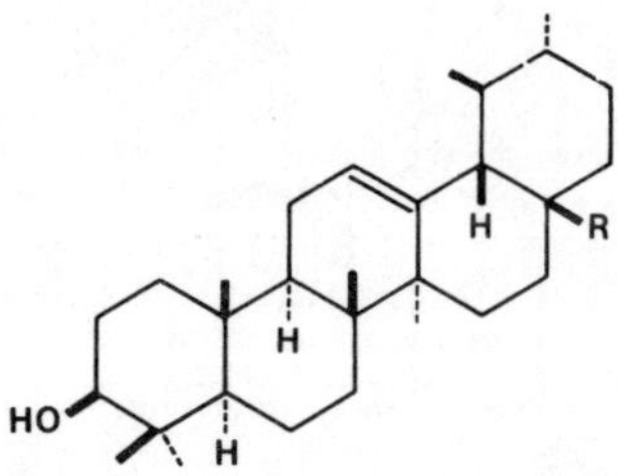

		R	ED_{50}
223	Uvaol	CH_2OH	PS(125)
224	Ursolic acid	COOH	PS(125)
225	α-Amyrin	CH_3	WA(37)

FIGURE 64. Uvaol,[212,213] ursolic acid,[212,213] and α-amyrin.[20,214]

600 mg/kg dose level,[210] and betulinic acid showed a T/C of 135 at 100 mg/kg and 140 at 50 mg/kg dose level in 3PS system.[211]

F. Ursanes

As a result of the continuing search for plants containing tumor inhibitors, Trumbull et al.[212] have shown that uvaol[212,213] **(223)** has a T/C 125 at 100 and 200 mg/kg dose level and ursolic acid[212,213] **(224)** has T/C 125 at 50 mg/kg dose level in 3PS. Another member of this class α-amyrin[20,214] **(225)** has been shown active in Walker carcinoma 256 (TWI[%] 63 at 50 to 400 mg/kg dose level). These are shown in Figure 64.

G. Oleananes

A number of glucosides of this class are known to possess activity against Walker carcinoma 256. The structure of aescin[215] **(226)**, isolated from the seeds of *Aesculus hippocastanum* L. is shown in Figure 65. It shows a T/C of 35 at 3 to 60 mg/kg dose level. The other compounds, myrsine saponin the active principle of *Myrsine africana* L.[216] and *Wallenia yunquensis* (Myrisinaceae)[217] (WA: T/C 27, 34, and 62 at 8, 6, and 4 mg/kg dose levels, respectively), saponin P from *Acer negundo* L. (active against S-180, WM),[218,219] and cyclamin isolated from *Cyclamen persicum* Mill. (WA: T/C 36 at 3 to 60 mg/kg dose levels)[219] have been shown active in vivo, but their complete structures are not known.

H. Taraxeranes

Only one compound of this class acetylaleuritolic acid **(227)**, (Figure 66) isolated from the roots of *Jatropha macrorhiza* Benth (Euphorbiaceae), has been shown to possess significant activity in the P-388 test systems (T/C 158 and 128 at 1.0 and 1.4 mg/kg dose levels, respectively).[220]

I. Quinonoid-Nortriterpenoids

Quinonoid-nortriterpenoids are quinone methides.[221] Iguesterin[222,223] **(228)** (Figure 67), isolated from the root bark of *Catha cassinoides*[222] and *Matenus canariensis,*[223] shows an ID_{50} of 0.6 μg/mℓ on HeLa cells. Its structure was determined spectroscopically and by synthesis from tingenone.[222] Pristimerin[223,226] **(229)** (Figure 67) isolated from *Pristimera indica* (Willd.) A. C. Smith, *Pristimera grahamii* (Wight) A. C.

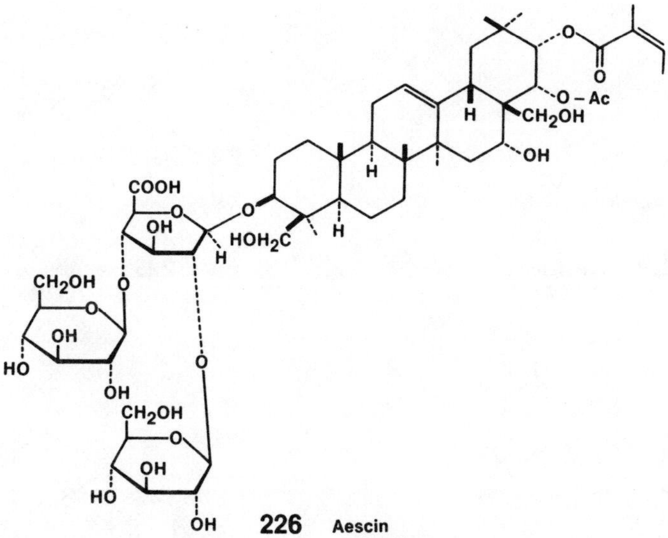

226 Aescin
WA (35)

FIGURE 65. Aescin.[215]

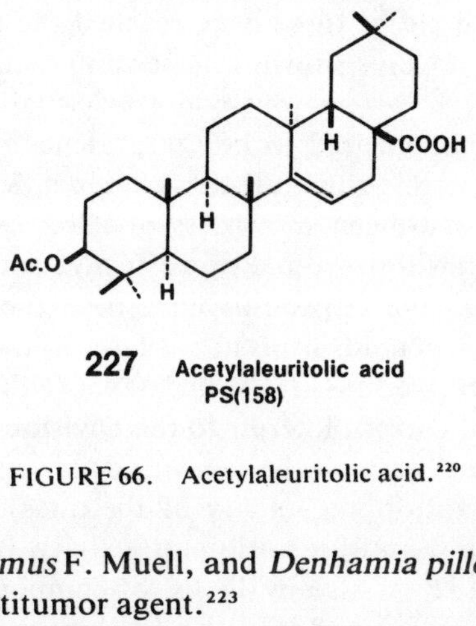

227 Acetylaleuritolic acid
PS(158)

FIGURE 66. Acetylaleuritolic acid.[220]

Smith,[224] *Celastrus dispermus* F. Muell, and *Denhamia pillosporides*[225] has been shown to be a potent but toxic antitumor agent.[223]

Two more compounds of this group tingenin A[223,226] (maitenin) **(230)** and tingenin B[226] **(231)** (Figure 67) have been isolated from the leaves of *Euonymus tingens* Wall (Celastraceae)[226] and *Maytenus* sp. (Celastraceae).[223] Besides cytotoxicity, tingenin A exhibits in vivo antitumor activity and significant inhibitory effect on protein and RNA synthesis.[223]

VII. CONCLUSIONS

As indicated earlier, only four monoterpenoids have been shown to have some antitumor activity. More work needs to be done to determine if these will be useful in cancer chemotherapy.

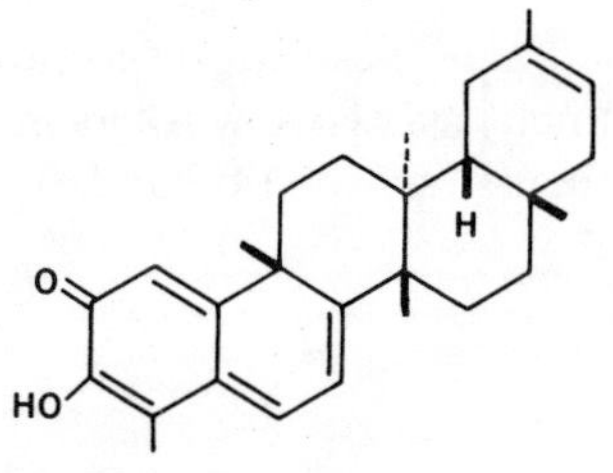

228 Iguesterin
ID_{50} 0.6

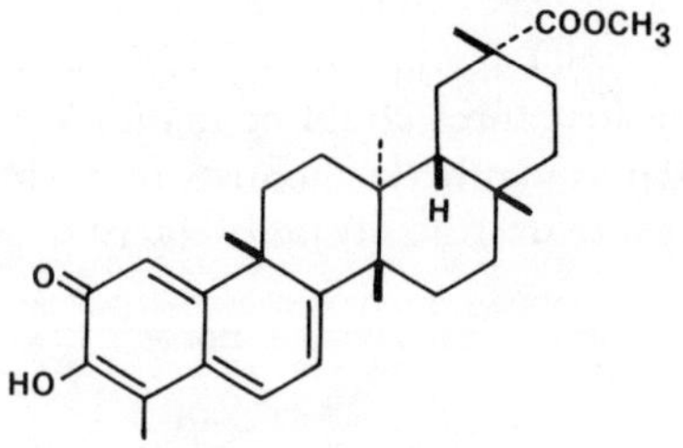

229 Pristimerin
ID_{50} 0.6

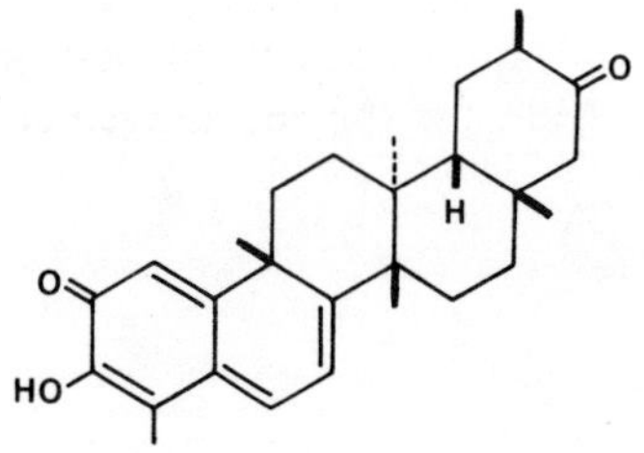

230 Tingenin A (Maitenin)
ID_{50} 0.3

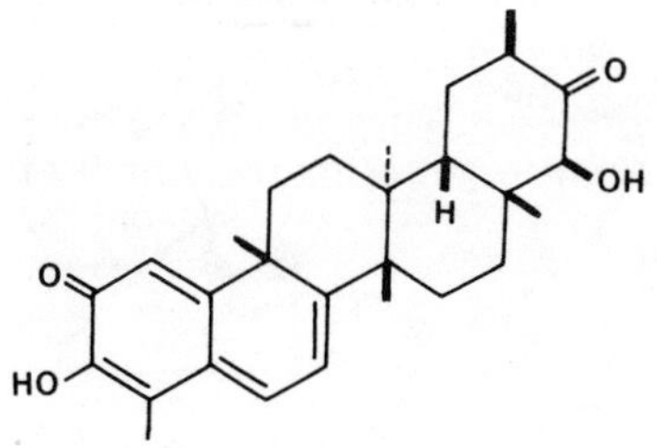

231 Tingenin B

FIGURE 67. Iguesterin,[222,223] pristimerin,[223,226] tingenin A,[223,226] and tingenin B.[226]

Although a large number of sesquiterpenoids have been shown to possess cytotoxic and/or antitumor activity, none of these have reached the clinical trial stage. Recent studies by Mitchell et al.[227-231] have shown that sesquiterpenoid lactones are the common denominators for allergic eczematous contact dermatitis. The immunochemical requisite for such dermatitis appeared to be the presence of an exocyclic methylene group conjugated to the γ-lactone ring in their structures. Since such a system is reactive to cysteine[4,21] and has also been shown essential for cytotoxicity,[22] it is possible that sesquiterpenoids as such will never make an effective anticancer agent.

The most important of all the terpenoids have been the diterpenoids and quassinoids. Though the active diterpenoids list is not as large as that of the sesquiterpenoids, two of these tripdiolide (NSC-163063) **(167)** and taxol (NSC-125973) **(170)** are in the list of chemical structures of current interest to the Division of Cancer Treatment and are being developed.

Following the report of antitumor activity of the quassinoid holacanthone **(187)**, several new quassinoids with promising antitumor activity were isolated. One of these, bruceantin (NSC-165563) **(182)**, is also in the list of compounds of current interest to the Division of Cancer Treatment and is in the development stage. Among triterpenoids, though cucurbitacins form the largest group and are highly cytotoxic, the chances of their making an anticancer drug are very slim because of their toxicity and low in vivo activity.

A few compounds from other classes of triterpenoids: dammaranes, lanostanes, limonoids, lupanes, ursanes, oleananes, taraxeranes, and quinonoid-nortriterpenoids have shown some activity but none of these are in the list of compounds of interest to the Division of Cancer Treatment. Perhaps more work is needed with this group of compounds.

In the past, work has been carried out on plant materials either by collecting a variety of samples from throughout the world, or by relying on a folklore history. This has resulted in the isolation of many known compounds with activity. We have attempted to classify and place all similar structural types together and compare their activities in a tabular form, so that a rapid evaluation of their activities with respect to changes in the structures could be made. We hope this will help the active workers in this field as well as synthetic chemists to design their experiments for isolation and/or synthesis of a more useful antitumor agent.

ACKNOWLEDGMENTS

This work was supported in part by the Public Health Service, National Cancer Institute under Contract No. NO-1-CO-75380. We are thankful to Dr. R. S. D. Mittal of Minerec Corporation, Baltimore, Md. and Dr. M. W. Slein of Frederick Cancer Research Center, Frederick, Md. for their assistance in collecting part of the literature. Our special thanks are due to Dr. R. J. White, Director, Chemotherapy Fermentation Program, Frederick Cancer Research Center, Frederick, Md., for the encouragement in finalizing this manuscript.

REFERENCES

1. **Ruzicka, L.,** *Experientia*, 9, 357, 1953.
2. **Devon, T. K. and Scott, A. I.,** *Handbook of Naturally Occurring Compounds*, Vol. II. Terpenes, Academic Press, New York, 1972.
3. **Kupchan, S. M.,** *Pure and Appl. Chem.*, 21, 227, 1970.
4. **Kupchan, S. M.,** *Fed. Proc.*, 33, 2288, 1974.
5. **Rodriguez, E., Towers, G. H. N., and Mitchell, J. C.,** *Phytochemistry*, 15, 1573, 1976.
6. **Hartwell, J. L.,** *Cancer Treat. Rep.*, 60, 1031, 1976.
7. **Pettit, G. R.,** *Biosynthetic Products for Cancer Chemotherapy*, Vol. 1., Plenum Press, New York, 1977.
8. **Cordell, G. A. and Farnsworth, N. R.,** *Lloydia*, 40, 1, 1977.
9. **Pettit, G. R.,** *Biosynthetic Products for Cancer Chemotherapy*, Vol. 2, Plenum Press, New York, 1978.
10. **Cordell, G. A.,** in *Progress in Phytochemistry*, Vol. 5, Reinhold, L., Harborne, J. B., and Swain, T., Eds., Pergamon Press, New York, 1978, 273.
11. **Schepartz, S. A.,** *Cancer Treat. Rep.*, 60, 975, 1976.
12. **Perdue, Jr., R. E.,** *Cancer Treat. Rep.*, 60, 987, 1976.
13. **Statz, D. and Coon, F. B.,** *Cancer Treat. Rep.*, 60, 999, 1976.
14. **Wall, M. E., Wani, M. C., and Taylor, H.,** *Cancer Treat. Rep.*, 60, 1011, 1976.
15. **Abbott, B. J.,** *Cancer Treat. Rep.*, 60, 1007, 1976.
16. **Spjut, R. W. and Perdue, Jr., R. E.,** *Cancer Treat. Rep.*, 60, 979, 1976.
17. **Geran, R. I., Greenberg, N. H., Macdonald, M. M., Schumacher, A. M., and Abbott, B. J.,** *Cancer Chemother. Rep.*, 3, 1, 1972.
18. **Barclay, A. S. and Perdue, Jr., R. E.,** *Cancer Treat. Rep.*, 60, 1081, 1976.
19. Instruction 14: Screening Data Summary Interpretation and Outline of Current Screen, Drug Evaluation Branch, Division of Cancer Treatment, National Cancer Institute, Bethesda, Md.
20. **Hartwell, J. L. and Abbott, B. J.,** *Ad. Pharmacol. Chemother.*, 7, 117, 1969.
21. **Fujita, E. and Nagao, Y.,** *Bioorg. Chem.*, 6, 287, 1977.
22. **Kupchan, S. M., Eakin, M. A., and Thomas, A. M.,** *J. Med. Chem.*, 14, 1147, 1971.
23. **Kupchan, S. M., Dessertine, A. L., Blaylock, B. T., and Bryan, R. F.,** *J. Org. Chem.*, 39, 2477, 1974.
24. **Jolad, S., Hoffmann, J. J., Wiedhopf, R. M., Cole, J. R., Bates, R. B., and Kriek, G. R.,** *Tetrahedron Lett.*, 4119, 1976.
25. **Rogers, D., Moss, G. P., and Neidle, S.,** *J. Chem. Soc. Chem. Commun.*, 142, 1972.
26. **Neidle, S. and Rogers, D.,** *J. Chem. Soc. Chem. Commun.*, 140, 1972.
27. **Rao, A. S., Kelkar, G. R., and Bhattacharyya, S. C.,** *Tetrahedron*, 9, 275, 1960.
28. **Doskotch, R. W. and El-Feraly, F. S.,** *J. Pharm. Sci.*, 58, 877, (1969).

29. Fischer, N. H., Mabry, T. J., and Kagan, H. B., *Tetrahedron,* 24, 4091, 1968.
30. Lee, K. H., Huang, H. C., Huang, E. S., and Furukawa, H., *J. Pharm. Sci.,* 61, 629, 1972.
31. Doskotch R. W. and El-Feraly, F. S., *J. Org. Chem.,* 35, 1928, 1970.
32. Kupchan, S. M., Fujita, T., Maruyama, M., and Britton, R. W., *J. Org. Chem.,* 38, 1260, 1973.
33. Kupchan, S. M., Ashmore, J. W., and Sneden, A. T., *Phytochemistry,* 16, 1834, 1977.
34. Gnecco, S., Poyser, J. P., Silva, M., Sammes, P. G., and Tyler, T. W., *Phytochemistry,* 12, 2469, 1973.
35. Bialecki, M., Hladon, B., Drozdz, B., Bloszyk, E., Szwemin, S., and Bobkiewicz, T., *Pol. J. Pharmacol. Pharm.,* 26, 511, 1974.
36. Herz, W., de Groote, R., Murari, R., and Kumar, N., *J. Org. Chem.,* 44, 2784, 1979.
37. Gonzalez, A. G., Darias, V., Alonso, G., Boada, J. N., and Feria, M., *Planta Med.,* 33, 356, 1978.
38. Govindachari, T. R., Joshi, B. S., and Kamat, V. N., *Tetrahedron,* 21, 1509, 1965.
39. Wiedhopf, R. M., Young, M., Bianchi, E., and Cole, J. R., *J. Pharm. Sci.,* 62, 345, 1973.
40. Ogura, M., Cordell, G. A., and Farnsworth, N. R., *Phytochemistry,* 17, 957, 1978.
41. Doskotch, R. W., Keely, Jr., S. L., Hufford, C. D., and El-Feraly, F. S., *Phytochemistry,* 14, 769, 1975.
42. Cassady, J. M., Ojima, N., Chang, C. J., and McLaughlin, J. L., *Phytochemistry,* 18, 1569, 1979.
43. Lee, K. H., Kimura, T., Okamoto, M., Cowherd, C. M., McPhail, A. T., and Onan, K. D., *Tetrahedron Lett.,* 1051, 1976.
44. Hall, I. H., Lee, K. H., Starnes, C. O., Eigebaly, S. A., Ibuka, T., Wu, Y. S., Kimura, T., and Haruna, M., *J. Pharm. Sci.,* 67, 1235, 1978.
45. Kupchan, S. M., Aynehchi, Y., Cassady, J. M., Schnoes, H. K. and Burlingame, A. L., *J. Org. Chem.,* 34, 3867, 1969.
46. Lee, K. H., Cowherd, C. M., and Wolo, M. T., *J. Pharm. Sci.,* 64, 1572, 1975.
47. Herz, W., Santhanam, P. S., Subramaniam, P. S., and Schmid, J. J., *Tetrahedron Lett.,* 3111, 1967.
48. Geissman, T. A., Turley, R. J., and Murayama, S., *J. Org. Chem.,* 31, 2269, 1966.
49. Pettit, G. R., Herald, C. L., Judd, G. F., Bolliger, G., Vanell, L. D., Lehto, E., and Pase, C. P., *Lloydia,* 41, 29, 1978.
50. Waddell, T. G. and Geissman, T. A., *Phytochemistry,* 8, 2371, 1969.
51. El-Feraly, F. S., Chan, Y.-M., Fairchild, E. H., and Doskotch, R. W., *Terahedron Lett.,* 1973, 1977.
52. Doskotch, R. W., El-Feraly, F. S., Fairchild, E. H., and Huang, C. T., *J. Org. Chem.,* 42, 3614, 1977.
53. Doskotch, R. W., El-Feraly, F. S., Fairchild, E. H., and Huang, C. T., *J. Chem. Soc. Chem. Commun.,* 402, 1976.
54. McPhail, A. T. and Onan, K. D., *J. Chem. Soc. Perkin Trans. II,* 578, 1976.
55. McPhail, A. T., Onan, K. D., Lee, K. H., Ibuka, T., and Huang, H. C., *Tetrahedron Lett.,* 3203, 1974.
56. Herz, W. and Wahlberg, I., *J. Org. Chem.,* 38, 2485, 1973.
57. Kupchan, S. M., Maruyama, M., Hemingway, R. J., Hemingway, J. C., Shibuya, S., Fujita, T., Cradwick, P. D., Hardy, A. D. U., and Sim, G. A., *J. Am. Chem. Soc.,* 93, 4914, 1971.
58. Kupchan, S. M., Maruyama, M., Hemingway, R. J., Hemingway, J. C., Shibuya, S., and Fujita, T., *J. Org. Chem.,* 38, 2189, 1973.
59. Lee, K. H., Kimura, T., Haruna, M., McPhail, A. T., Onan, K. D., and Huang, H. C., *Phytochemistry,* 16, 1068, 1977.
60. Torrance, S. J., Geissman, T. A., and Chedekel, M. R., *Phytochemistry,* 8, 2381, 1969.
61. Herz, W. and Sharma, R. P., *Phytochemistry,* 14, 1561, 1975.
62. Herz, W. and Sharma, R. P., *J. Org. Chem.,* 41, 1248, 1976.
63. Lee, K. H., Ibuka, T., Huang, H. C., and Harris, D. L., *J. Pharm. Sci.,* 64, 1077, 1975.
64. McPhail, A. T., Onan, K. D., Lee, K. H., Ibuak, T., Kozuka, M., Shingu, T., and Huang, H. C., *Tetrahedron Lett.,* 2739, 1974.
65. Lee, K. H., Furukawa, H., Kozuka, M., Huang, H. C., Luhan, P. A., and McPhail, A. T., *J. Chem. Soc. Chem. Commun.,* 476, 1973.
66. Kupchan, S. M., Davies, V. H., Fujita, T., Cox, M. R., and Bryan, R. F., *J. Am. Chem. Soc.,* 93, 4916, 1971.
67. Kupchan, S. M., Davies, V. H., Fujita, T., Cox, M. R., Restivo, R. J., and Bryan, R. F., *J. Org. Chem.,* 38, 1853, 1973.
68. Pal, R., Kulshreshtha, D. K., and Rastogi, R. P., *J. Pharm. Sci.,* 65, 918, 1976.
69. Baruah, N. C., Sharma, R. P., Madhusudanan, K. P., Thyagarajan, G., Herz, W., and Murari, R., *J. Org. Chem.,* 44, 1831, 1979.
70. Raffauf, R. F., Huang, P. K. C., Quesne, P. W. L., Levery, S. B., and Brennan, T. F., *J. Am. Chem. Soc.,* 97, 6884, 1975.

71. Quesne, P. W. L., Levery, S. B., Menachery, M. D., Brennan, T. F., and Raffauf, R. F., *J. Chem. Soc. Perkin Trans. I*, 1572, 1978.
72. Herz, W., de Groote, R., Murari, R., and Blount, J. F., *J. Org. Chem.*, 43, 3559, 1978.
73. Herz, W., Murari, R., and Govindan, S. V., *Phytochemistry*, 18, 1337, 1979.
74. Lee, K. H., Ibuka, T., Kozuka, M., McPhail, A. T., and Onan, K. D., *Tetrahedron Lett.*, 2287, 1974.
75. Herz, W., de Vivar, A. R., and Lakshmikantham, M. V., *J. Org. Chem.*, 30, 118, 1965.
76. Pettit, G. R., Herald, C. L., Judd, G. F., Bolliger, G., and Thayer, P. S., *J. Pharm. Sci.*, 64, 2023, 1975.
77. Kupchan, S. M., Cassady, J. M., Kelsey, J. E., Schnoes, H. K., Smith, D. H., and Burlingame, A. L., *J. Am. Chem. Soc.*, 88, 5292, 1966.
78. Dullforce, T. A., Sim, G. A., White, D. N. J., Kelsey, J. E., and Kupchan, S. M., *Tetrahedron Lett.*, 973, 1969.
79. Kupchan, S. M., Cassady, J. M., Bailey, J., and Knox, J. R., *J. Pharm. Sci.*, 54, 1703, 1965.
80. Jolad, S. D., Wiedhopf, R. M., and Cole, J. R., *J. Pharm. Sci.*, 63, 1321, 1974.
81. de Vivar, A. R., Cabrera, A., Ortega, A., and Romo, J., *Tetrahedron*, 23, 3903, 1967.
82. Corbella, A., Gariboldi, P., Jommi, G., Samek, Z., Holub, M., Drozdz, B., and Bloszyk, E., *J. Chem. Soc. Chem. Commun.*, 386, 1972.
83. Samek, Z., Holub, M., Drozdz, B., Iommi, G., Corbella, A., and Gariboldi, P., *Tetrahedron Lett.*, 4775, 1971.
84. Gonzalez, A. G., Bermejo, J., Cabrera, I., Massanet, G. M., Mansilla, H., and Galindo, A., *Phytochemistry*, 17, 955, 1978.
85. Kupchan, S. M., Kelsey, J. E., Maruyama, M., Cassady, J. M., Hemingway, J. C., and Knox, J. R., *J. Org. Chem.*, 34, 3876, 1969.
86. Kupchan, S. M., Hemingway, J. C., Cassady, J. M., Knox, J. R., McPhail, A. T., and Sim, G. A, *J. Am. Chem. Soc.*, 89, 465, 1967.
87. Herz, W., Poplawski, J., and Sharma, R. P., *J. Org. Chem.*, 40, 199, 1975.
88. Kupchan, S. M., Kelsey, J. E., Maruyama, M., and Cassady, J. M., *Tetrahedron Lett.*, 3517, 1968.
89. Cassady, J. M., Abramson, D., Cowall, P., Chang, C. J., and McLaughlin J. L., *J. Nat. Prod.*, 42, 427, 1979.
90. Fayos, L. J., Blanco, S. G., and Ripoll, M. M., *Acta Crystallogr.*, B34, 2669, 1978.
91. Silva, M., Wiesenfeld, A., Sammes, P. G., and Tyler, T. W., *Phytochemistry*, 16, 379, 1977.
92. Doskotch, R. W. and Hufford, C. D., *J. Pharm. Sci.*, 58, 186, 1969.
93. Quallich, G. J. and Schlessinger, R. H., *J. Am. Chem. Soc.*, 101, 7627, 1979.
94. Miller, H. E. and Mabry, T. J., *J. Org. Chem.*, 32, 2929, 1967.
95. Fischer, N. H. and Mabry, T. J., *Tetrahedron*, 23, 2529, 1967.
96. Herz, W. and Hogenauer, G., *J. Org. Chem.*, 26, 5011, 1961.
97. Torrance, S. J., Wiedhopf, R. M., and Cole, J. R., *J. Pharm. Sci.*, 64, 887, 1975.
98. Herz, W., Watanabe, H., Miyazaki, M., and Kishida, Y., *J. Am. Chem. Soc.*, 84, 2601, 1962.
99. Romo, J., Nathan, P. J., and Fernando Diaz, A., *Tetrahedron*, 20, 79, 1964.
100. Dominguez, E. and Romo, J., *Tetrahedron*, 19, 1415, 1963.
101. Grieco, P. A., Ohfune, Y., and Majetich, G., *Tetrahedron Lett.*, 3265, 1979.
102. Hall, I. H., Lee, K. H., and Eigebaly, S. A., *J. Pharm. Sci.*, 67, 552, 1978.
103. Pettit, G. R., Budzinski, J. C., Cragg, G. M., Brown, P., and Johnston, L. D., *J. Med. Chem.*, 17, 1013, 1974.
104. Ohfune, Y., Grieco, P. A., Wang, C.-L. J., and Majetich, G., *J. Am. Chem. Soc.*, 100, 5946, 1978.
105. Roberts, M. R. and Schlessinger, R. H., *J. Am. Chem. Soc.*, 101, 7626, 1979.
106. Herz, W., Gast, C. M., and Subramaniam, P. S., *J. Org. Chem.*, 33, 2780, 1968.
107. Herz, W., Rajappa, S., Roy, S. K., Schmid, J. J., and Mirrington, R. N., *Tetrahedron*, 22, 1907, 1966.
108. Pettit, G. R., Herald, C. L., Gust, D., Herald, D. L., and Vanell, L. D., *J. Org. Chem.*, 43, 1092, 1978.
109. Lee, K. H., Ibuka, T., McPhail, A. T., Onan, K. D., Geissman, T. A., and Waddell, T. G., *Tetrahedron Lett.*, 1149, 1974.
110. Einck, J. J., Herald, C. L., Pettit, G. R., and Von Dreele, R. B., *J. Am. Chem. Soc.*, 100, 3544, 1978.
111. Herz, W., Aota, K., Hall, A. L., and Srinivasan, A., *J. Org. Chem.*, 39, 2013, 1974.
112. Lee, K. H., Imakura, Y., Sims, D., McPhail, A. T., and Onan, K. D., *Phytochemistry*, 16, 393, 1977.
113. Sims, D., Lee, K. H., Wu, R. Y., Furukawa, H., Itoigawa, M., and Yonaha, K., *J. Nat. Prod.*, 42, 282, 1979.

114. Aota, K., Caughlan, C. N., Emerson, M. T., Herz, W., Inayama, S., and Haque, M., *J. Org. Chem.*, 35, 1448, 1970.
115. Herz, W., Ueda, K., and Inayama, S., *Tetrahedron*, 19, 483, 1963.
116. Herz, W. and Roy, S. K., *Phytochemistry*, 8, 661, 1969.
117. Von Dreele, R. B., Pettit, G. R., Cragg, G. M., and Ode, R. H., *J. Am. Chem. Soc.*, 97, 5256, 1975.
118. Herz, W., Rohde, W. A., Rabindran, K., Jayaraman, P., and Viswanathan, N., *J. Am. Chem. Soc.*, 84, 3857, 1962.
119. Ivie, G. W., Witzel, D. A., and Rushing, D. D., *J. Agric. Food Chem.*, 23, 845, 1975.
120. Lucas, R. A., Rovinski, S., Kiesel, R. J., Dorfman, L., and MacPhillamy, H. B., *J. Org. Chem.*, 29, 1549, 1964.
121. Herz, W., Aota, K., Holub, M., and Samek, Z., *J. Org. Chem.*, 35, 2611, 1970.
122. Lee, K. H., Imakura, Y., Sims, D., McPhail, A. T., and Onan, K. D., *J. Chem. Soc. Chem. Commun.*, 341, 1976.
123. Imakura, Y., Lee, K. H., Sims, D., and Hall, I. H., *J. Pharm. Sci.*, 67, 1228, 1978.
124. Kupchan, S. M., Hemingway, R. J., Werner, D., Karim, A., McPhail, A. T., and Sim, G. A., *J. Am. Chem. Soc.*, 90, 3596, 1968.
125. Kupchan, S. M., Hemingway, R. J., Werner, D., and Karim, A., *J. Org. Chem.*, 34, 3903, 1969.
126. Iio, H., Isobe, M., Kawai, T., and Goto, T., *J. Am. Chem. Soc.*, 101, 6076, 1979.
127. Kieczykowski, G. R. and Schlessinger, R. H., *J. Am. Chem. Soc.*, 100, 1938, 1978.
128. Kupchan, S. M., Hemingway, R. J., Karim, A., and Werner, D., *J. Org. Chem.*, 34, 3908, 1969.
129. Herz, W. and Hogenauer, G., *J. Org. Chem.*, 27, 905, 1962.
130. Herz, W. and Viswanathan, N., *J. Org. Chem.*, 29, 1022, 1964.
131. Marshall, J. A. and Cohen, N., *J. Org. Chem.*, 29, 3727, 1964.
132. Herz, W., Mitra., R. B., Rabindran, K., and Viswanathan, N., *J. Org. Chem.*, 27, 4041, 1962.
133. Kupchan, S. M., Baxter, R. L., Chiang, C.-K., Gilmore, C. J., and Bryan, R. F., *J. Chem. Soc., Chem. Commun.*, 842, 1973.
134. Toubiana, R. and Gaudemer, A., *Tetrahedron Lett.*, 1333, 1967.
135. Bianchi, E., Culvenor, C. C. J., and Loder, J. W., *Aust. J. Chem.*, 21, 1109, 1968.
136. Jamieson, G. R., Reid, E. H., Turner, B. P., and Jamieson, A. T., *Phytochemistry*, 15, 1713, 1976.
137. Bermejo, J., Breton, J. L., Fajardo, M., nd Gonzalez, A. G., *Tetrahedron Lett.*, 3475, 1967.
138. Kupchan, S. M., LaVoie, E. J., Branfman, A. R., Fei, B. Y., Bright, W. M., and Bryan, R. F., *J. Am. Chem. Soc.*, 99, 3199, 1977.
139. Jolad, S. D., Wiedhopf, R. M., and Cole, J. R., *J. Pharm. Sci.*, 64, 1889, 1975.
140. Arai, T., Koyama, Y., Suenaga, T., and Morita, T., *J. Antibiot. Ser. A*, 16, 132, 1963.
141. Fujita, E., Nagao, Y., Kaneko, K., Nakazawa, S., and Kuroda, H., *Chem. Pharm. Bull.*, 24, 2118, 1976.
142. Fujita, E., Nagao, Y., Node, M., Kaneko, K., Nakazawa, S., and Kuroda, H., *Experientia*, 32, 203, 1976.
143. Kubo, I., Miura, I., Nakanishi, K., Kamikawa, T., Isobe, T., and Kubota, T., *J. Chem. Soc., Chem. Commun.*, 555, 1977.
144. Kubo, I., Pettei, M. J., Hirotsu, K., Tsuji, H., and Kubota, T., *J. Am. Chem. Soc.*, 100, 628, 1978.
145. Kupchan, S. M., Sweeny, J. G., Baxter, R. L., Murae, T., Zimmerly, V. A., and Sickles, B. R., *J. Am. Chem. Soc.*, 97, 672, 1975.
146. Kupchan, S. M. and Baxter, R. L., *Science*, 187, 652, 1974.
147. Kupchan, S. M., Shizuri, Y., Sumner, Jr., W. C., Haynes, H. R., Leighton, A. P., and Sickles, B. R., *J. Org. Chem.*, 41, 3850, 1976.
148. Ogura, M., Koike, K., Cordell, G. A., and Farnsworth, N. R., *Planta Med.*, 33, 128, 1978.
149. Sakata, K., Kawazu, K., and Mitsui, T., *Agric. Biol. Chem.*, 35, 2113, 1971.
150. Kupchan, S. M., Shizuri, Y., Murae, T., Sweeny, J. G., Haynes, H. R., Shen, M. S., Barrick, J. C., Bryan, R. F., Helm, D., and Wu, K. K., *J. Am. Chem. Soc.*, 98, 5719, 1976.
151. Hecker, E., *Cancer Res.*, 28, 2338, 1968.
152. Kupchan, S. M., Uchida, I., Branfman, A. R., Dailey, Jr., R. G., and Fei, B. Y., *Science*, 191, 571, 1976.
153. Evans, F. J. and Soper, C. J., *Lloydia*, 41, 193, 1978.
154. Kupchan, S. M., Baxter, R. L., Ziegler, M. F., Smith, P. M., and Bryan, R. F., *Experientia*, 31, 137, 1975.
155. Hayashi, Y., Sakan, T., Sakurai, Y., and Tashiro, T., *Gann*, 66, 587, 1975.
156. Ito, S. and Kodama, M., *Heterocycles*, 4, 595, 1976.
157. Kupchan, S. M., Karim, A., and Marcks, C., *J. Org. Chem.*, 34, 3912, 1969.
158. Marletti, F., Monache, F. D., Bettolo, G. B. M., Araujo, M. D. C. M. D., Cavalcanti, M. D. S. B., D'Albuquerque, I. L., and Lima, O. G. D., *Gazz. Chim. Ital.*, 106, 119, 1976.

159. Zelnik, R., Lavie, D., Levy, E. C., Wang, A. H.-J., and Paul, I. C., *Tetrahedron*, 33, 1457, 1977.
160. Loder, J. W., Culvenor, C. C. J., Nearn, R. H., Russell, G. B., and Stanton, D. W., *Aust. J. Chem.*, 27, 179, 1974.
161. Loder, J. W. and Nearn, R. H., *Tetrahedron Lett.*, 2497, 1975.
162. Loder, J. W. and Nearn, R. H., *Aust. J. Chem.*, 28, 651, 1975.
163. Kupchan, S. M., Court, W. A., Dailey, Jr., R. G., Gilmore, C. J., Bryan, R. F., *J. Am. Chem. Soc.*, 94, 7194, 1972.
164. Kupchan, S. M. and Schubert, R. M., *Science*, 185, 791, 1974.
165. Sher, F. T. and Berchtold, G. A., *J. Org. Chem.*, 42, 2569, 1977.
166. Uemura, D. and Hirata, Y., *Tetrahedron Lett.*, 1387, 1972.
167. Uemura, D., Katayama, C., and Hirata, Y., *Tetrahedron Lett.*, 283, 1977.
168. Wani, M. C., Taylor, H. L., Wall, M. E., Coggon, P., and McPhail, A. T., *J. Am. Chem. Soc.*, 93, 2325, 1971.
169. Powell, R. G., Miller, R. W., and Smith, Jr., C. R., *J. Chem. Soc. Chem. Commun.*, 102, 1979.
170. Kupchan, S. M., Sigel, C. W., Matz, M. J., Gilmore, C. J., and Bryan, R. F., *J. Am. Chem. Soc.*, 98, 2295, 1976.
171. Torrance, S. J., Wiedhopf, R. M., Cole, J. R., Arora, S. K., Bates, R. B., Beavers, W. A., and Cutler, R. S., *J. Org. Chem.*, 41, 1855, 1976.
172. Wall, M. E. and Wani, M. C., *7th Int. Symp. Chem. Natl. Prod.*, Riga, U.S.S.R., Abstr., E138, 614, 1970.
173. Wall, M. E. and Wani, M. C., *J. Med. Chem.*, 21, 1186, 1978.
174. Ghosh, P. C., Larrahondo, J. E., LeQuesne, P. W., and Raffauf, R. F., *Lloydia*, 40, 364, 1977.
175. Wani, M. C., Taylor, H. L., Wall, M. E., McPhail, A. T., and Onan, K. D., *J. Chem. Soc. Chem. Commun.*, 295, 1977.
176. Kupchan, S. M., Lacadie, J. A., Howie, G. A., and Sickles, B. R., *J. Med. Chem.*, 19, 1130, 1976.
177. Kupchan, S. M., Britton, R. W., Lacadie, J. A., Ziegler, M. F., and Sigel, C. W., *J. Org. Chem.*, 40, 648, 1975.
178. Eigebaly, S. A., Hall, I. H., Lee, K. H., Sumida, Y., Imakura, Y., and Wu., R. Y., *J. Pharm. Sci.*, 68, 887, 1979.
179. Hall, I. H., Lee, K. H., Eigebaly, S. A., Imakura, Y., Sumida, Y., and Wu, R. Y., *J. Pharm. Sci.*, 68, 883, 1979.
180. Sim, K. Y., Sims, J. J., and Geissman, T. A., *J. Org. Chem.*, 33, 429, 1968.
181. Kupchan, S. M., Britton, R. W., Ziegler, M. F., and Siegel, C. W., *J. Org. Chem.*, 38, 178, 1973.
182. Kupchan, S. M. and Streelman, D. R., *J. Org. Chem.*, 41, 3481, 1976.
183. Polonsky, J., and Zylber, N. B., *Bull. Soc. Chim. Fr.*, 2793, 1965.
184. Seida, A. A., Kinghorn, A. D., Cordell, G. A., and Farnsworth, N. R., *Lloydia*, 41, 584, 1978.
185. Polonsky, J. and Fourrey, J. L., *Tetrahedron Lett.*, 3983, 1964.
186. Kupchan, S. M. and Lacadie, J. A., *J. Org. Chem.*, 40, 654, 1975.
187. Polonsky, J., Varon, Z., Jacquemin, H., and Pettit, G. R., *Experientia*, 34, 1122, 1978.
188. Wani, M. C., Taylor, H. L., Thompson, J. B., and Wall, M. E., *Lloydia*, 41, 578, 1978.
189. Wani, M. C., Taylor, H. L., Thompson, J. B., Wall, M. E., McPhail, A. T. and Onan, K. D., *Tetrahedron*, 35, 17, 1979.
190. Lee, K. H., Imakura, Y., and Huang, H. C., *J. Chem. Soc. Chem. Commun.*, 69, 1977.
191. Konopa, J., Matuszkiewicz, A., Hrabowska, M., and Onoszka, K., *Arzneim Forsch.*, 24, 1741, 1974.
192. Kupchan, S. M., Sigal, C. W., Guttman, L. J., Restive, R. J., and Bryan, R. F., *J. Am. Chem. Soc.*, 94, 1353, 1972.
193. Lavie, D., Shvo, Y., Willner, D., Enslin, P. R., Hugo, J. M., and Norton, K. B., *Chem. Ind. (London)*, 951, 1959.
194. Kupchan, S. M., Gray, A. H., and Grove, M. D., *J. Med. Chem.*, 10, 337, 1967.
195. Doskotch, R. W., Malik, M. Y., and Beal, J. L., *Lloydia*, 32, 115, 1969.
196. Kupchan, S. M. and Tsou, G., *J. Org. Chem.*, 38, 1055, 1973.
197. Enslin, P. R., Hugo, J. M., Norton, K. B., and Rivett, D. E. A., *J. Chem. Soc.*, 4779, 1960.
198. Kupchan, S. M., Tsou, G., and Siegel, C. W., *J. Org. Chem.*, 38, 1420, 1973.
199. Kupchan, S. M., Meshulam, H., and Sneden, A. T., *Phytochemistry*, 17, 767, 1978.
200. Enslin, P. R., Hugo, J. M., Norton, K. B., and Rivett, D. E. A., *J. Chem. Soc.*, 4787, 1960.
201. Kupchan, S. M., Smith, R. M., Aynehchi, Y., and Maruyama, M., *J. Org. Chem.*, 35, 2891, 1970.
202. Merwe, K. J., Enslin, P. R., and Pachler, K., *J. Chem. Soc.*, 4275, 1963.
203. Enslin, P. R. and Norton, K. B., *J. Chem. Soc.*, 529, 1964.
204. Anisimov, M. M., Shentsova, E. B., Shcheglov, V. V., Strigina, L. I., Uvarova, N. I., Levina, E. V., Oshitok, G. I., and Elyakov, G. B., *Toxicon*, 13 (Abstr.), 86, 1975.
205. Tanaka, O., Nagai, M., and Shibata, S., *Chem. Pharm. Bull.*, 14, 1150, 1966.
206. Chan, W. R., Taylor, D. R., and Yee, T., *J. Chem. Soc. C*, 311, 1970.

207. **Jolad, S. D., Wiedhopf, R. M., and Cole, J. R.,** *J. Pharm. Sci.*, 66, 889, 1977.
208. **Polonsky, J., Varon, Z., Arnoux, B., Pascard, C., Pettit, G. R., Schmidt, J. H., and Lange, L. M.,** *J. Am. Chem. Soc.*, 100, 2575, 1978.
209. **Polonsky, J., Varon, Z., Arnoux, B., Pascard, C., Pettit, G. R., and Schmidt, J. M.,** *J. Am. Chem. Soc.*, 100, 7731, 1978.
210. **Sheth, K., Bianchi, E., Wiedhopf, R., and Cole, J. R.,** *J. Pharm. Sci.*, 62, 139, 1973.
211. **Sheth, K., Jolad, S., Wiedhopf, R., and Cole, J. R.,** *J. Pharm. Sci.*, 61, 1819, 1972.
212. **Trumbull, E. R., Bianchi, E., Eckert, D. J., Wiedhopf, R. M., and Cole, J. R.,** *J. Pharm. Sci.*, 65, 1407, 1976.
213. **Zurcher, A., Jeger, O., and Ruzicka, L.,** *Helv. Chim. Acta.*, 37, 2145, 1954.
214. **Allan, G. G., Fayez, M. B. E., Spring, F. S., and Stevenson, R.,** *J. Chem. Soc.*, 456, 1956.
215. **Wulff, G. and Tschesche, R.,** *Tetrahedron*, 25, 415, 1969.
216. **Kupchan, S. M., Steyn, P. S., Grove, M. D., Horsfield, S. M., and Meitner, S. W.,** *J. Med. Chem.*, 12, 167, 1969.
217. **Kim, H. K., Farnsworth, N. R., Fong, H. H. S., Blomster, R. N., and Persinos, G. J.,** *Lloydia*, 33, 30, 1970.
218. **Kupchan, S. M., Takasugi, M., Smith, R. M., and Steyn, P. S.,** *J. Org. Chem.*, 36, 1972, 1971.
219. **Kupchan, S. M., Hemingway, R. J., Knox, J. R., Barboutis, S. J., Werner, D., and Barboutis, M. A.,** *J. Pharm. Sci.*, 56, 603, 1967.
220. **Torrance, S. J., Wiedhopf, R. M. and Cole, J. R.,** *J. Pharm. Sci.*, 66, 1348, 1977.
221. **Turner, A. B.,** *Prog. Chem. Organic Nat. Prod.*, 24, 288, 1966.
222. **Gonzalez, A. G., Francisco, C. G., Freire, R., Hernandez, R., Salazar, J. A., and Suarez, E.,** *Phytochemistry*, 14, 1067, 1975.
223. **Gonzalez, A. G., Darias, V., Boada, J., and Alonso, G.,** *Planta Med.* 32, 282, 1977.
224. **Bhatnager, S. S., and Divekar, P. V.,** *J. Sci. Ind. Res.*, 10B, 56, 1951.
225. **Grant, P. K.and Johnson, A. W.,** *J. Chem. Soc.*, 4079, 1957.
226. **Nakanishi, K., Gullo, V. P., Miura, I., Govindachari, T. R., and Viswanathan, N.,** *J. Am. Chem. Soc.* 95, 6473, 1973.
227. **Mitchell, J. C., Fritig, B., Singh, B., and Towers, G. H. N.,** *J. Invest. Derm.*, 54, 233, 1970.
228. **Mitchell, J. C. and Dupuis, G.,** *Br. J. Derm.*, 84, 139, 1971.
229. **Mitchell, J. C., Geissman, T. A., Dupuis, G., and Towers, G. H. N.,** *J. Invest. Derm.*, 56, 98, 1971.
230. **Mitchell, J. C., Dupuis, G., and Towers, G. H. N.,** *Br. J. Derm.*, 86, 568, 1972.
231. **Mitchell, J. C., Dupuis, G., and Geissman, T. A.,** *Br. J. Derm.*, 87, 235, 1972.

INDEX

A

B

C

D

E

F

G

N

O

P

Q

R

S

T

U

V

W

X

Y

Z